METAL IONS IN
BIOLOGICAL SYSTEMS

VOLUME 44

**Biogeochemisty, Availability,
and Transport of Metals
in the Environment**

METAL IONS IN BIOLOGICAL SYSTEMS

Edited by

Astrid Sigel
Helmut Sigel

Department of Chemistry
Inorganic Chemistry
University of Basel
CH-4056 Basel, Switzerland

and ## Roland K. O. Sigel

Institute of Inorganic Chemistry
University of Zürich
CH-8057 Zürich, Switzerland

VOLUME 44

Biogeochemistry, Availability, and Transport of Metals in the Environment

CRC Press
Taylor & Francis Group
Boca Raton London New York

CRC Press is an imprint of the
Taylor & Francis Group, an **informa** business
A TAYLOR & FRANCIS BOOK

The figure on the cover is Figure 5 of Chapter 9 by R.E.M. Rickaby and D.P. Schrag.

First published 2005 by Taylor & Francis

Published 2019 by CRC Press
Taylor & Francis Group
6000 Broken Sound Parkway NW, Suite 300
Boca Raton, FL 33487-2742

© 2005 by Taylor & Francis Group, LLC
CRC Press is an imprint of Taylor & Francis Group, an Informa business

First issued in paperback 2019

No claim to original U.S. Government works

ISBN 13: 978-0-367-45421-0 (pbk)
ISBN 13: 978-0-8493-3820-5 (hbk)

Library of Congress Cataloging-in-Publication Data

Biogeochemistry, availability, and transport of metal in the environment / editors, Astrid Sigel, Helmut Sigel, Roland K.O. Sigel.
    p. cm. -- (Metal ions in biological systems ; v. 44)
    Includes bibliographical references.
    ISBN 0-8493-3820-4 (alk. paper)
    1. Biogeochemical cycles. 2. Nonmetals. I. Sigel, Astrid. II. Sigel, Helmut. III. Sigel
Roland K. O. IV. Series.

QP532.M47 vol. 44
[QH344]
572'.51--dc22
[577'.14]   2004061452

Library of Congress Card Number 2004061452

Preface to the Series

Recently, the importance of metal ions to the vital functions of living organisms, hence their health and well-being, has become increasingly apparent. As a result, the long-neglected field of "bioinorganic chemistry" is now developing at a rapid pace. The research centers on the synthesis, stability, formation, structure, and reactivity of biological metal ion-containing compounds of low and high molecular weight. The metabolism and transport of metal ions and their complexes are being studied, and new models for complicated natural structures and processes are being devised and tested. The focal point of our attention is the connection between the chemistry of metal ions and their role for life.

No doubt, we are only at the brink of this process. Thus, it is with the intention of linking coordination chemistry and biochemistry in their widest sense that the *Metal Ions in Biological Systems* series reflects the growing field of "bioinorganic chemistry". We hope, also, that this series will help to break down the barriers among the historically separate spheres of chemistry, biochemistry, biology, medicine, and physics, with the expectation that a good deal of future outstanding discoveries will be made in the interdisciplinary areas of science.

Should this series prove a stimulus for new activities in this fascinating "field", it would serve its purpose and would be a satisfactory result for the efforts spent by the authors.

Fall 1973   *Helmut Sigel*

Preface to Volume 44

The preceding Volume 43, entitled *Biogeochemical Cycles of the Elements* is closely related to the present Volume 44 which is devoted to the *Biogeochemistry, Availability, and Transport of Metals in the Environment.*

The book opens with a summary about the role of the atmosphere in metal cycling, the importance of long-range metal transport in aerosols, and the factors which govern the atmospheric transport; several elements are individually considered. The atmospheric transport of iron, one of the limiting nutrients for phytoplankton, is of high significance, since a third to half of the world's fixation of carbon dioxide is estimated to occur in the oceans as a result of the photosynthetic activity by phytoplankton, having thus a large impact on the global carbon cycle. As outlined in Chapter 2, it is clear now that biologically derived iron binding ligands play an integral role in the cycling of this element in the oceans. In this context marine bacterial siderophores and mediated iron uptake are discussed.

The interrelations between speciation and bioavailability of trace metals in freshwater environments and in soils are emphasized in Chapters 3 and 4, respectively. Next, heavy metal uptake by higher plants, algae, and cyanobacteria is critically evaluated in particular regarding cadmium, copper, nickel, and zinc. On the one hand, these metals are, in part, micronutrients but on the other hand, at higher concentrations they are toxic by inhibiting plant metabolism. Indeed, plants have developed strategies to resist the toxicity of heavy metals; there are so-called "excluders", which actively prevent metal accumulation inside the cells, as well as "hyperaccumulators", which bind heavy metals with high affinity. The latter offer promising approaches for cleaning up contaminated soils (phytoremediation) as well as for mining metals (phytomining).

The geochemical cycle of arsenic, an element poisonous at elevated concentrations, has come into focus due to its dispersal by mining but even more so due to its natural occurrence in Bangladeshi groundwater in the Ganges delta, where the role of human activities on arsenic levels is subtle and indirect. Antimony, a close relative of arsenic, is also a potentially toxic trace element whose environmental significance clearly outweights the attention it has received to date. The anthropogenic impacts on its biogeochemistry and cycling are addressed in Chapter 7; no biological function has yet been attributed to antimony, but its toxicity to animals is comparable to that of arsenic and lead (see Volume 43).

The release of radionuclides from nuclear sites and their subsequent mobility in the environment are subjects of intense public concern. Chapter 8 provides an overview on the biogeochemistry of key radionuclides, like americium, uranium, plutonium, cesium, technetium, etc., summarizes what is known about their interactions with microorganisms, and discusses how such interactions have an impact on the mobility of radionuclides in the environment. The terminating chapter considers the biogeochemistry of carbonates, the biomineralization processes of different organisms, and it shows how trace metal proxies can be used to reconstruct past ocean conditions, providing thus information about past climates.

Astrid Sigel
Helmut Sigel
Roland K. O. Sigel

Contents of Volume 44

Contributors

Numbers in parentheses indicate the pages on which the authors' contributions begin.

Roger D. Beckie *Department of Earth and Ocean Sciences, University of British Columbia, 6339 Stores Road, Vancouver, BC, Canada V6T 1Z4,* E-mail: rbeckie@eos.ubc.ca (145)

Renata Behra *Swiss Federal Institute for Environmental Science and Technology EAWAG, P.O. Box 611, CH-8600 Dübendorf, Switzerland* (47)

Torunn Berg *Norwegian Institute for Air Research, P.O. Box 100, NO-2027 Kjeller, Norway,* Fax: +47-63-898050, E-mail: tbe@nilu.no (1)

Alison Butler *Department of Chemistry and Biochemistry, University of California, Santa Barbara, CA 93106-9510, USA,* Fax: +1-805-893-4120, E-mail: butler@chem.ucsb.edu (21)

Bin Chen *Institute of Environmental Geochemistry, University of Heidelberg, Im Neuenheimer Feld 236, D-69120 Heidelberg, Germany,* E-mail: b.chen@ ugc.uni-heidelberg.de (171)

Charles F. Harvey *Parsons Laboratory, Department of Civil and Environmental Engineering, Massachusetts Institute of Technology 48-321, Cambridge, MA 02139, USA,* Fax: +1-617-258-8850, E-mail: charvey@mit.edu (145)

Michael Krachler *Institute of Environmental Geochemistry, University of Heidelberg, Im Neuenheimer Feld 236, D-69120 Heidelberg, Germany,* E-mail: krachler@ugc.uni-heidelberg.de (171)

Peter M. H. Kroneck *Mathematisch-Naturwissenschaftliche Sektion, Fachbereich Biologie, Universität Konstanz, Postfach M665, D-78457 Konstanz, Germany,* E-mail: peter.kroneck@uni-konstanz.de (97)

Hendrik Küpper   *Mathematisch-Naturwissenschaftliche Sektion, Fachbereich Biologie, Room M702, Universität Konstanz, D-78457 Konstanz, Germany,* E-mail: hendrik.kuepper@uni-konstanz.de (97)

Jon R. Lloyd   *Williamson Research Centre for Molecular Environmental Sciences and Department of Earth Sciences, The University of Manchester, Manchester M13 9PL, UK,* Fax: +44-161-275-3947, E-mail: jon.lloyd@man.ac.uk (205)

Bernard Ludwig   *Department of Environmental Chemistry, University of Kassel, Nordbahnhofstrasse 1a, D-37213 Witzenhausen, Germany,* E-mail: bludwig@uni-kassel.de (75)

Jessica D. Martin   *Department of Chemistry and Biochemistry, University of California, Santa Barbara, CA 93106-9510, USA* (21)

Kerstin Michel   *Department of Environmental Chemistry, University of Kassel, Nordbahnhofstrasse 1a, D-37213 Witzenhausen, Germany,* E-mail: kerstin.michel@uni-kassel.de (75)

Joanna C. Renshaw   *Williamson Research Centre for Molecular Environmental Sciences and Department of Earth Sciences, The University of Manchester, Manchester M13 9PL, UK* (205)

Rosalind E. M. Rickaby   *Department of Earth Sciences, University of Oxford, Parks Road, Oxford OX1 3PR, UK,* E-mail: rosr@earth.ox.ac.uk (241)

Daniel P. Schrag   *Laboratory for Geochemical Oceanography, Department of Earth and Planetary Sciences, Harvard University, 20 Oxford Street, Cambridge, MA 02138, USA,* Fax: +1-617-496-4387; E-mail: schrag@eps.harvard.edu (241)

William Shotyk   *Institute of Environmental Geochemistry, University of Heidelberg, Im Neuenheimer Feld 236, D-69120 Heidelberg, Germany,* E-mail: shotyk@ugc.uni-heidelberg.de (171)

Laura Sigg   *Swiss Federal Institute for Environmental Science and Technology EAWAG, P.O. Box 611, CH-8600 Dübendorf, Switzerland,* Fax: +41-1-823-5311, E-mail: laura.sigg@eawag.ch (47)

Eiliv Steinnes   *Department of Chemistry, Norwegian University of Science and Technology, NO-7491 Trondheim, Norway,* E-mail: eiliv.steinnes@chem.ntnu.no (1)

Contents of Previous Volumes

---

*Out of print

---

[*]Out of print

Volume 27. Electron Transfer Reactions in Metalloproteins

Volume 36. Interrelations Between Free Radicals and Metal Ions in Life Processes

[†]Deceased

[†]Deceased

Comments and suggestions with regard to contents, topics, and the like for future
volumes of the series are welcome.

The following Marcel Dekker, Inc., books are also of interest for any reader involved with bioinorganic chemistry or who is dealing with metals or other inorganic compounds:

HANDBOOK ON TOXICITY OF INORGANIC COMPOUNDS

Edited by Hans G. Seiler and Helmut Sigel, with Astrid Sigel

In 74 chapters, written by 84 international authorities, this book covers the physiology, toxicity, and levels of tolerance, including prescriptions for detoxification, for all elements of the Periodic Table (up to atomic number 103). The book also contains short summary sections for each element, dealing with the distribution of the elements, their chemistry, technological uses, and ecotoxicity as well as their analytical chemistry.

HANDBOOK ON METALS IN CLINICAL AND ANALYTICAL CHEMISTRY

Edited by Hans G. Seiler, Astrid Sigel, and Helmut Sigel

This book is written by 80 international authorities and covers over 3500 references. The first part (15 chapters) focuses on sample treatment, quality control, etc., and on the detailed description of the analytical procedures relevant for clinical chemistry. The second part (43 chapters) is devoted to a total of 61 metals and metalloids; all these contributions are identically organized covering the clinical relevance and analytical determination of each element as well as, in short summary sections, its chemistry, distribution, and technical uses.

HANDBOOK ON METALLOPROTEINS

Edited by Ivano Bertini, Astrid Sigel, and Helmut Sigel

The book consists of 23 chapters written by 43 international authorities. It summarizes a large part of today's knowledge on metalloproteins, emphasizing their structure–function relationships, and it encompasses the metal ions of life: sodium, potassium, magnesium, calcium, vanadium, chromium, manganese, iron, cobalt, nickel, copper, zinc, molybdenum, and tungsten.

1

Atmospheric Transport of Metals

Torunn Berg[1] and Eiliv Steinnes[2]

[1]Norwegian Institute of Air Research, P.O. Box 100, NO-2027 Kjeller, Norway
[2]Department of Chemistry, Norwegian University of Science
and Technology, NO-7491 Trondheim, Norway

1. INTRODUCTION: THE ROLE OF THE ATMOSPHERE IN METAL CYCLING

While atmospheric transport of metals was hardly mentioned in the scientific literature 50 years ago, it has become a frequently discussed theme in more recent years. The main reason for the increased interest is undoubtedly the growing concern about contamination of the environment by toxic metals. It has become evident that atmospheric transport and deposition is an important pathway of these elements to the terrestrial environment, and in some cases also to fresh water systems. Moreover, the interest has gradually shifted from only considering the impact of local sources to include transport and effects on the regional and global scale.

The importance of the atmosphere as a pathway of metals depends very much on the physical and chemical properties of the metal concerned. Metals that may occur in a chemical form that is volatile at the temperature of emission are likely to be carried over greater distances in the atmosphere than those contained in emitted particulate matter (PM). Most volatile metal species will tend to condense on particles in the atmosphere. Since smaller particles have a higher specific surface than larger ones, the result of this condensation is a relative enrichment on the smaller particles. Smaller particles have a lower deposition velocity in the air than larger ones, their residence time in the atmosphere is therefore longer, and they may be transported over hundreds and even thousands of kilometers before they reach the ground. In the specific case of mercury the elemental form is sufficiently volatile even at room temperature that emissions in the form of Hg^0 are dispersed throughout the entire hemisphere where they occur [1].

The importance of long-range atmospheric transport of metals in aerosols was first recognized in the case of lead by Patterson and his group [2], who observed a significant increase of Pb in surface layers of Greenland ice relative to deeper horizons. More recently significant long-range transport has also been shown for elements such as Zn, As, Cd, and Se [3–5]. Other frequently discussed toxic metals such as Cr, Mn, Co, Ni, and Cu tend to be associated with larger particles and are hence preferentially deposited closer to the source, like the major crustal metals such as Al, Ca, Ti, and Fe [5].

By combination of chemical data and meteorological models it may be possible to calculate fairly accurately from where a given polluted air mass originates. This tool has been important so far in assessing the large-scale pathways of metal transport in the atmosphere [6], and may also be significant in assessing where future problems may arise as a result of the present large-scale escalation of industrial activities in many countries where the previous activity was small.

In addition to the emissions from industrial and other anthropogenic activities, some natural processes also provide significant contributions to the content of metals in the atmosphere [7,8]. These processes include wind erosion, volcanic activity, biogenic emissions, forest fires, sea-salt spray, and in the case of mercury general outgassing of Hg^0 from rocks and soils [7,8]. In this chapter, present and past contributions to metals in the atmosphere from natural and anthropogenic sources will be evaluated, and factors governing the atmospheric transport and deposition of metals will be reviewed. In addition, some factors related to sampling and analysis will be briefly discussed.

2. FACTORS GOVERNING ATMOSPHERIC TRANSPORT

2.1. Source Term

Elements are termed atmophile when their mass transport through the atmosphere is greater than that in streams. Many atmophile elements are volatile and have metal oxides of relatively low boiling point. It is known that some of these metals, i.e., Hg, As, Se, Sn, and Pb, can be methylated [9] and/or released into the atmosphere as vapors and that Hg and probably As and Se are released mainly as inorganic vapor from burning of coal [9]. Elemental Hg has usually a long residence time in the atmosphere. Divalent species of Hg are also emitted such as reactive gaseous mercury (RGM) and particle-bound Hg. Divalent mercury has moderate residence times and a higher affinity to condensed phases [1].

Larger particles emitted from coal-fired furnaces are primarily oxides of Al, Si, Ca, Fe, Na, Mg, and K, while smaller particles are highly enriched in volatile trace metals such as As, Sb, Se, Cd, Pb, and Zn. An intermediate behavior is found for Ba, Be, Cr, Sn, Ni, Sr, U, and V showing a slight enrichment in the fine particle fraction [9,10].

The multimodal shape of trace element size distributions is also found in the ambient atmosphere. As, Sb, Se, Cd, Pb, and Zn are present predominantly in the accumulation mode [aerodynamic diameter (AD) 0.3–0.8 μm], Al, Si, Ca, Fe, Na, Mg, and K generally follow the shape of the coarse mode (AD > 3 μm) whereas Ba, Be, Cr, Cu, Sn, Mn, Ni, Sr, U, and V exhibit an intermediate behavior with ADs of ~1–5 μm [9]. This is important not only from a health viewpoint since fine particles (aerodynamic ratio <2.5 μm) are respirable, but also because fine particles tend to persist in the atmosphere where they can undergo chemical reactions and be transported from their sources over long-distances to pristine areas of the environment [9]. Based on studies of the long-range transport of several trace elements in Western Europe, and on literature data Pacyna et al. [11] inferred the local deposition within a 150 km grid to be only 5% of the metal emissions from all sources.

In between release to and removal from the atmosphere, the species may be transported horizontally and vertically [12]. The source-transport-sink sequence defines the physical lifetime of the constituents. However, chemical changes may occur on a time-scale comparable with, or smaller than this lifetime.

2.2. Wet/Dry Deposition

Metals are removed from the atmosphere by dry deposition and wet deposition. Dry deposition denotes the direct transfer of species, both gaseous and particulate, to the Earth's surface and proceeds without the aid of precipitation. Wet deposition, on the other hand, encompasses all processes by which airborne species are transferred to the Earth's surface in aqueous form (i.e., rain, snow, or fog): (1) dissolution of atmospheric gases in airborne droplets (cloud drops, rain, or fog), (2) removal of atmospheric particles when they serve as nuclei for the condensation of atmospheric water to form a cloud or fog droplet, and (3) removal of atmospheric particles when the particle collides with a droplet both within and below clouds [13]. For dry deposition particle size, several meteorological parameters, and the surface structure of the receptor are important parameters determining the deposition mechanism [14]. For wet deposition rainfall rate and the concentration of the component of interest in the liquid phase determine the wet deposition flux.

The relative importance of dry vs. wet deposition of the various compounds depends on the chemical state, physical parameters and the surface characteristics. Since particle size is a major factor influencing the deposition velocity, the relative contribution of dry to total deposition is higher near the sources (e.g., [15]). There seems to be a general agreement that wet deposition is the main removal mechanism for trace elements associated with particles below 1 μm at remote sites, except in some outlying dry areas. Dry deposition may possibly dominate for aerosols associated with particles in the size range above 1 μm radius (e.g., Mn, Fe) at remote sites. In general, for atmospheric long range transport to remote areas in Scandinavia and the Arctic, dry deposition appears to be more important for the removal of pollutants close to the emission source, while wet deposition is more important to remove particles "en route" [16].

In areas where topographical features cause large variations in deposition, use of measured concentrations and precipitation amounts makes it possible to determine the wet deposition more directly and with more detailed spatial resolution than what is possible from dispersion models. Similarly, dry deposition of individual species may be interfered as the product of the ambient airborne concentrations and their respective dry deposition velocities [17]. In this case, it is essential to take into account seasonal variations and differences in ground cover.

The first major studies of the deposition of trace elements from the atmosphere started in the United Kingdom in the beginning of the 1970s [18]. Since then several national and international programs are monitoring heavy metals in deposition. Figure 1 shows monitoring results for Pb in precipitation from the European Monitoring and Assessment Programme (EMEP). The lowest concentrations of Pb during 2001 were found in northern Scandinavia, Iceland,

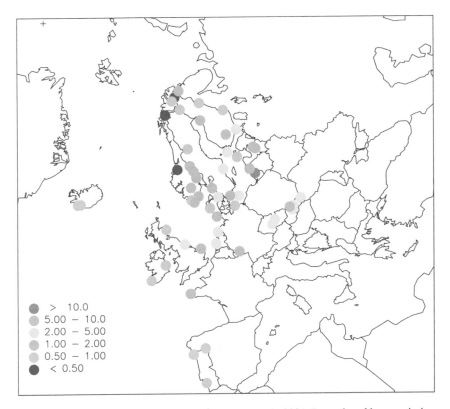

Figure 1   Lead in precipitation (μg Pb/L) in Europe in 2001. Reproduced by permission from Ref. [19].

Ireland, and Portugal, where the annual averages were below 1 μg Pb/L. An exception was one of the Norwegian stations which is located close to the large heavy metal emission sources at the Kola Peninsula in Russia. The highest concentrations were seen in the Czech Republic, Slovakia, and Lithuania with annual concentration means between 3.5 and 4.5 μg Pb/L.

3. SOURCES OF METALS IN AIR

3.1. Natural Sources

The most comprehensive global assessment of natural sources to metals in the atmosphere is that of Ref. [7], and the following text essentially follows this evaluation. The most predominant natural sources of trace elements in the atmosphere are biogenic processes, volcanic dust, rock and soil-derived dust, forest fires, and sea-salt spray. The contributions from natural processes are much more difficult to assess than those from air pollution because the sources are in most cases much more dispersed and the fluxes are low. Nevertheless their contribution to the global inventory of metals in the air is in most cases probably at least of the same order as that from anthropogenic processes. A comparison of contributions from natural and anthropogenic sources of some key metals to the atmosphere along with an indication of the most important natural sources for each of the metals is presented in Table 1.

3.1.1. Biogenic Emissions

Contributions from biogenic processes include volatile products from biologically mediated processes both in the terrestrial and marine environment, and contribute a significant portion of the total Hg, As, and Se emitted to the atmosphere from natural processes [7,20]. Biogenic contributions of other elements are primarily associated with particles of pollen, spores, waxes, fungi, algae, etc. [7,20].

3.1.2. Marine Aerosols

Marine aerosols are formed when bubbles burst on the sea surface and are brought into the atmosphere by wind action. In addition to solutes in seawater these aerosols may also contain fragments of the organic surface microlayer on the sea, which is known to concentrate some trace elements [7,20]. For most metals the marine aerosols is an insignificant source due to their low concentrations in seawater.

3.1.3. Windblown Mineral Dust

This is the overall most important natural source of trace elements, in particular for elements that are abundant in the Earth's crust [7,20]. Contributions are particularly high from areas with sparse vegetation cover, such as deserts [7,20]. The

Table 1 Comparison of Estimated Annual Global Emissions for Selected Metals from Natural and Anthropogenic Sources (10^3 tonnes) and Indication of the Dominant Natural Sources for Each Element

Metal	Anthropogenic sources		Natural sources	Dominant natural source
	1983	1995		
As	18.8	5.0	12	VOL, WBM, BIO
Cd	7.6	3.0	1.3	VOL
Cr	30.5	14.7	44	WBM, VOL
Cu	35.4	25.9	28	VOL, WBM
Hg	3.6	2.2	2.5	BIO, VOL
Mn	38.3	11.0	317	WBS
Mo	3.3	2.6	3.0	WBM, FF, BIO
Ni	55.7	95.3	30	VOL, WBM
Pb	332	119	12	WBM, VOL
Sb	3.5	1.6	2.4	WBM, VOL, SSP
Se	3.8	4.6	9.3	BIO
V	86.0	240	28	WBM
Zn	132	45.0	45	WBM, VOL, FF, BIO

Abbreviations: VOL, volcanoes; BIO, biogenic emissions; WBM, windblown minerogenic dust; FF, wild forest fires; SSP, sea-salt particles.
Source: Natural emissions are median values from Nriagu [7]. The anthropogenic emissions in 1983 are assessments by Nriagu and Pacyna [8] and the 1995 ones by Pacyna and Pacyna [20].

corresponding anthropogenic component from open agricultural land is difficult to distinguish from the natural one.

3.1.4. Forest Fires

Wild forest fires occur all over the world where forests exist, and they are a significant source of trace metals enriched in plants, such as Zn and Mo [7,20].

3.1.5. Volcanic Emissions

Volcanic emissions are probably the second most important natural source of metals in the atmosphere, and this includes both gaseous emissions and dust particles [7,20]. Due to the episodicity of major events and difficult conditions for measurements during such events this is probably the most uncertain of all the factors mentioned here.

3.2. Anthropogenic Sources

A comparison of median global emission rates of trace elements from natural vs. anthropogenic sources (Table 1) indicates that the current anthropogenic emissions of Cd, Ni, Pb, and V clearly exceed the corresponding fluxes from natural sources. For As, Cr, Cu, Hg, Mo, Sb, Se, and Zn the natural and anthropogenic contributions are probably of similar magnitude, whereas the natural emissions dominate in the case of Mn. Within Europe, however, the anthropogenic sources are more important in relative terms because of the relatively low occurrence of volcanoes and deserts [21], and the same probably applies for eastern North America. Moreover, metals primarily emitted from high-temperature anthropogenic sources are more likely to be found in the intermediate (1–5 μm) and the fine (0.3–0.8 μm) particle mode than metals associated with wind-blown mineral dust which is the dominant natural source for a majority of trace metals [7,20].

The generally most important anthropogenic source categories of trace metal emission to the atmosphere are stationary fossil fuel combustion and non-ferrous metal production. In addition, vehicular traffic, iron and steel production, cement production, and waste incineration are significant sources for some metals. The relative contributions from different source categories to the estimated worldwide emission of 16 trace metals in 1995 are shown in Table 2. In the following, the importance of the various source categories to the anthropogenic emission of different metals are briefly discussed, mainly on the basis of information given in Ref. [20].

Table 2 Estimates of the Relative Contributions (%) from Different Source Categories to the Worldwide Emission of Trace Metals to the Atmosphere in 1995

Element	SFF	VT	NFM	ISP	CP	WI	GP
As	16.1		69.0	7.0	5.3	2.5	
Cd	23.2		72.8	2.1	0.6	1.3	
Cr	68.9			19.2	9.1	2.8	
Cu	27.3		69.7	0.5		2.4	
Hg	66.0		7.3	1.3	6.0	4.9	14.5
In			100				
Mn	85.2		0.5	9.6		4.6	
Mo	100						
Ni	90.4		9.3		0.1	0.1	
Pb	9.8	74.4	12.4	2.5	0.2	0.7	
Sb	46.8		35.4	0.4		17.4	
Se	89.1		10.1	0.2		0.5	
Sn	89.0		8.1			2.9	
Tl	100						
V	99.9		0.03	0.03		0.01	
Zn	16.5		71.7	3.7	4.7	3.4	

Abbreviations: SFF, stationary fossil fuel combustion; VT, vehicular traffic; NFM, non-ferrous metal production; ISP, iron and steel production; CP, cement production; WI, waste incineration; GP, gold production.
Source: Data are from Ref. [20].

3.2.1. Stationary Fossil Fuel Combustion

The materials predominantly burned in power plants and heat-producing units are coal and fuel oil. Coal burning is the predominant source to atmospheric emissions of Cr, Hg, Mn, Mo, Sb, Se, Sn, and Tl, and is also a significant source to the emissions of As, Cd, Cu, Pb, and Zn [7,20]. Combustion of fuel oils on the other hand is responsible for a major part of the Ni emission worldwide and almost 100% of the V emission. According to the figures in Table 1 there has been a tendency of reduced metal emissions from coal burning over the last 20 years. The emissions of Ni and particularly V, however, increased considerably between 1983 and 1995, mainly because of a strong increase in Asian countries where emission controls are generally less efficient than in Europe and North America. Asian countries, predominantly China and India, were also responsible for 60% of the total Hg emissions from stationary fossil fuel combustion in 1995 [7,20].

3.2.2. Non-ferrous Metal Production

Production of non-ferrous metals is the most important emission source for As, Cd, Cu, In, and Sn, and is also significant for Hg, Ni, Pb, Sb, Se, and Sn [7,20]. The main part of these emissions is associated with copper production, except for Pb and Zn where the largest emissions are from the lead and zinc production, respectively. Emissions of Cd, Hg, and Pb from the zinc production are also significant [7,20].

3.2.3. Vehicular Traffic

A major part of the lead emissions to the atmosphere during the last 50 years has been from the use of leaded petrol, and it still was in 1995 [7,20]. In Europe and North America, however, this source is currently contributing substantially less to air pollution than in the past due to the out-phasing of leaded gasoline in many countries. Vehicular traffic is also a source to emissions of other metals, such as platinum group metals (Rh, Pd, Pt) from exhaust catalysts, but no inventory on the global emissions of these metals seems to be available. From their chemical properties, these metals are not expected to be subject to long-range atmospheric transport but rather to be deposited mainly in the vicinity of the road [7,20].

3.2.4. Iron and Steel Production

The production of iron and ferro-alloys is not the major source of emission for any of the trace elements discussed here, but appreciable emissions are encountered for As, Cr, and Mn [7,20]. Significant amounts of Pb and Zn are also emitted, although constituting only 3–4% of the total global emissions of these metals [7,20].

3.2.5. Cement Production

Although this industry is a large source of aerosols in general, it is less important than some of the above categories as a source of trace metals to the atmosphere. In the case of As, Cr, Hg, and Zn 5–10% of the global emissions are ascribed [7,20] to cement production, for other metals it is relatively less important.

3.2.6. Waste Incineration

This is a source of metal emission that is likely to be gradually more important in the future. So far it is relatively most significant for Sb (17%), but also for a number of other metals (As, Cr, Cu, Hg, Sn, Zn) for which 2–5% of the global anthropogenic emissions in 1995 were assumed to originate from this source.

3.2.7. Gold Production

This source is important for Hg since metallic mercury is used to extract grains of gold metal from sediments. In the refining of gold from this mixture, the residual Hg metal is frequently removed by evaporation. Gold production, mostly in Africa and South America, is assumed to be responsible for $\sim$15% of the global anthropogenic Hg emissions [7,20].

3.3. Re-emission of Mercury

In comparison to the other heavy metals, one of the major distinguishing features of Hg is the great extent to which it is re-circulated in the environment via the atmosphere. Since 1890 more than 2×10^5 tonnes of Hg have been deposited to terrestrial soils [1]. The re-emission of mercury from this large pool may represent a significant global flux to the atmosphere. During the last 15 years a considerable number of significant studies has focused on air-surface exchange of Hg [1].

3.4. Methods for Source Analysis

3.4.1. Enrichment Factors

Use of enrichment factors may provide useful evidence of the source of elements in aerosols and precipitation in remote areas [22]. The enrichment factor EF_{crust} relates the concentration of a given element in air to the concentration (X) of a crustal element (e.g., Al, Sc, Ti, or Fe) in air, normalized to the ratio of concentration in the crust related to the reference element in the crust:

$$EF_{crust} = \frac{(X/Fe)_{air}}{(X/Fe)_{crust}}$$

An EF of approximately unity suggests that crustal erosion is the primary source of the observed element in the atmosphere. Values much higher than unity imply sources other than crustal dust. These sources may include anthropogenic activities and natural processes (e.g., biogenic processes, forest wildfires, volcanic eruptions, sea salts). Potential problems in the use of EFs are the variation in crustal composition from one site to another, and the possibility of fractionating during formation of the soil-derived aerosols [3]. Another drawback of this method is the inability to discriminate between anthropogenic contributions and other natural processes that produce an enrichment of the abundance of the observed element as compared to crustal abundances [3]. A delicate problem when using EFs, especially in polluted regions, is the similarity of the matrix of major elements in coal fly ash and the crustal composition.

In Fig. 2, enrichment factors with respect to average crustal rock [23] are reported from six background sites in Norway using Fe as crustal indicator [3].

3.4.2. Multivariate Statistical Methods

One commonly used method for source analysis is the principal component analysis (PCA) [24,25]. PCA is a special case of factor analysis, which transforms the original set of inter-correlated variables into a set of uncorrelated variables that are linear combinations of the original variables. The first principal component is the linear combination of the variables that accounts for a maximum of the total variability of the data set. The second principal component explains a maximum of the variability not accounted for by the first component, and so on. The objective is to find a minimum number of principal components that explain most of the variance in the data set. The principal components are

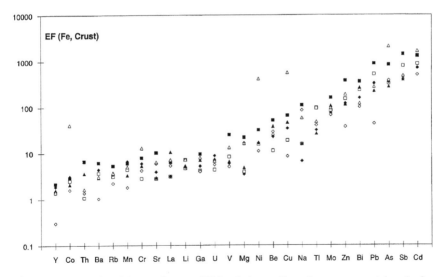

Figure 2 Crustal enrichment factors (EFs) relative to Fe and average crustal rock, for 27 elements at six rural and remote sites in Norway. Reproduced by permission from Ref. [3].

statistically independent and, typically, the first few components explain almost all the variability of the whole data set. The minor principal components, which explain only a minor part of the data, can be eliminated, thus simplifying the analysis.

However, these minor components contain most of the random error, so eliminating them tends to remove extraneous variability from the analysis [24,25]. The main advantage of PCA is to identify unusual sources that may not have been considered previously and to find the major contributing source classes [24,25]. In remote areas factor analysis may tend to uncover obviously influencing sources such as marine, crustal, and mixed anthropogenic ones, and do not always allow one to obtain a fine resolution of the contributions from various distant source regions to the chemical composition of the remote aerosol.

3.4.3. Dispersion Models

Atmospheric dispersion models capable of reproducing the observed deposition and air concentration patterns of heavy metals are sophistical tools for analyzing the extent of the antropogenic impact of emissions. Several models have been successful to simulate local and regional scale transport of heavy metals by coupling meteorological and chemical processes. The first model of long-range transport of heavy metals in Europe [11] was used to verify emission data through a comparison of measured and calculated concentrations of heavy metals in ambient air.

Models for evaluation of heavy metal transport in the atmosphere have traditionally been divided in two large groups—models of Eulerian and Lagrangian types [6,26]. In the first group the transport is calculated in the fixed co-ordinate system strictly linked with geographical grid. In the second group the co-ordinate system is connected with an air particle moving in the air flux. Each type of models has its advantages and shortcomings [6]. Models have been developed and used both on local/regional and hemispheric scales. Atmospheric processes that may be included in the models are emissions, dispersion, advection, and wet and dry deposition.

Inclusion of particle size spectrum to the parameterization of the model has been an important step [27]. Particle size greatly affects the removal of heavy metals from the atmosphere. An Eulerian model with particle scheme has been applied to study long-range transport of Pb, Cd, Zn, and As from sources in Eurasia and North America to the Arctic [27]. It was concluded that the simulation of the deposition of studied metals in the Arctic demonstrated an appreciable impact of emissions from sources in these regions. In contrast to other heavy metals, atmospheric mercury is represented by a great number of different reactive compounds [22]. In the atmosphere oxidation reactions can take place both in the gaseous and liquid phase. The key question of atmospheric mercury chemistry lies in the evaluation of a relative importance of oxidation and reduction in the cloud liquid phase [26]. Modeling of mercury is therefore much more complex than modeling of other heavy metals and inclusion of a chemical reaction scheme is also required in the models.

4. SAMPLING AND DETERMINATION OF METALS

4.1. Air Concentrations

Most heavy metals in air are sampled with a filter using either high- or low-volume samplers. The choice depends on the sampling period [28]. For short sampling periods (daily), a high volume sampler is usually needed. With a high volume sampler it is possible to collect samples of 1600 m^3 per 24 hours. Low volume samplers are in comparison in the range of $1-3 \text{ m}^3$ per hour. The filter pack method [28], used for sampling of the main air components is a low volume sampler and can be used for heavy metal sampling if the sampling period is longer, i.e., for weekly sampling. Another important issue is which particle size should be sampled, fine (PM2.5), coarse + fine (PM10) or total suspended particulate matter (TSP).

Most of the high or low volume samplers provide the possibility to collect either total suspended particle matter or a fraction with a defined cut-off; which to prefer depends on the aim of the measurements [28]. In order to obtain the best estimate of the deposition of heavy metals, a size distribution is preferred. However, several studies indicate that the mass distribution is different for the elements, and to obtain an informative size distribution 7–8 fractions should be sampled. This is very expensive and for monitoring purposes it is generally sufficient to sample one fraction. Sampling of particles $<10 \text{ } \mu\text{m}$ will to a large extent contain the main fraction of long range transported heavy metals and it is recommended that this fraction is collected [28].

The sampling of elemental mercury in air is usually done using gold traps with gas-meters or mass flow controllers for air volume measurements. During the last 10 years, automated instruments for the sampling and analysis of mercury in air have been made available. The automated method applies the same basic principles as the manual method and has been shown to generate comparable results [29,30].

4.2. Atmospheric Deposition

It is recommended to use a wet-only collector, which is open only during precipitation events, for precipitation sampling [28]. When choosing which wet-only sampler to use, it is important that no parts of the precipitation collector are made from metal. Further, it is essential that all parts can easily be cleaned and are of known composition. As for material, high-density polyethylene collectors are recommended.

Bulk collectors that are open at all times may also be used for deposition measurements. Bulk collectors tend to give too high metal concentrations due to dry deposition, but in some areas with little dry deposition there is practically no difference between the two types of collector, i.e., in the Nordic countries [28]. When a bulk collector can be used, it is recommended to use a sampler with separate collective funnel and collection bottle for easy cleaning. In addition, the funnel should have high walls. In order to prevent insects, leaves, etc. to enter the collection bottle one can use a sieve made of, i.e., polycarbonate with larger grid size.

Mercury is collected in special precipitation samplers [30]. Two alternative materials may be used for funnels and collection bottles: borosilicate glass and a halocarbon such as Teflon or PFA. As for the other heavy metals, the sampling vessels can be bulk samplers or wet-only. For monitoring purposes, bulk sampling using funnels and bottles is normally adequate. Wet-only samplers are used by the German national monitoring program as well as by research groups working in the Great Lakes area and in the US National Atmospheric Deposition Programme. Wet-only samplers have the advantage that they avoid particle dry deposition, although the contribution of gaseous or particulate mercury species to the wet deposition fluxes in non-industrialized or non-urban areas is probably not large [30].

Detailed knowledge about national atmospheric deposition patterns and locations of significant trace element emission sources requires a very extensive sampling network, which is not possible using precipitation samplers. In such cases, mosses used as biomonitors may give additional information. Biomonitors may be defined as organisms that can be used for the recognition and quantitative determinations of anthropogenically induced environmental factors. The moss technique as a mean of surveying atmospheric heavy metal deposition was developed in the late 1960s [31]. Mosses, especially the carpet-forming species, obtain most of their supply of chemical substances directly from precipitation and from the impaction and sedimentation of airborne particles. A number of epigeic mosses have been proposed [5] as biomonitors of atmospheric heavy metals. The feather mosses *Hylocomium splendens* and *Pleurozium schreberi* have been most commonly used in Scandinavia [32,33]. *Sphagnum* mosses from peat bogs, and epiphytic *Hypnum cupressiforme* have also been used in moss surveys [5].

4.3.   Retrospective Sampling

In some cases, it is possible to study the past trends in atmospheric deposition of metals by proper selection of media where the metals are being stored after their arrival. The most promising candidates for such retrospective studies of past metal deposition appear to be snow and ice cores from glaciers, lake sediments, and peat cores from ombrotrophic bogs. In all cases, dating of individual layers is a prerequisite for these studies, e.g., by ^{210}Pb, ^{14}C, anthropogenic radionuclides from known events, or tephra from volcanic eruptions [34]. One important result of these retrospective studies is the elucidation of the deposition history of Pb in

Europe. It has been generally believed that the air pollution problem associated with lead emissions is mainly a result of leaded petrol, but recent studies [35,36] have shown that the air pollution with Pb was substantial long before the advent of leaded petrol.

The use of snow and ice cores for this purpose requires that no re-distribution of deposited material by melting occurs. For this reason, the use of snow and ice cores is restricted to areas where the temperature is below 0°C around the year, i.e., in the polar regions and occasionally on high mountains else-where. Ice coring has been successfully applied to study temporal trends of metal deposition in Greenland [37] and Antarctica.

Lake sediments are generally more available geographically, but there are limitations since atmospheric deposition is not the only source of metals to lake sediments, and sometimes surface sediment mixing may be a problem [36].

In ombrotrophic bogs, the surface peat layer is elevated above the influence of the groundwater and hence is supplied by chemical substances only from the atmosphere. This approach has in some cases been successfully used to study atmospheric deposition trends of metals over the last several thousand years [38]. A basic condition for this method is that the metals do not move vertically in the peat profile. Ombrotrophic bogs occur mainly in the boreal zone, but may also be found at high-altitude sites closer to the equator.

4.4. Analytical Techniques

Neutron activation analysis (NAA), X-ray fluorescence (XRF) and particle-induced X-ray emission analysis (PIXE) allow non-destructive, multi-element analysis of solid samples, and are therefore well established techniques in the analysis of atmospheric aerosols [39]. Up to 30–40 elements may be determined by instrumental neutron activation analysis (INAA), compared to only 15–20 by XRF or PIXE. The latter two techniques offer high speed, low cost (XRF), and require small samples (PIXE), so that they have often been preferred in programs where large sample loads are encountered [39].

Atomic absorption spectrometry (AAS) is perhaps the most widely used technique for atmospheric samples [39]. AAS is a single element technique where the analyte element has to be brought into solution before determination. The technique is, however, inexpensive, simple and has reasonably high sensi-tivity of several elements of environmental interest, particularly when changing from ordinary flame (flame atomic absorption spectrometry, FAAS) to electro-thermal atomization AAS (graphite furnace atomic absorption spectrometry, GF-AAS). For these reasons, it is also often used to complement INAA or XRF analyses.

Inductively coupled plasma atomic emission spectrometry (ICP-AES) has in common with AAS that it is necessary to have the analyte elements in solution. ICP-AES is a suitable technique for many elements, and the sensitivity is between FAAS and GF-AAS [39].

Inductively coupled plasma-mass spectrometry (ICP-MS) is a powerful analytical technique for multi-element investigations [5], particularly those involving a large number of analyses such as environmental studies. Since the first commercial ICP-MS instrument became available in 1983, the technique has been described in a number of publications and textbooks [40]. The most outstanding features that make ICP-MS an attractive technique for environmental trace element studies are:

- Good multi-element capability for determining a large number of elemental concentrations over the range of μg/mL to sub ng/mL.
- High sensitivity for ultra trace element determinations with detection limits in aqueous solution generally below 0.1 ng/mL (Fig. 1). Such low detection limits are necessary for determining most trace elements in, e.g., precipitation samples.
- Rapid analysis with typical sample throughput times of 2–5 min: This is important in environmental studies where large sample numbers have to be processed.
- Low detection limits for many elements not readily determined by other analytical techniques (Li, Be, As, Sb, Bi, rare-earth elements, noble metals).
- Several available sample introduction systems including conventional pneumatic, ultrasonic, high-solids nebulizers, and low sample consumption nebulizers as well as other injection techniques such as flow-injection, electrothermal volatilization, and laser-ablation.
- The ability to determine isotope ratios provides the opportunity to undertake tracer and speciation studies.

However, the full potential of ICP-MS cannot be exploited by conventional quadrupole-based instrumentation because of spectral overlap. Background species generated by the ICP ion source produce isobaric or inter-element spectral overlaps, arising primarily from molecular species formed by a combination of sample, plasma gas and matrix, or solvent constituents [40]. A very effective approach from the improvement of ICP-MS analysis is the pairing of a high-resolution sector field mass analyzer with ICP ion source. This combination allows interpretation of spectral interferences in elemental analysis and also permits the operation of the plasma in its normal "hot" mode so that levels of oxides are not preferentially raised as in the cold plasma [40]. The high-resolution ICP-MS technique is able to determine more metals at lower detection limits and has therefore a growing popularity among those who are measuring heavy metals [40].

The most reliable technique for the analysis of mercury is atomic fluorescence spectrometry (AFS). The analysis of mercury in air samples is generally made using double amalgamation cold vapor atomic fluorescence spectroscopy (CVAFS). In this procedure, the gold trap is mounted in series with a second analytical trap in a gas stream (Hg-free argon) leading to the CVAFS detector.

The most common procedure for the determination of mercury in precipitation is by reduction of the aqueous Hg to Hg^0, purging onto gold traps and thermal desorption and analysis using CVAFS.

5. CONCLUDING REMARKS

The atmosphere has proved to be an important pathway for the re-distribution of metals on the Earth's surface, both by natural processes and by emissions from anthropogenic sources. In most cases, a major part of the metals will be deposited within the region where the emission took place. In some cases, however, the terrestrial surface can be significantly contaminated from metals emitted by sources located as much as 1000 km away, and this may lead to appreciable uptake in food chains [32]. In the case of Hg where most of the emissions are as gaseous Hg^0, a large part of the emissions are being distributed over the entire hemisphere. Thus, metal emissions from anthropogenic sources may have impacts very far away from where they occurred. This makes metal emissions anywhere on the Earth a potential global concern.

In recent years, the metal emissions from industries in Europe and North America have been substantially reduced. At the same time, however, there has been a strong industrial growth in many countries of the Third World, in particular in Southeast Asia. In 1995, about 50% of the global trace metal emissions from anthropogenic sources came from Asia [20], and with the current development this fraction may become even greater. Transfer of technology to these countries for cleaner production of energy and industrial goods appears to be an urgent matter in order to avoid large-scale damage associated with the emission of trace metals and other air pollutants.

ABBREVIATIONS

AAS	atomic absorption spectrometry
AD	aerodynamic diameter
AFS	atomic fluorescence spectrometry
CVAFS	cold vapor atomic fluorescence spectroscopy
EF	enrichment factor
EMEP	European Monitoring and Assessment Programme
FAAS	flame atomic absorption spectrometry
GF-AAS	graphite furnace atomic absorption spectrometry
ICP-AES	inductively coupled plasma-mass atomic emission spectrometry
ICP-MS	inductively coupled plasma-mass spectrometry
INAA	instrumental neutron activation analysis
NAA	neutron activation analysis
PCA	principal component analysis
PFA	tetrafluoroethylene perfluoroalkoxy vinyl ether copolymer resins
PIXE	particle-induced X-ray emission analysis

PM   particulate matter
RGM   reactive gaseous mercury
TSP   total suspended particulate matter
XRF   X-ray fluorescence

REFERENCES

1. Schroeder WH, Munthe J. Atmos Environ 1998; 32:809–822.
2. Murozumi M, Chow TJ, Patterson CC. Geochim Cosmochim Acta 1969; 33:
   1247–1294.
3. Berg T, Røyset O, Steinnes E. Atmos Environ 1994; 28:3519–3536.
4. Steinnes E, Rambæk JP, Hanssen JE. Chemosphere 1992; 25:735–752.
5. Berg T. Atmospheric Trace Element Deposition in Norway Studied by ICP-MS.
   Dr.scient thesis, Department of Chemistry, University of Trondheim, AVH,
   Norway, 1993.
6. Gusev A, Ilyin I, Petersen G, van Pul A. Long-range Transport Model Intercomparison Studies. Model Intercomparison Study for Cadmium, EMEP Meteorological
   Synthesizing Centre-East, Moscow, Russia. Technical Report 2/2001.
7. Nriagu JO. Nature 1989; 338:47–49.
8. Nriagu JO, Pacyna JM. Nature 1988; 333:134–139.
9. Merian E, ed. Metals and Their Compounds in the Environment. Occurrence Analysis
   and Biological Relevance. Weinheim: VCH, 1991:1–1438.
10. Coles DG, Ragaini RC, Ondov JM, Fisher GL, Silberman D, Prentice B. Environ Sci
    Technol 1979; 13:455–459.
11. Pacyna JM, Semb A, Hanssen JE. Tellus 1984; 36B:163–178.
12. Wayne RP. Chemistry of Atmospheres. 3rd ed. Padstow, Cornwall: Oxford University
    Press, 2000.
13. Seinfelt JH, Pandis SN. Atmospheric Chemistry and Physics. From Air Pollution to
    Climate Change. New York: John Wiley & Sons. Inc., 1998:1–1324.
14. Puxbaum H. In: Merian E, ed. Metals and Their Compounds in the Environment.
    Occurrence, Analysis and Biological Relevance. Weinheim, Germany: VCH, 1991:
    257–286.
15. Galloway JN, Thornton JD, Norton SA, Volchoc HL, McLean RAN. Atmos Environ
    1982; 16:1677–1700.
16. Pacyna JM. In: Pacyna JM, Ottar B, eds. Control and Fate of Atmospheric Trace
    Metals. NATO ASI Series 268. Dordrecht, Holland: Kluwer, 1989:15–31.
17. Tørseth K. The Atmospheric Chemistry and Deposition of Major Inorganic Com-
    pounds in Rural Areas of Norway. Dr.scient thesis, Faculty of Chemistry and
    Biology, Norwegian University of Science and Technology, Trondheim, Norway,
    1999.
18. Cawse PA. A Survey of Atmospheric Trace Elements in the U.K. (1972–73). AERE
    Harwell, England. AERE Report R 7669, 1974.
19. Aas W, Hjellbrekke AG. Heavy Metals and POP measurements, 2001. Norwegian
    Institute for Air Research, Kjeller, Norway. NILU EMEP/CCC-Report 1/2003.
20. Pacyna JM, Pacyna EG. Environ Rev 2001; 9:269–298.
21. Pacyna JM. Adv Environ Sci Technol 1986; 17:33–52.
22. Zoller WH, Gladney ES, Duce RA. Science 1974; 183:198–200.

23. Mason B. Principles of Geochemistry. New York: John Wiley, 1966:1–329.
24. Hopke PK. Atmos Environ 1988; 22:1777–1792.
25. Berg T, Røyset O, Steinnes E. Environ Monit Assess 1994; 31:259–273.
26. Ryaboshapko A, Ilyin I, Bullock R, Ebinghaus R, Lohman K, Munthe J, Petersen G, Segneur C, Wangberg I. Intercomparison Study of Numerical Models for Long-Range Atmospheric Transport of Mercury. Stage I. Comparison of Chemical Modules for Mercury Transformations in a Cloud/Fog Environment, EMEP Meteorological Synthesizing Centre-East, Moscow, Russia. Technical Report 2/2001.
27. Galperin MV, Erdman LK, Subbotin SR, Sofiev MA, Afinogenova OG. Modelling Experience of the Arctic Pollution with Sulphur and Nitrogen Compounds, Heavy Metals from Sources of the Northern Hemisphere, EMEP Meteorological Synthesizing Centre-East, Moscow, Russia. Technical Report 2/1994.
28. EMEP Co-operative Programme for Monitoring and Evaluation of the Long-range Transmission of Air Pollutants in Europe. EMEP Manual for Sampling and Chemical Analysis. Norwegian Institute for Air Research, Kjeller, Norway. EMEP/CCC-Report 1/95.
29. Ebinghaus R, Jennings SG, Schroeder WH, Berg T, Donaghy T, Guentzel J, Kenny C, Kock HH, Kvietkus K, Landing W, Munthe J, Prestbo EM, Schneeberger D, Slemr F, Sommar J, Urba A, Wallschläger D, Xiao Z. Atmos Environ 1999; 33:3063–3073.
30. Munthe J. Guidelines for the Sampling and Analysis of Mercury in Air and Precipitation. Swedish Environmental Research Institute, Gothenburg, Sweden. IVL-report L 96/204, 1996.
31. Rühling Å, Tyler G. Bot Notiser 1968; 121:321–342.
32. Steinnes E. Water Air Soil Pollut: Foc 2001; 1:449–460.
33. Steinnes E, Berg T, Sjøbakk TE. J Phys IV France 2003; 107:1271–1273.
34. Langdon PG, Barber KE. Holocene 2004; 14:21–33.
35. Dunlap CE, Steinnes E, Flegal AR. Earth Planet Sci Lett 1999; 167:81–88.
36. Eades LJ, Farmer JG, MacKenzie AB, Kirika A, Bailey-Watts AE. Sci Total Environ 2002; 292:55–67.
37. Boutron C, Candelone JP, Hong SM. Sci Total Environ 1995; 161:233–241.
38. Shotyk W, Weiss D, Appleby P. Science 1998; 281:1635–1640.
39. Maenhaut W. In: Pacyna JM, Ottar B, eds. Control and Fate of Atmospheric Trace Metals. NATO ASI Series 268. Dordrecht, Holland: Kluwer, 1989:259–302.
40. Montaser A. Inductively Coupled Plasma Mass Spectrometry. New York: Wiley-VCH Inc., 1998.

2

The Marine Biogeochemistry of Iron

Alison Butler and Jessica D. Martin

*Department of Chemistry and Biochemistry,
University of California, Santa Barbara,
California 93106-9510, USA*

1. BACKGROUND

1.1. The Iron Hypothesis

A third to a half of the world's fixation of carbon dioxide is estimated to occur in the oceans as a result of photosynthetic activity by phytoplankton [1]. Specifically, the majority of this marine carbon dioxide fixation has been shown to occur in coastal environments. In vast regions of the world's oceans, however, chlorophyll levels from photosynthetic microorganisms are unusually low, leading to low levels of primary production, despite the fact that these waters are replete in major nutrients like nitrate, phosphate, and silicate. These high-nitrate-low chlorophyll (HNLC) regions also coincide with very low iron levels, ranging from 20 pM to 1 nM in surface seawater [2–10].

The recognition that the HNLC regions correlated with low iron concentration led to the development of the "Iron Hypothesis" by Martin [4,11], which states that primary productivity in large areas of the world's oceans is limited by low iron concentrations. The assertion is that photosynthetic microorganisms are unable to utilize the available nitrate, phosphate, and silicate due to a lack of iron, and that addition of iron to the HNLC regions would result in an increase in growth of photosynthetic microorganisms and primary productivity. This increase in growth could efficiently consume carbon dioxide from the atmosphere, leading to removal of carbon dioxide to the deep oceans as the remains of the phytoplankton sink away from the ocean surface. Consequently, as suggested by the iron hypothesis, atmospheric carbon dioxide levels could be significantly decreased if iron addition to a large area of the ocean could stimulate sufficient primary production and if this carbon could be exported to the deep oceans.

The region of high primary productivity to the west of the Galapagos Islands bolstered the iron hypothesis. The conjecture was that an influx of iron from volcanic sources such as rock or ash near the Galapagos Islands carried by the prevailing winds and currents was fertilizing these waters. The hypothesis that increasing iron concentrations could allow photosynthetic microorganisms to proliferate was tantalizing and quickly led to the early *in vitro* supplementation studies.

The iron hypothesis has now been tested at least eight times on large ($\sim$70–100 km^2) patches of ocean water in the equatorial Pacific, the eastern subarctic Pacific, and the Southern Oceans. The first open-ocean iron enrichment experiment (IronEx I) was completed in 1993 in the equatorial Pacific Ocean [6,12–15]. In this experiment, a single 4 nM iron enrichment was made to a patch of open surface water, resulting in a threefold increase in chlorophyll. Within 4 days, the added iron was no longer detectable. The magnitude and longevity of the biological and geochemical responses were much smaller than predicted, most likely due to the rapid loss of iron from the system [6,16].

A second series of experiments (IronEx II, 1996) tested the idea that iron loss decreased efficacy in IronEx I and was completed in the equatorial Pacific

Ocean, near the Galapagos Islands [17–26]. Multiple additions of iron over several days stimulated massive phytoplankton blooms reaching a maximum 2 days after the last addition of iron, resulting in a 90 μatm drawdown of CO_2 and a 5 μM drawdown of nitrate. The iron(III) present in surface ocean waters has been shown to be essentially fully complexed by an organic ligand or class of ligands, "L" [25,27–30]. An intriguing result of IronEx II was that the concentration of the organic ligand, "L", increased in a short time-span, meeting the concentration of added iron; thus, it has been proposed that "L" is biologically derived [25]. The results from IronEx II unequivocally support the iron hypothesis for the limitation of primary productivity by iron concentration.

In addition to IronEx I and IronEx II, mesoscale additions of ferrous sulfate have been made in the subarctic Pacific Ocean, during the Subarctic Pacific Iron Experiment for Ecosystems Dynamics Study (SEEDS, 2001) [31] and the Subarctic Ecosystem Response to Iron Enrichment Study (SERIES, 2002) [32]. SEEDS resulted in a 40-fold increase of chlorophyll and a 13 μM drawdown of nitrate; SERIES resulted in a greater than 10-fold increase in chlorophyll and >5 μM drawdown of nitrate.

Three large-scale experiments in the Southern Ocean have been completed and a fourth is in progress. These include the Southern Ocean Iron Release Experiment (SOIREE, 1999) [33–53], the Southern Ocean Iron Experiment (SOFeX) [54–56], an experiment in the Atlantic sector of the Southern Ocean called EisenEx (named for the German word for iron, Eisen; 2000) [57,58], and the European Iron Fertilization Experiment (Eifex, 2004) which took place in early 2004; the results from Eifex have not yet been reported. The completed experiments have all demonstrated significant increases in primary productivity (e.g., 40-fold increase in chlorophyll) and biomass of phytoplankton in response to iron addition. Corresponding reductions in atmospheric carbon dioxide and nitrate concentrations also occurred. The bloom stimulated by SOIREE lasted for over 50 days. The unexpected persistence of this bloom led to the proposal that phytoplankton release compounds to keep iron in a useable form, although the nature of this compound or compounds is not yet known. Results of SOFeX, which were only recently reported [56], demonstrated that carbon could be exported to the deep oceans, below 100 m, however, instead of the predicted sequestration of 100,000 tonnes carbon per 1000 tonnes Fe added, only 1000 tonnes of carbon was exported [56]. Thus, the results of these studies support the iron hypothesis and also provide further evidence for the presence of biologically derived iron binding ligands which play an integral role in iron cycling in the oceans.

In addition to phytoplankton such as diatoms and cyanobacteria, heterotrophic bacteria constitute an important class of microorganisms in the ocean that are also limited by the low iron levels in HNLC regions. Heterotrophic bacteria comprise up to half the total particulate organic carbon in ocean waters [59] and in some regions, such as the subarctic Pacific, heterotrophic bacteria can even contain higher cellular concentrations of iron than phytoplankton [60]. During iron supplementation-induced phytoplankton blooms, heterotrophic bacteria also increase in numbers. Thus, heterotrophic bacteria compete successfully for iron against phytoplankton and cyanobacteria.

1.2. Mechanism of Microbial Iron Acquisition

Iron is required for growth of the vast majority of all bacteria [61,62]. While iron is the fourth most abundant element by weight in the Earth's surface, the insolubility of iron at the neutral pH conditions in which most bacteria grow severely limits the availability of this essential nutrient [K_{SP} of Fe(OH)$_3$ = 10^{-39}]. To overcome this apparent lack of iron, microorganisms have evolved an elaborate mechanism to acquire iron. Under aerobic growth conditions, bacteria and other microorganisms produce siderophores which are low molecular weight, ferric-ion-specific compounds for the solubilization and sequestration of iron(III). Siderophores are generally produced under conditions of low iron availability and secreted into the surrounding environment where they can complex ferric ion. Iron-bound siderophores are returned to the cell through an active transport system, based on siderophore-specific outer membrane receptor proteins. The synthesis of both the siderophores and the outer membrane receptor proteins is tightly regulated by iron concentrations within the cell. Once the iron concentration inside the cell is sufficient, the biosynthesis of siderophores and receptor proteins is repressed.

1.2.1. Siderophores

The structures of hundreds of siderophores produced by terrestrial and freshwater bacteria are known. Two common iron(III)-binding motifs found among the known structures are catechols and hydroxamic acids but other iron-binding functional groups are also present. Several species of *Enterobacteriaceae* such as *E. coli* produce enterobactin (Fig. 1), an example of a *tris*-catecholate-containing siderophore [63]. Desferrioxamine B (Fig. 1), produced by *Streptomyces* species, is an example of a *tris*-hydroxamate siderophore [64]. While these two examples each only contain a single type of iron-binding group, many siderophores coordinate iron(III) with a combination of ligand types.

Hexadentate siderophores such as enterobactin and desferrioxamine B generally bind iron with one-to-one stoichiometry. Enterobactin is the cyclic triester of 2,3-dihydroxy-*N*-benzoyl-*S*-serine and forms an octahedral complex with ferric ion [65,66]. While many hexadentate siderophores are known, others are only tetradentate, such as rhodotorulic acid [67], alcaligin [68], putrebactin [69], and bisucaberin [70] (Fig. 2). Tetradentate siderophores often form Fe$_2$(siderophore)$_3$ complexes to achieve octahedral coordination of iron(III). The crystal structure of the Fe$_2$(alcaligin)$_3$ complex revealed that only one alcaligin acts to bridge the iron(III) ions as opposed to a triple-bridged helicate structure [71].

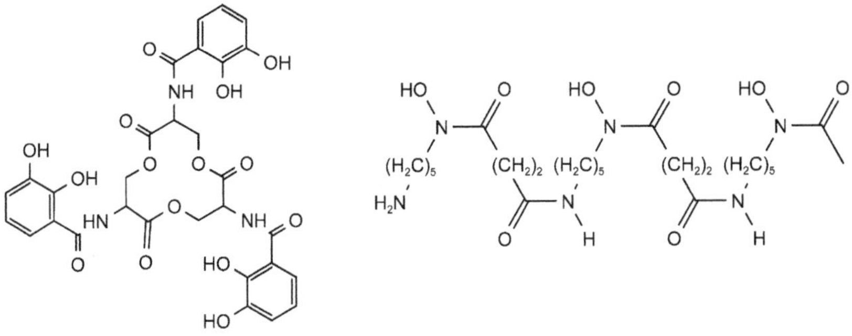

Enterobactin   Desferrioxamine B

Figure 1 Structures of enterobactin and desferrioxamine B, examples of catecholate and hydroxamate siderophores, respectively.

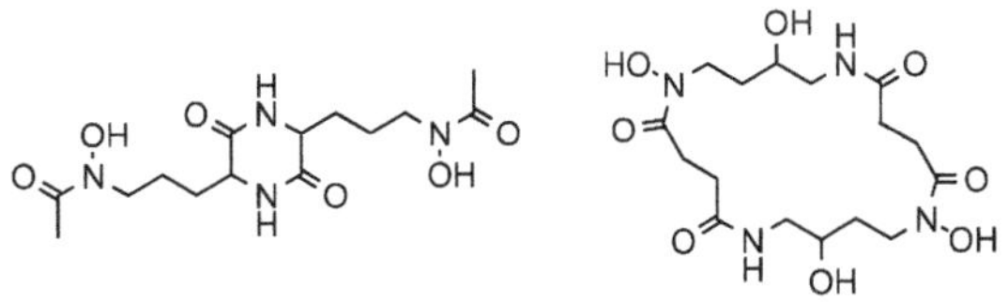

Rhodotorulic Acid   Alcaligin

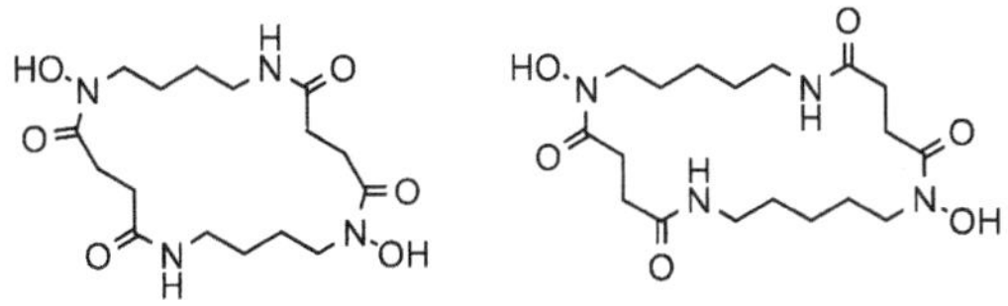

Putrebactin   Bisucaberin

Figure 2 Structures of examples of tetradentate siderophores: rhodotorulic acid, alcaligin, putrebactin, and bisucaberin.

Siderophores complex ferric ion with particularly high affinity constants. The stability constant is often reported as the complex formation constant, β_{FeLH} between hydrated ferric ion, Fe_{aq}^{3+}, and the fully deprotonated siderophore ligand (Sid^{n-}):

$$Fe_{aq}^{3+} + Sid^{n-} \xrightleftharpoons{\beta_{110}^{\text{siderophore}}} FeSid^{(3-n)}$$

$$\beta_{110}^{\text{siderophore}} = \frac{[FeSid^{(3-n)}]}{[Fe_{aq}^{3+}][Sid^{n-}]}$$

The values of the formation constants span a wide range, from $10^{22.5}$ for ferric aerobactin to 10^{49} for the ferric enterobactin complex. Determination of these proton-independent formation constants requires knowledge of the pK_a value of each coordinating group. Because little fully deprotonated siderophore or hexaqua iron(III) species would be present in solution at physiological pH, as well as to avoid problems associated with hydrolysis of ferric ion, the stability constant is often determined by competing the siderophore against another ligand with well defined thermodynamics of interaction with ferric ion such as ethylenediaminetetraacetate (EDTA):

$$\text{Fe(EDTA)}^{1-} + \text{Sid}^{n-} \; \underset{}{\overset{K_{overall}}{\rightleftharpoons}} \; \text{FeSid}^{(3-n)} + \text{EDTA}^{4-}$$

$$K_{overall} = \frac{[\text{FeSid}^{(3-n)}][\text{EDTA}^{4-}]}{[\text{FeEDTA}^-][\text{Sid}^{n-}]}$$

$$= \left(\frac{[\text{FeSid}^{(3-n)}]}{[\text{Fe}_{aq}^{3+}][\text{Sid}^{n-}]} \right) \left(\frac{[\text{EDTA}^{4-}][\text{Fe}_{aq}^{3+}]}{[\text{FeEDTA}^-]} \right) = \frac{\beta_{110}^{siderophore}}{\beta_{110}^{EDTA}}$$

Note that β values are comparable only for ligands of the same denticity.

1.2.2. Outer Membrane Receptor Proteins

Bacteria have evolved a class of high affinity outer membrane receptor proteins which recognize specific iron(III)–siderophore complexes and are involved in energy-dependent active transport of the ferric siderophore complex across the outer membrane. Three receptors from *E. coli* whose crystal structures have recently been determined are FhuA (i.e., ferric hydroxamate uptake; [72]) which recognizes ferric ferrichrome siderophores, FepA (i.e., ferric enterobactin permease; [73]) which recognizes ferric enterobactin, and FecA (i.e., ferric citrate protein; [74]) which recognizes ferric citrate (Fig. 3).

The structural core for these receptors is a membrane-spanning, 22-strand, anti-parallel β-barrel similar in overall structure to other membrane porins, although each of the siderophore receptors have a segment of residues at the N-terminus which folds inside the β-barrel, effectively corking the barrel from the periplasmic side of the outer membrane. When the ferric siderophore complex binds to the receptor, FhuA, FepA, and FecA undergo conformational changes, which through interactions with the TonB (i.e., transport of iron) protein lead to the transport and release of the ferric siderophore complex into the periplasmic space.

TonB is a cytoplasmic membrane protein which spans the periplasmic space binding the siderophore receptor proteins and also binding the cytoplasmic membrane proteins, ExbB and ExbD (Fig. 4) [75]. The TonB/ExbB/ExbD ternary complex mediates the signal transduction of the electrochemical potential of the cytoplasmic membrane to the outer membrane, allowing transport and release of the ferric siderophore complex. Once the ferric siderophore is released

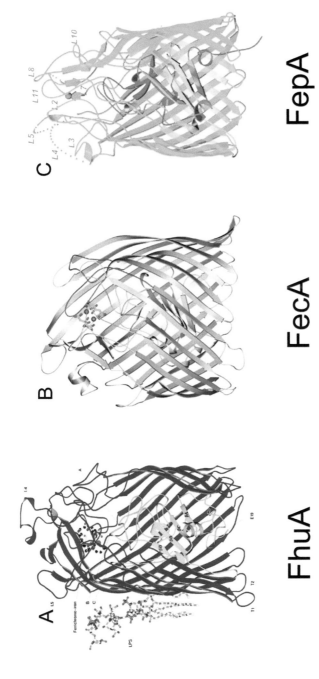

Figure 3 Crystal structures of outer membrane receptor proteins FhuA, FecA, and FepA. Reproduced with permission from Refs. [72–74], respectively.

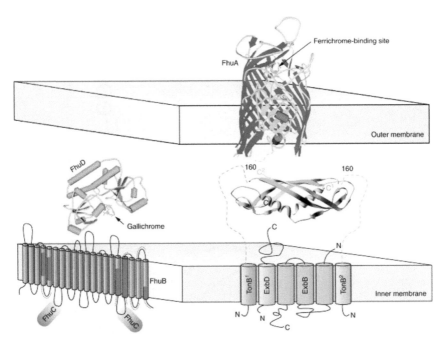

Figure 4 Diagram of the proteins involved in ferrichrome transport, i.e., the TonB complex. Reproduced with permission from Ref. [75].

to the periplasmic space, it is bound by a high affinity periplasmic binding protein (e.g., FhuD K_d 0.1 μM for the ferric ferrichrome siderophore complex), preventing the reverse transport of the iron-siderophore complex across the outer membrane.

The binding and transport of a ferric siderophore complex is usually highly specific. The kinetics of ferric siderophore uptake show saturation behavior with an apparent K_d value in the range of 0.1–100 nM. For example, FepA binds $Fe^{(III)}(enterobactin)^{3-}$ with a K_d of <0.1 nM. Other complexes, such as $Rh^{(III)}(catecholate)_3^{3-}$ (a kinetically inert complex) do not affect the rate of uptake of $Fe^{(III)}(enterobactin)^{3-}$, however, $Rh^{(III)}(N,N\text{-dimethyl-2,3-dihydroxybenzamide})_3$ does block uptake of $Fe^{(III)}(enterobactin)^{3-}$ at ~100 μM. Thus, while the trilactone backbone of enterobactin is not required for recognition and uptake, the catecholamide moiety is an essential feature of enterobactin for iron uptake to occur.

Several types of systems exist for the transport of iron across the cytoplasmic membrane. These include the translocation of ionic ferric ion, ferrous iron, and the ferric siderophore complexes. Transport of the ferric-ion and ferric-siderophore complexes typically requires use of an ATP binding cassette (ABC) transporter system in an energy-dependent process.

2. MARINE BACTERIAL SIDEROPHORES

2.1. Structures

Many marine cyanobacteria and heterotrophic bacteria have been shown to produce siderophores in culture in response to iron stress [76–79]. Those marine siderophores that have been structurally characterized are shown in Fig. 5. Alterobactin A is a cyclic depsipeptide produced by *Pseudoalteromonas leuteoviolacea* [80]. In the absence of a bound metal ion [e.g., Fe(III), Ga(III)], alterobactin A hydrolyzes to the linear form, alterobactin B. Pseudoalterobactins A and B are produced by *Pseudoalteromonas* sp. KP20-4 and are chemically related to the alterobactins [81]. Petrobactin is a citrate-derived marine sidero-phore produced by the hydrocarbon degrading bacterium *Marinobacter hydrocarbonoclasticus* [82,83]. Aerobactin and desferrioxamine G are known terrestrial siderophores that have also been isolated from marine bacteria (*Vibrio* sp. DS40M5 and *Vibrio* sp. BLI-41, respectively [84,85]).

The aquachelins and marinobactins are two distinct families of amphiphilic peptidic siderophores (Fig. 6) produced by two different genera of bacteria [86]. The aquachelins and marinobactins are distinguished by a unique peptidic head group that coordinates iron(III) and the presence of one of a series of different fatty acid appendages at the amine terminus [86]. The amphibactins (Fig. 6) form another distinct family of amphiphilic siderophores, characterized by a shorter peptide head group of only four amino acids and a series of at least 10 different fatty acid tails [87].

The characteristic features of these marine siderophores are (i) the presence of the α-hydroxy acid group in the alterobactins, aerobactin, petrobactin, the aquachelins, and the marinobactins, all of which were isolated from oceanic bacteria and (ii) the presence of suites of amphiphilic siderophores that vary only by the fatty acid chain length.

2.2. Photoreactivity

The presence of the α-hydroxy acid is of direct importance to this review because Fe(III) coordinated to α-hydroxy acids undergoes photoinduced reduction of Fe(III) and oxidation of the ligand. The prime example of this reaction is the actinometer, ferrioxalate [i.e., *tris*-oxalato-iron(III)].

Photolysis of the Fe(III)−aquachelin complexes in natural sunlight results in ligand oxidation and truncation as well as reduction of Fe(III) to Fe(II), as shown in Fig. 7 [88]. The same product with an m/z of 780 is formed whether each of the aquachelins is photolyzed separately or as the physiologically produced mixture of aquachelins A–D. The only amino acid lost in the photo-reaction is β-hydroxy-aspartate. The peptide photoproduct retains the two hydroxamate groups and is capable of binding iron(III), as evinced by detection of an iron adduct (m/z 833) in ESI-MS [88].

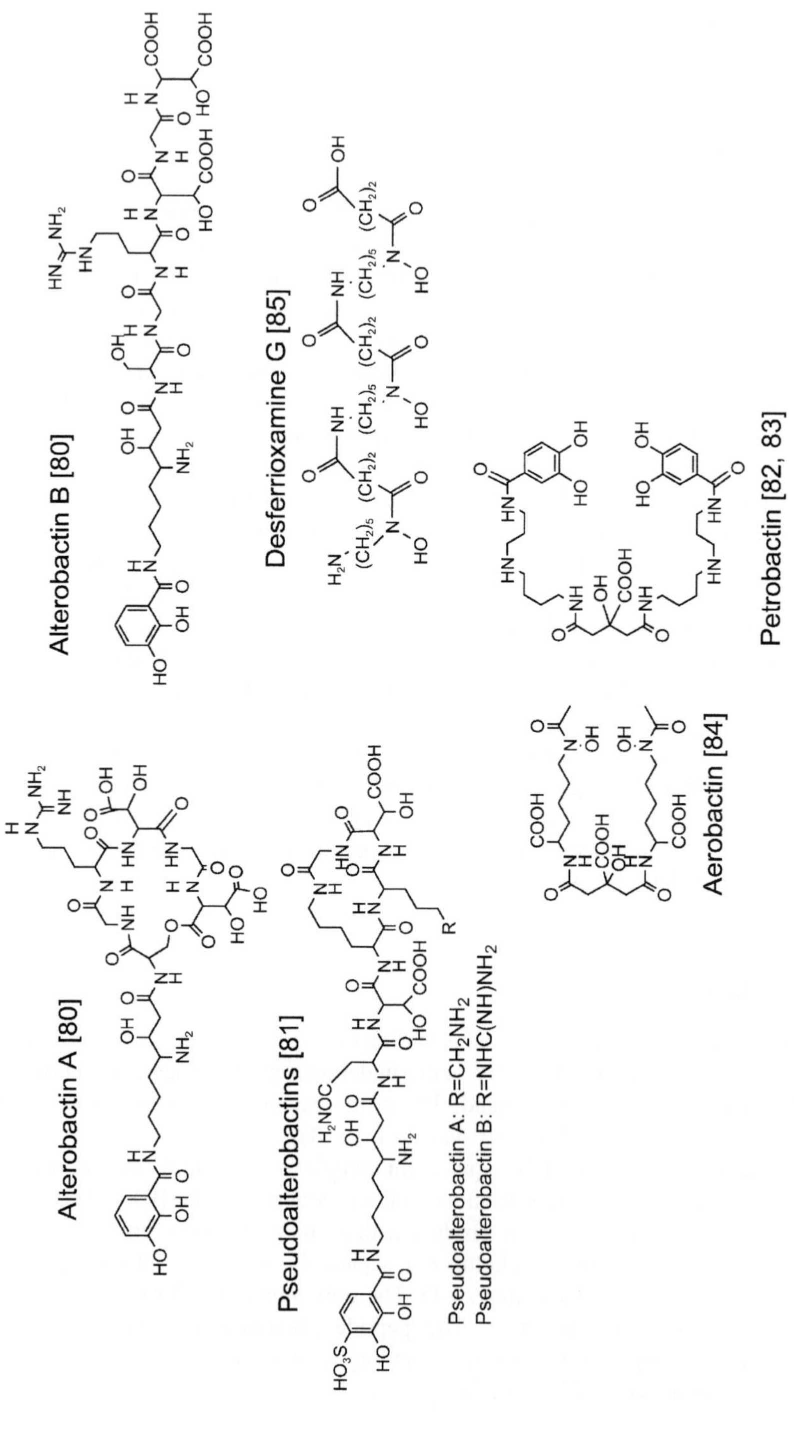

Figure 5   Structures of known marine siderophores.

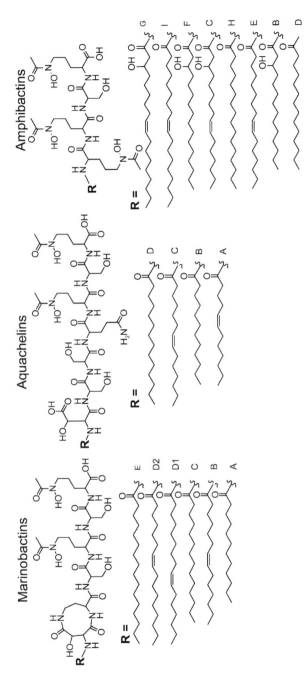

Figure 6 Structures of amphiphilic marine siderophores.

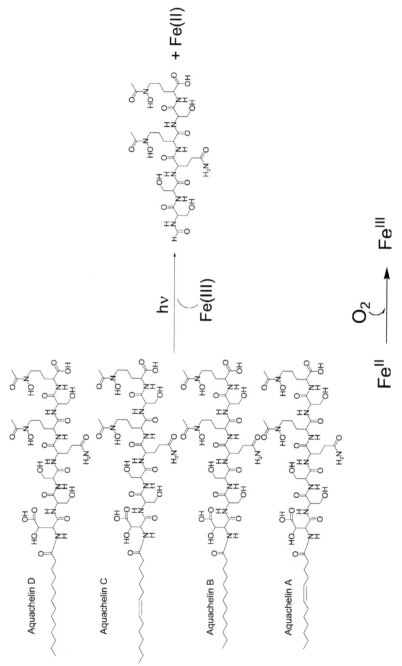

Figure 7 Photolysis of the Fe(III)–aquachelin siderophores. Under conditions where oxygen is present, Fe(II) will be rapidly oxidized to Fe(III) and will be available for chelation by the aquachelin photoproduct or native aquachelin siderophore.

The conditional stability constant ($K_{\text{FeL,Fe'}}^{\text{cond}}$) of the Fe(III)–aquachelin peptide photoproduct was determined by competitive ligand equilibration-adsorptive cathodic stripping voltammetry (CLE-ACSV) which demonstrated a significant drop in Fe(III)-binding strength relative to the native aquachelin side-rophore: i.e., $K_{\text{FeL,Fe'}}^{\text{cond}}$ of $10^{12.2}$ M^{-1} for aquachelin B vs. $K_{\text{FeL,Fe'}}^{\text{cond}}$ of $10^{11.5}$ M^{-1} for the peptide photoproduct [88]. This decrease in Fe(III)-binding strength is consistent with partial loss of the ligating ability of the siderophore ligand, due to loss of the β-hydroxy-aspartate moiety as determined by structural analysis of the peptide photoproduct. The reduced conditional stability constant of the photoproduct compared to the native siderophores is tantalizing given field observations of a strong Fe(III)-binding ligand L_1 in seawater (average $K_{\text{FeL,Fe'}}^{\text{cond}}$ $10^{12.5 \pm 0.3}$ M^{-1}) in which L_1 is present at lower concentration than the weaker Fe(III)-binding ligand L_2 (average $K_{\text{FeL,Fe'}}^{\text{cond}}$ $10^{11.6 \pm 0.2}$ M^{-1}) which is present at greater concentration [89].

The photoreactivity of the Fe(III)–aquachelin complexes is initiated by a β-hydroxy-aspartate-to-iron(III) ligand-to-metal charge transfer reaction [88]. The UV-visible spectrum of photolyzed Fe(III)–aquachelin C shows loss of an electronic transition in the near UV (300 nm) which likely corresponds to the charge transfer from β-hydroxy-aspartate to iron(III). The photoreactivity of iron(III) complexes of simple polycarboxylate molecules containing α-hydroxy acids (e.g., citrate, oxalate) is well known, leading to ligand oxidation and iron(III) reduction [90,91].

The availability of Fe(III) to microorganisms has also been found to be differentially affected by complexation to siderophore vs. the siderophore photoproduct in the case of the aquachelins [88]. ^{59}Fe uptake experiments with a natural assemblage of planktonic organisms from the oligotrophic Atlantic Ocean showed that photolysis increased the bioavailability of ^{59}Fe(III) added as the aquachelin B complex [88]. As expected the intact ^{59}Fe(III)–aquachelin B complex was largely unavailable to the natural assemblage. In contrast, uptake of ^{59}Fe(III) from previously photolyzed ^{59}Fe(III)–aquachelin B was indistinguishable from incubations in which inorganic ^{59}FeCl$_3$ was added [88]. Iron uptake experiments mediated by other siderophores and their photoproducts are in progress.

In addition to the photoreactivity of the Fe(III)–aquachelins, the iron(III)–marinobactins are also photoreactive, although the photoreaction has not been studied as extensively as the aquachelins.

Petrobactin is a citrate-derived marine siderophore with unique 3,4-dihydroxy catecholate appendages. The α-hydroxy acid functionality of citrate siderophores is also photochemically active. Photolysis of the Fe(III)–petrobactin complex results in photodecarboxylation and oxidation of siderophore and reduction of the iron [82] (Fig. 8).

These early results show that microbial Fe(III) chelates facilitate photochemical cycling of iron in ocean surface waters [88]. The key features of this cycle involve photolysis of Fe(III)–ligand complexes, with reduction of iron(III) to iron(II) and oxidation of the ligand.

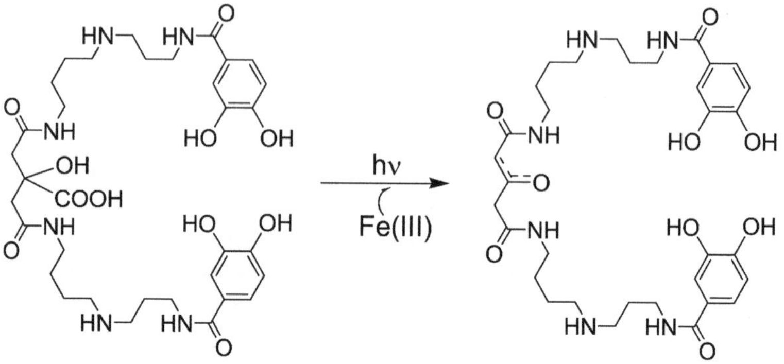

Figure 8 Photolysis of the Fe(III)–petrobactin siderophore, illustrating the photodecarboxylation reaction to produce the petrobactin photoproduct.

The fate of photolytically produced iron(II) is uncertain, but could include direct biological uptake, oxidation to iron(III), or oxidation with subsequent complexation by another ligand. The photooxidized siderophore retains the ability to coordinate Fe(III) with appreciable affinity and could therefore continue to contribute to the upper ocean iron(III) ligand pool, possibly as part of a weaker ligand class. Iron bound by the oxidized siderophore ligand may be more available for biological uptake, since the conditional stability constant of Fe(III)-photoproduct is reduced relative to that of the original ferric ligand. The potential effects of photolysis on Fe(III)–ligand complexes in recently upwelled water masses could be another important component of this cycle, operative over longer timescales [88].

2.3. Amphiphilic Characteristics

Very few of the hundreds of structurally characterized siderophores are amphiphilic. Known terrestrial or pathogenic amphiphilic siderophores include the ornibactins (Fig. 9) produced by *Burkholderia* (*Pseudomonas*) *cepacia* [92], corrugatin produced by *Pseudomonas corrugata* [93], acinetoferrin produced by *Acinetobacter haemolyticus* [94], rhizobactin 1021 produced by *Sinorhizobium meliloti* [95], and the carboxymycobactins (Fig. 9) produced by *Mycobacteria* [96]. Of these, only the ornibactins and the carboxymycobactins are produced as suites of siderophores differing by length of fatty acid appendage or by minor structural modifications.

By contrast, more than half of the structurally characterized marine siderophores are amphiphilic in nature. Moreover, these amphiphilic siderophores have been isolated from diverse marine bacteria which belong to distinct clades of bacteria spanning the α and β subgroups of the *Proteobacteria*, suggesting that amphiphilic siderophores may constitute a common iron acquisition strategy for marine bacteria [87] (Fig. 10).

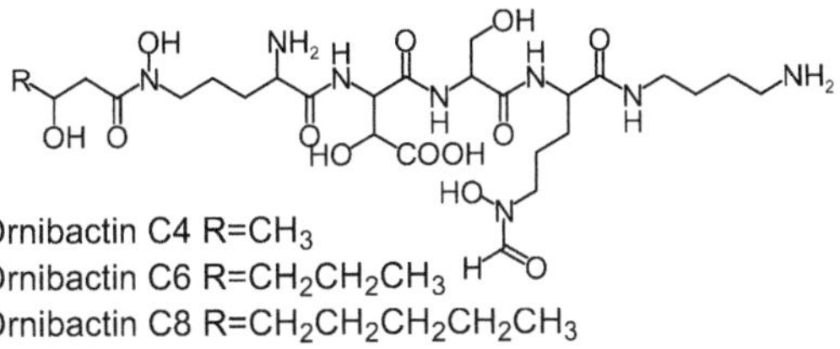

Ornibactin C4 R=CH₃
Ornibactin C6 R=CH₂CH₂CH₃
Ornibactin C8 R=CH₂CH₂CH₂CH₂CH₃

Ornibactins

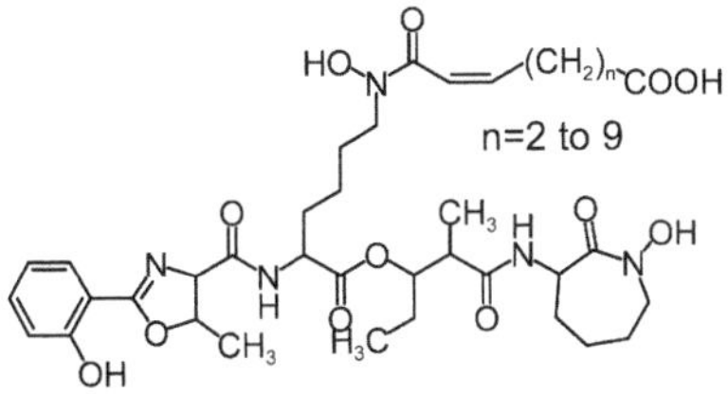

n=2 to 9

Carboxymycobactins

Figure 9 Structures of the ornibactins and carboxymycobactins, examples of terrestrial amphiphilic siderophores.

The marinobactins produced by *Marinobacter* sp. strain DS40M6 are composed of a six amino acid peptidic headgroup and one of a series of saturated or unsaturated acyl appendages ranging from C_{12} to C_{16} [86]. From the carboxy terminus, the marinobactin headgroup is composed of L-*N*-acetyl, *N*-hydroxy-ornithine, L-serine, D-*N*-acetyl, *N*-hydroxy-ornithine, D-serine, and an unusual nine membered ring formed by condensation of L-diaminobutyric acid and D-*threo*-β-hydroxy-aspartate. These siderophores, while significantly lipidic in character, are present after centrifugation of the bacterial culture in the supernatant and not associated with the cell membranes, emphasizing the truly amphiphilic character of these molecules.

The aquachelins are another suite of amphiphilic siderophores produced by the marine bacterium *Halomonas aquamarina* strain DS40M3. The aquachelins are composed of a seven amino acid headgroup and one of a series of acyl appendages including saturated and unsaturated C_{12} and C_{14} fatty acids [86]. From the carboxy terminus, the aquachelin headgroup consists of L-*N*-acetyl, *N*-hydroxy-ornithine, D-serine, D-*N*-acetyl, *N*-hydroxy-ornithine, D-glutamine, L-serine, D-serine, and L-*threo*-β-hydroxy-aspartate. Like the marinobactins, the aquachelins are found in the supernatant after centrifugation of the bacterial culture.

In contrast to the supernatant-localized siderophores from *Marinobacter* sp. strain DS40M6 and *H. aquamarina* strain DS40M3, the amphibactins, produced by marine *Vibrio* sp. strain R-10, are membrane-associated marine

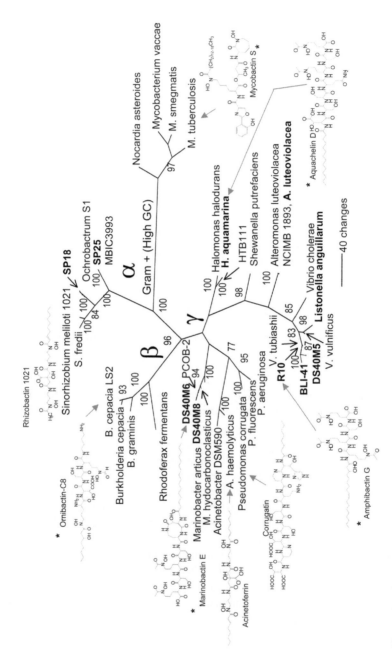

Figure 10 Phylogenetic tree demonstrating the diversity of bacterial species known to produce amphiphilic siderophores. Siderophores produced in suites varying by length of fatty acid appendage are indicated with an asterisk.

siderophores [87]. The only other known membrane bound siderophores are the mycobactins which are lipidic siderophores produced by *Mycobacteria* [96] and the structurally related nocobactins [97], formobactins [98] and amamistatins [99,100]. The amphibactins contain a peptide headgroup of only four amino acid residues and one of a series of acyl appendages ranging from C_{14} to C_{18}, including saturated, unsaturated, and hydroxylated moieties. The amphibactin headgroup is composed of *N*-acetyl, *N*-hydroxy-ornithine, L-serine, *N*-acetyl, *N*-hydroxy-ornithine, and *N*-acetyl, *N*-hydroxy-ornithine, beginning at the carboxy terminus. These siderophores contain the shortest peptide headgroup and the longest series of acyl appendages, increasing the hydrophobicity of these siderophores and contributing to the membrane association under culture conditions.

The amphiphilic character of these marine siderophores suggests their surface activity in solution. The marinobactin siderophores are known to have unusually low critical micelle concentrations (CMC) [86]. For the apo-siderophores, the CMCs range from 25 μM for marinobactin E (M_E) to 150 μM for the shorter tailed siderophores [86]. The ferri-siderophores have slightly higher CMCs. Dynamic light scattering (DLS) results and cryoelectron microscopy of Fe(III)–M_E and Fe(III)–M_D established the presence of spherical particles ranging from 140 to 180 nm, indicative of vesicles [86]. On the other hand, DLS and cryoelectron microscopy of apo-M_E suggest that the iron-free siderophore is limited to micellar formations that are small and out of the range of detection by the DLS instrument used [86]. Spontaneous self-assembly of vesicles upon addition of iron to M_E solutions represents the first example of metal-induced micellar-to-vesicular phase change (Fig. 11) of a biologically produced compound.

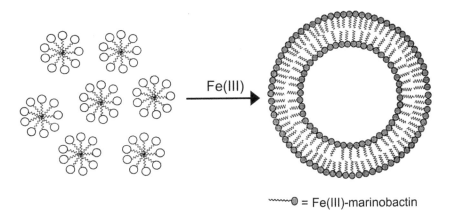

⌇⌇⌇● = Fe(III)-marinobactin

Figure 11   Illustration of the micelle to vesicle transition induced by the addition of Fe(III) to a solution of apo-marinobactin at a concentration above the CMC.

2.4. Membrane and Bacterial Partitioning

Microbial lipopeptides comprise a class of biomolecules produced by microorganisms to modify the immediate environment. Lipopeptides and related compounds are known to interact with surfaces and to modify microbial activities. Specifically, bacterial lipopeptides facilitate growth on hydrocarbons [101], influence bacterial cell surface hydrophobicity [102], induce or facilitate bacterial adhesion to surfaces promoting biofilm formation [103], alter cell membrane fluidity [104], and exhibit antimicrobial activities [104]. The similarities between the structural components of lipopeptides and the known amphiphilic marine siderophores suggest that these functional aspects may be shared. It is therefore of great interest to investigate the potential interactions between amphiphilic siderophores and bacterial membranes.

The marinobactins are the most extensively studied of the known marine amphiphilic siderophores. Siderophore-membrane interactions have been monitored with small and large unilamellar vesicles (SUVs, 40 nm, and LUVs, 200 nm, respectively) using NMR line-broadening, fluorescence quenching, and ultracentrifugation techniques [105]. Purified samples of M_E were utilized for evaluation of the membrane affinity of these compounds. Apo-M_E was found to have significant affinity for 1,2-dimyristoyl-*sn*-3-glycero-phosphocholine (DMPC) vesicles with molar partition coefficients $K_x^{\text{apo-ME}}$ of 6.3×10^5 for SUVs and 3.6×10^5 for LUVs. The iron(III) complex, Fe(III)–M_E, exhibited molar partition coefficients of $K_x^{\text{Fe-ME}}$ of 1.3×10^4 for SUVs and 9.6×10^3 for LUVs, indicating a 50-fold decrease in membrane affinity upon chelation of Fe(III). Both the apo-M_E and Fe-M_E partition coefficients are on the same order as other membrane-active detergents such as octyl glucoside, Triton X-100, and sodium cholate [106,107].

The cell surfaces of gram-negative bacteria have net negative charges due to the presence of lipopolysaccharide (LPS). The presence of LPS in the outer membrane of gram-negative bacteria could potentially influence cell-surface interactions. The similar membrane affinity of apo-M_E and Fe(III)–M_E with DMPC, a zwitterionic lipid, and 1,2-dimyristoyl-*sn*-3-glycero-phosphate, an anionic lipid, indicates that the partitioning of these lipopeptides is dominated by hydrophobic interactions [105]. Additionally, Fe–M_E was found to rapidly associate with vesicles made of LPS derived from *E. coli*, in a manner similar to the interaction observed with DMPC indicating that the significant membrane affinities demonstrated in these experiments may correlate with *in vivo* bacterial cell interactions.

Perhaps the most biologically relevant partitioning experiments for the marinobactins involved the physiological mixture of siderophores (Fig. 12). While all of the marinobactin siderophores demonstrate significant membrane affinity for DMPC LUVs, apo-M_E partitions the most extensively [105]. This result is as would be expected, since this siderophore has the longest fatty acid appendage of the marinobactins, a saturated C_{16} moiety. For each subsequent

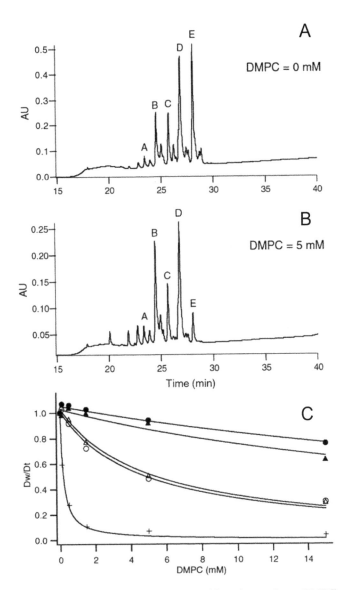

Figure 12 Partitioning of apo-marinobactin siderophores into DMPC vesicles. (A) Physiological mixture of marinobactin siderophores incubated with 0 mM DMPC. (B) Physiological mixture of marinobactin siderophores incubated with 5 mM DMPC. Note the difference in the y-axis. (C) Plot of the ratio of siderophore remaining in aqueous phase to the total siderophore concentration vs. the concentration of DMPC in each sample.

marinobactin siderophore, reduction of the fatty acid appendage by two methylene carbons or the introduction of one unit of unsaturation results in a decrease in partition coefficient of approximately one order of magnitude [105]. Within experimental error, the partition coefficient for apo-M_E determined from the physiological mixture agrees with that determined for purified samples.

3.   THOUGHTS ON BIOLOGICAL SIGNIFICANCE

3.1.   Marine Photoreactive Siderophores

The majority of the pelagic marine siderophores that have been structurally characterized contain α-hydroxy acid moieties, either in the form of citrate or β-hydroxy-aspartate that participate in iron(III) binding. When coordinated to Fe(III), these siderophores are photoreactive as observed for the Fe(III)–aquachelins [88] and Fe(III)–petrobactin [82] (see above), producing Fe(II) and an oxidized siderophore. The oxidized siderophore retains the ability to coordinate iron(III) through the remaining hydroxamate ligands, in the case of aquachelin, or the remaining catecholate ligands, in the case of petrobactin. The largest input of iron to the oceans comes in the form of windborne dust. Thus, a potentially attractive feature of the α-hydroxy acid moiety is the photoreactivity when coordinated to Fe(III). This photoinduced reduction of Fe(III) could contribute to the reductive dissolution of colloidal iron minerals, if the α-hydroxy acid coordinates to the iron oxide surface. Thus photoreactivity may be a common characteristic of oceanic Fe(III)–siderophore complexes, although the physiological significance is still under investigation.

Given the well-known photoreactivity of ferric citrate and the actinometer ferric oxalate, $Fe(Ox)_3^{3-}$ [91,108], it is surprising that the photoreactivity of Fe(III) coordinated to siderophores containing the citrate backbone was not realized earlier. Several siderophores isolated from terrestrial bacterial sources contain citrate or β-hydroxy aspartate, however, the bacteria from which these siderophores are isolated are largely enteric bacteria (e.g., γ-proteobacteria) and not likely to be exposed to the ultraviolet radiation required to affect the photooxidation of the ligand and reduction of Fe(III).

3.2.   Marine Amphiphilic Siderophores

Marine microbial production of suites of amphiphilic siderophores differing in the fatty acid appendage is intriguing. The membrane affinity, surface activity, and physical characteristics of these siderophores suggest that variations in the amphiphilic character within the suites of the marinobactins, aquachelins, and amphibactins confer important physiological functions. While the results summarized earlier are beginning to reveal some of the functional characteristics of these molecules, the physiological effect of these features has not been fully elucidated. Clearly, the composition of the outer membranes of bacteria and the environment immediately surrounding the cells is important. Small forces affect the equilibrium distribution of the siderophores, between cell-association and dissociation. Iron(III) coordination also affects the partitioning, as well as the nature of the self-assembled particles (e.g., micelle, vesicle) in the cell-released state. Multiple interactions occur between these amphiphilic molecules, the bacterial cells, and their environment which are intriguing and suggest tantalizing areas of future investigation.

4. CONCLUDING REMARKS

Structural characterization of siderophores produced by marine bacteria and investigations of the iron acquisition processes employed by these bacteria has been initiated only relatively recently. For the most part the siderophore structures that have been elucidated within this short timeframe are defined by the general characteristic features of (a) α-hydroxy acid moieties which when coordinated to Fe(III) are photoreactive and (b) suites of siderophores with varying fatty acids attached to a constant head group that coordinates Fe(III).

The next stage is to determine if marine bacteria make use of these features in a physiologically advantageous manner, as well as to ask if the evolutionary development of these structural features can be traced. Given the enormous diversity of marine bacteria [109], the discovery of new classes of siderophores and iron acquisition strategies can certainly be anticipated. Developments in molecular biology techniques for cloning the biosynthetic gene clusters for siderophores point to the possibility that novel siderophores may be obtained by recombinant means from bacteria that cannot actually be brought into culture.

Clearly, the mesoscale addition of iron to the HNLC regions of the oceans to test the iron hypothesis emphasizes the global importance of iron. To understand the global cycling of many elements, including most importantly carbon, a molecular level understanding of the ways that microorganisms sequester iron is essential.

ACKNOWLEDGMENTS

Funding from NIH GM38130 and The Center for Environmental BioInorganic Chemistry (CEBIC), an NSF Environmental Molecular Science Institute (NSF CHE-0221978) is gratefully acknowledged, as is the long-standing collaboration with Prof. Margo G. Haygood (Scripps Institution of Oceanography, University of California San Diego).

ABBREVIATIONS AND DEFINITIONS

ABC	ATP binding cassette
ATP	adenosine 5′ triphosphate

CLE-ACSV	competitive ligand equilibration-adsorptive cathodic stripping voltammetry
CMC	critical micelle concentration
DLS	dynamic light scattering
DMPC	1,2-dimyristoyl-*sn*-3-glycero-phosphocholine
EDTA	ethylenediaminetetraacetate
Eifex	European Iron Fertilization Experiment
ESI-MS	electrospray ionization-mass spectrometry
HNLC	high-nitrate-low chlorophyll
IronEx	open-ocean iron enrichment experiment
K^{cond}	conditional stability constant
LPS	lipopolysaccharide
LUV	large unilamellar vesicle
M_E	marinobactin E
SEEDS	Subarctic Pacific Iron Experiment for Ecosystems Dynamics Study
SERIES	Subarctic Ecosystem Response to Iron Enrichment Study
SOFex	Southern Ocean Iron Experiment
SOIREE	Southern Ocean Iron Release Experiment
SP	solubility product
SUV	small unilamellar vesicle

REFERENCES

1. Field CB, Behrenfeld MJ, Randerson JT, Falkowski P. Science 1998; 281:237–240.
2. Johnson KS, Coale KH, Elrod VA, Tindale NW. Mar Chem 1994; 46:319–334.
3. Johnson KS, Gordon RM, Coale KH. Mar Chem 1997; 57:137–161.
4. Martin JH. Paleoceanography 1990; 5:1–13.
5. Martin JH, Fitzwater SE. Nature 1988; 331:341–343.
6. Martin JH, Coale KH, Johnson KS, Fitzwater SE, Gordon RM, Tanner SJ, Hunter CN, Elrod VA, Nowicki JL, Coley TL, Barber RT, Lindley S, Watson AJ, Vanscoy K, Law CS, Liddicoat MI, Ling R, Stanton T, Stockel J, Collins C, Anderson A, Bidigare R, Ondrusek M, Latasa M, Millero FJ, Lee K, Yao W, Zhang JZ, Friederich G, Sakamoto C, Chavez F, Buck K, Kolber Z, Greene R, Falkowski P, Chisholm SW, Hoge F, Swift R, Yungel J, Turner S, Nightingale P, Hatton A, Liss P, Tindale NW. Nature 1994; 371:123–129.
7. O'Sullivan DW, Hanson AK, Miller WL, Kester DR. Limnol Oceanogr 1991; 36:1727–1741.
8. Moore JK, Doney SC, Glover DM, Fung IY. Deep-Sea Res Part II-Top Stud Oceanogr 2002; 49:463–507.
9. Morel FMM, Price NM. Science 2003; 300:944–947.
10. Morel FMM, Milligan AJ, Saito MA. In: Turekian KK, Holland HD, eds. Treatise on Geochemistry. Cambridge, UK: Elsevier Science Ltd., 2003:113–143.
11. Martin JH, Gordon RM, Fitzwater SE. Limnol Oceanogr 1991; 36:1793–1802.

12. Boyd PW, Goldblatt RH, Harrison PJ. Deep-Sea Res Part II-Top Stud Oceanogr 1999; 46:2645–2668.
13. Coale KH, Johnson KS, Fitzwater SE, Blain SPG, Stanton TP, Coley TL. Deep-Sea Res Part II-Top Stud Oceanogr 1998; 45:919–945.
14. Stanton TP, Law CS, Watson AJ. Deep-Sea Res Part II-Top Stud Oceanogr 1998; 45:947–975.
15. Gordon RM, Johnson KS, Coale KH. Deep-Sea Res Part II-Top Stud Oceanogr 1998; 45:995–1041.
16. Cullen JJ. Limnol Oceanogr 1995; 40:1336–1343.
17. Wells ML. Mar Chem 2003; 82:101–114.
18. Cavender-Bares KK, Rinaldo A, Chisholm SW. Limnol Oceanogr 2001; 46:778–789.
19. Landry MR, Ondrusek ME, Tanner SJ, Brown SL, Constantinou J, Bidigare RR, Coale KH, Fitzwater S. Mar Ecol: Prog Ser 2000; 201:27–42.
20. Bollens GCR, Landry MR. Mar Ecol: Prog Ser 2000; 201:43–56.
21. Mann EL, Chisholm SW. Limnol Oceanogr 2000; 45:1067–1076.
22. Erdner DL, Anderson DM. Limnol Oceanogr 1999; 44:1609–1615.
23. Steinberg PA, Millero FJ, Zhu XR. Mar Chem 1998; 62:31–43.
24. Behrenfeld MJ, Bale AJ, Kolber ZS, Aiken J, Falkowski PG. Nature 1996; 383:508–511.
25. Rue EL, Bruland KW. Limnol Oceanogr 1997; 42:901–910.
26. Coale KH, Johnson KS, Fitzwater SE, Gordon RM, Tanner S, Chavez FP, Ferioli L, Sakamoto C, Rogers P, Millero F, Steinberg P, Nightingale P, Cooper D, Cochlan WP, Landry MR, Constantinou J, Rollwagen G, Trasvina A, Kudela R. Nature 1996; 383:495–501.
27. Debaar HJW, Dejong JTM, Bakker DCE, Loscher BM, Veth C, Bathmann U, Smetacek V. Nature, 1995; 373:412–415.
28. Wu J, Luther GW. Mar Chem 1995; 50:159–177.
29. Rue EL, Bruland KW. Mar Chem 1995; 50:117–138.
30. Gledhill M, Vandenberg CMG. Mar Chem 1994; 47:41–54.
31. Tsuda A, Takeda S, Saito H, Nishioka J, Nojiri Y, Kudo I, Kiyosawa H, Shiomoto A, Imai K, Ono T, Shimamoto A, Tsumune D, Yoshimura T, Aono T, Hinuma A, Kinugasa M, Suzuki K, Sohrin Y, Noiri Y, Tani H, Deguchi Y, Tsurushima N, Ogawa H, Fukami K, Kuma K, Saino T. Science 2003; 300:958–961.
32. Boyd PW, Law CS, Wong CS, Nojiri Y, Tsuda A, Levasseur M, Takeda S, Rivkin R, Harrison PJ, Strzepek R, Gower J, McKay RM, Abraham E, Arychuk M, Barwell-Clarke J, Crawford W, Crawford D, Hale M, Harada K, Johnson K, Kiyosawa H, Kudo I, Marchetti A, Miller W, Needoba J, Nishioka J, Ogawa H, Page J, Robert M, Saito H, Sastri A, Sherry N, Soutar T, Sutherland N, Taira Y, Whitney F, Wong SKE, Yoshimura T. Nature 2004; 428:549–553.
33. Boyd PW, Watson AJ, Law CS, Abraham ER, Trull T, Murdoch R, Bakker DCE, Bowie AR, Buesseler KO, Chang H, Charette M, Croot P, Downing K, Frew R, Gall M, Hadfield M, Hall J, Harvey M, Jameson G, LaRoche J, Liddicoat M, Ling R, Maldonado MT, McKay RM, Nodder S, Pickmere S, Pridmore R, Rintoul S, Safi K, Sutton P, Strzepek R, Tanneberger K, Turner S, Waite A, Zeldis J. Nature 2000; 407:695–702.
34. Bakker DCE, Watson AJ, Law CS. Deep-Sea Res Part II-Top Stud Oceanogr 2001; 48:2483–2507.

35. Bowie AR, Maldonado MT, Frew RD, Croot PL, Achterberg EP, Mantoura RFC, Worsfold PJ, Law CS, Boyd PW. Deep-Sea Res Part II-Top Stud Oceanogr 2001; 48:2703–2743.
36. Boyd PW, Law CS. Deep-Sea Res Part II-Top Stud Oceanogr 2001; 48:2425–2438.
37. Boyd PW, Abraham ER. Deep-Sea Res Part II-Top Stud Oceanogr 2001; 48: 2529–2550.
38. Croot PL, Bowie AR, Frew RD, Maldonado MT, Hall JA, Safi KA, Roche JL, Boyd PW, Law CS. Geophys Res Lett 2001; 28:3425–3428.
39. Frew R, Bowie A, Croot P, Pickmere S. Deep-Sea Res Part II-Top Stud Oceanogr 2001; 48:2467–2481.
40. Gall MP, Boyd PW, Hall J, Safi KA, Chang H. Deep-Sea Res Part II-Top Stud Oceanogr 2001; 48:2551–2570.
41. Gall MP, Strzepek R, Maldonado M, Boyd PW. Deep-Sea Res Part II-Top Stud Oceanogr 2001; 48:2571–2590.
42. Hall JA, Safi K. Deep-Sea Res Part II-Top Stud Oceanogr 2001; 48:2591–2613.
43. Karsh KL, Trull TW, Lourey AJ, Sigman DM. Limnol Oceanogr 2003; 48: 1058–1068.
44. Law CS, Ling RD. Deep-Sea Res Part II-Top Stud Oceanogr 2001; 48:2509–2527.
45. Law CS, Abraham ER, Watson AJ, Liddicoat MI. J Geophys Res 2003; 108.
46. Maldonado MT, Boyd PW, LaRoche J, Strzepek R, Waite A, Bowie AR, Croot PL, Frew RD, Price NM. Limnol Oceanogr 2001; 46:1802–1808.
47. Nodder SD, Charette MA, Waite AM, Trull TW, Boyd PW, Zeldis J, Buesseler KO. Geophys Res Lett 2001; 28:2409–2412.
48. Nodder SD, Waite AM. Deep-Sea Res Part II-Top Stud Oceanogr 2001; 48: 2681–2701.
49. Trull T, Rintoul SR, Hadfield M, Abraham ER. Deep-Sea Res Part II-Top Stud Oceanogr 2001; 48:2439–2466.
50. Trull TW, Armand L. Deep-Sea Res Part II-Top Stud Oceanogr 2001; 48: 2655–2680.
51. Waite AM, Nodder SD. Deep-Sea Res Part II-Top Stud Oceanogr 2001; 48: 2635–2654.
52. Watson AJ, Bakker DCE, Ridgwell AJ, Boyd PW, Law CS. Nature 2000; 407: 730–733.
53. Zeldis J. Deep-Sea Res Part II-Top Stud Oceanogr 2001; 48:2615–2634.
54. Buesseler KO, Andrews JE, Pike SM, Charette MA. Science 2004; 304:414–417.
55. Cassar N, Laws EA, Bidigare RR, Popp BN. Global Biogeochem Cycles 2004; 18.
56. Coale KH, Johnson KS, Chavez FP, Buesseler KO, Barber RT, Brzezinski MA, Cochlan WP, Millero FJ, Falkowski PG, Bauer JE, Wanninkhof RH, Kudela RM, Altabet MA, Hales BE, Takahashi T, Landry MR, Bidigare RR, Wang XJ, Chase Z, Strutton PG, Friederich GE, Gorbunov MY, Lance VP, Hilting AK, Hiscock MR, Demarest M, Hiscock WT, Sullivan KF, Tanner SJ, Gordon RM, Hunter CN, Elrod VA, Fitzwater SE, Jones JL, Tozzi S, Koblizek M, Roberts AE, Herndon J, Brewster J, Ladizinsky N, Smith G, Cooper D, Timothy D, Brown SL, Selph KE, Sheridan CC, Twining BS, Johnson ZI. Science 2004; 304:408–414.
57. Gervais F, Riebesell U, Gorbunov MY. Limnol Oceanogr 2002; 47:1324–1335.
58. Croot PL, Laan P. Anal Chim Acta 2002; 466:261–273.
59. Pakulski JD, Coffin RB, Kelley CA, Holder SL, Downer R, Aas P, Lyons MM, Jeffrey WH. Nature 1996; 383:133–134.

60. Tortell PD, Maldonado MT, Price NM. Nature 1996; 383:330–332.
61. Templeton DM. Molecular and Cellular Iron Transport. New York: Marcel Dekker, 2002.
62. Sigel A, Sigel H, eds. Iron Transport and Storage in Microorganisms, Plants and Animals, Vol. 35 of Metal Ions in Biological Systems. New York: Marcel Dekker, 1998.
63. O'Brien IG, Gibson F. Biochim Biophys Acta 1970; 215:393–402.
64. Bickel H, Bosshardt R, Gaumann E, Reusser P, Vischer E, Voser W, Wettstein A, Zahner H. Helv Chim Acta 1960; 43:2118–2128.
65. McArdle JV, Sofen SR, Cooper RS, Raymond KN. Inorg Chem 1978; 17: 3075–3078.
66. Isied SS, Kuo G, Raymond KN. J Am Chem Soc 1976; 98:1763–1767.
67. Atkin CL, Neilands JB. Biochemistry 1968; 7:3734–3739.
68. Nishio T, Tanaka N, Hiratake J, Katsube Y, Ishida Y, Oda J. J Am Chem Soc 1988; 110:8733–8734.
69. Ledyard KM, Butler A. J Biol Inorg Chem 1997; 2:93–97.
70. Takahashi A, Nakamura H, Kameyama T, Kurasawa S, Naganawa H, Okami Y, Takeuchi T, Umezawa H. J Antibiot 1987; 40:1671–1676.
71. Hou ZG, Raymond KN, O'Sullivan B, Esker TW, Nishio T. Inorg Chem 1998; 37:6630–6637.
72. Ferguson AD, Hofmann E, Coulton JW, Diederichs K, Welte W. Science 1998; 282:2215–2220.
73. Buchanan SK, Smith BS, Venkatramani L, Xia D, Esser L, Palnitkar M, Chakraborty R, van der Helm D, Deisenhofer J. Nat Struct Biol 1999; 6:56–63.
74. Ferguson AD, Chakraborty R, Smith BS, Esser L, van der Helm D, Deisenhofer J. Science 2002; 295:1715–1719.
75. Braun V, Braun M. Curr Opin Microbiol 2002; 5:194–201.
76. Granger J, Price NM. Limnol Oceanogr 1999; 44:541–555.
77. Wilhelm SW, Trick CG. Limnol Oceanogr 1994; 39:1979–1984.
78. Trick CG. Curr Microbiol 1989; 18:375–378.
79. Hutchins DA, Witter AE, Butler A, Luther GW. Nature 1999; 400:858–861.
80. Reid RT, Live DH, Faulkner DJ, Butler A. Nature 1993; 366:455–458.
81. Kanoh K, Kamino K, Leleo G, Adachi K, Shizuri Y. J Antibiot 2003; 56: 871–875.
82. Barbeau K, Zhang G, Live DH, Butler A. J Am Chem Soc 2002; 124:378–379.
83. Bergeron RJ, Huang G, Smith RE, Bharti N, McManis JS, Butler A. Tetrahedron 2003; 59:2007–2014.
84. Haygood MG, Holt PD, Butler A. Limnol Oceanogr 1993; 38:1091–1097.
85. Martinez JS, Haygood MG, Butler A. Limnol Oceanogr 2001; 46:420–424.
86. Martinez JS, Zhang GP, Holt PD, Jung H-T, Carrano CJ, Haygood MG, Butler A. Science, 2000; 287:1245–1247.
87. Martinez JS, Carter-Franklin JN, Mann EL, Martin JD, Haygood MG, Butler A. Proc Natl Acad Sci USA 2003; 100:3754–3759.
88. Barbeau K, Rue EL, Bruland KW, Butler A. Nature 2001; 413:409–413.
89. Macrellis HM, Trick CG, Rue EL, Smith G, Bruland KW. Mar Chem 2001; 76: 175–187.
90. Zuo YG, Holgne J. Environ Sci Technol 1992; 26:1014–1022.
91. Faust BC, Zepp RG. Environ Sci Technol 1993; 27:2517–2522.

92. Stephan H, Freund S, Beck W, Jung G, Meyer JM, Winkelmann G. Biometals 1993; 6:93–100.
93. Risse D, Beiderbeck H, Taraz K, Budzikiewicz H, Gustine D. Z Naturforsch, C: Biosci 1998; 53:295–304.
94. Okujo N, Sakakibara Y, Yoshida T, Yamamoto S. Biometals 1994; 7:170–176.
95. Persmark M, Pittman P, Buyer JS, Schwyn B, Gill JPR, Neilands JB. J Am Chem Soc 1993; 115:3950–3956.
96. Ratledge C, Dale J. Mycobacteria: Molecular Biology and Virulence. Oxford, UK: Blackwell Science, Ltd., 1999.
97. Ratledge C, Patel PV. J Gen Microbiol 1976; 93:141–152.
98. Murakami Y, Kato S, Nakajima M, Matsuoka M, Kawai H, Shinya K, Seto H. J Antibiot 1996; 49:839–845.
99. Suenaga K, Kokubo S, Shinohara C, Tsuji T, Uemura D. Tetrahedron Lett 1999; 40:1945–1948.
100. Kokubo S, Suenaga K, Shinohara C, Tsuji T, Uemura D. Tetrahedron 2000; 56:6435–6440.
101. Kim HS, Kim SB, Park SH, Oh HM, Park YI, Kim CK, Katsuragi T, Tani Y, Yoon BD. Biotechnol Lett 2000; 22:1431–1436.
102. Ahimou F, Jacques P, Deleu M. Enzyme Microb Technol 2000; 27:749–754.
103. Neu TR. Microbiol Rev 1996; 60:151–166.
104. Wilkinson SG. In: Ratledge C, Wilkinson SG, eds. Microbial Lipids. San Diego, CA: Academic Press Ltd., 1988:299–488.
105. Xu G, Martinez JS, Groves JT, Butler A. J Am Chem Soc 2002; 124:13408–13415.
106. Tsao HK, Tseng WL. Chem Phys 2001; 115:8125–8132.
107. Inoue T. In: Rosoff M, ed. Vesicles. New York, NY: Marcel Dekker, 1996:151–193.
108. Balzani V, Carassiti V. Photochemistry of Coordination Compounds. London: Academic Press, 1970.
109. Venter JC, Remington K, Heidelberg JF, Halpern AL, Rusch D, Eisen JA, Wu D, Paulsen I, Nelson KE, Nelson W, Fouts DE, Levy S, Knap AH, Lomas MW, Nealson K, White O, Peterson J, Hoffman J, Parsons R, Baden-Tillson H, Pfannkoch C, Rogers Y-H, Smith HO. Science 2004; 304:66–74.

3

Speciation and Bioavailability of Trace Metals in Freshwater Environments

Laura Sigg and Renata Behra

EAWAG, Swiss Federal Institute for Environmental Science and Technology, P.O. Box 611, CH-8600 Dübendorf, Switzerland

1. INTRODUCTION: SPECIATION AND BIOAVAILABILITY OF TRACE METALS

A number of trace elements, both essential and non-essential, are present at elevated concentrations over the background levels in numerous freshwaters. Elements with important anthropogenic inputs into freshwaters comprise in particular copper, zinc, nickel, chromium, lead, cadmium, mercury, and silver [1]. Essential elements, which are required at trace concentrations to sustain vital functions, such as copper, zinc, nickel, and chromium, become toxic at higher concentrations, whereas non-essential metals such as cadmium, lead, mercury, and silver have mostly detrimental effects. Measurements of elevated total concentrations are not sufficient to evaluate the impact of these metals to aquatic organisms.

The impact of metals on aquatic organisms presupposes direct or indirect interactions between metal and organism, such as binding to biological ligands and transfer through the biological membranes. These interactions strongly depend on the chemical speciation of the metals. Knowledge on the speciation of metals as related to bioavailability is thus essential to the assessment of metal impacts and effects on aquatic organisms [2]. The key questions to be asked with respect to the bioavailability of metals are

- In which chemical species are the metals present in freshwaters?
- Can these species be taken up by the organisms, either directly or indirectly?
- Which measurable parameters can be used to predict the bioavailability of metals?

The complex overall interactions between metals, ligands, and organism receptors have to be assessed to understand the bioavailability of metals to aquatic organisms in freshwater systems [3]. The free ion activity of metals has been shown to be a key parameter for the evaluation of the bioavailability of metals [4,5]. However, some limitations of the free ion activity model have been discussed [6,7], indicating that additional factors have to be considered.

Bioavailability of metals needs to be discussed as a function of the interactions of metals with ligands forming strong and weak complexes in solution, with solid phases of different solubility, by precipitation, co-precipitation, or adsorption of the metals. Factors to be considered to evaluate the bioavailability of metals comprise the thermodynamic stability of complexes and the resulting free ion activity, the kinetic lability of complexes, the distribution over various size ranges (from low molecular weight to macromolecular and colloidal size ranges), and the hydrophilic or hydrophobic properties of the complexes. Furthermore, uptake of metals depends on biological factors such as growth rates of organisms and the cellular metabolism of absorbed metals.

In this chapter, bioavailability of trace metals in freshwater environments will be discussed in relationship to speciation. To this end, the speciation of trace metals in freshwater, as related to the presence of various types of ligands will first be considered. Some examples of experimentally determined speciation of metals in natural freshwaters will be shown. The significance of speciation with respect to bioavailability of metals to aquatic organisms, in particular to algae will be examined. Some examples of experimental work relating speciation and bioavailability of metals in natural waters are shown. The role of sediments in providing bioavailable metals will also be discussed. The focus of this chapter is on elements that are of ecotoxicological significance and are often present at elevated levels in freshwater, namely copper, zinc, nickel, cobalt, cadmium, and lead.

2. SPECIATION OF TRACE METALS IN FRESHWATERS

2.1. Ligands in Freshwater

Various ligands in freshwater, which bind metals in various types of complexes, include inorganic ligands, organic ligands over a wide range of sizes and chemical properties, in particular humic and fulvic acids (HA and FA), organic compounds released by organisms or introduced into water by anthropogenic activities, macromolecules and particles in the colloidal size range, and surface functional groups of particles and of organisms. Metals in freshwater are thus typically distributed among complexes of different complexing properties, with respect to thermodynamic stability, kinetic lability, and size range.

Inorganic ligands comprise first of all hydroxide and carbonate, and depending on the particular conditions, chloride, phosphate, sulfate, and sulfide (Table 1). The inorganic speciation of metals depends on the relative concentrations of these ligands under various conditions and on pH. The pH of natural freshwater may vary from typically neutral or slightly alkaline 7.0–8.5 in carbonate buffered hardwaters, to less than 7.0 in softwaters with little carbonate, and down to around 4 in freshwaters influenced by acidic atmospheric precipitation.

Table 1   Typical Concentrations of Ligands in Freshwater

Components	Typical range in freshwater (M)
HCO_3^-	10^{-4} to 5×10^{-3}
CO_3^{2-}	10^{-6} to 10^{-4}
Cl^-	10^{-5} to 10^{-3}
SO_4^{2-}	10^{-5} to 10^{-3}
HS^-	10^{-6} to 10^{-3} (anoxic)
	10^{-9} to 10^{-7} (oxic)[a]
DOM[b] as C	10^{-4} to 10^{-3}
DOM functional groups	10^{-6} to 10^{-4}
Amino acids	10^{-7} to 10^{-5}
Thiols[c]	10^{-9} to 10^{-7}

[a]Ref. [10].
[b]DOM: dissolved organic matter, concentration given either as organic carbon or as concentration of functional groups.
[c]Refs. [28,30].

Alkalinity is typically in the range 1–5 mM in hardwaters, whereas much smaller alkalinity <1 mM occurs in softwaters. The relative importance of carbonate and hydroxide as ligands depends thus on the pH and concentration conditions. Chloride is in freshwater usually present in the range of 10^{-4} to 10^{-3} mol L^{-1} and may be an important ligand for some metals, especially for silver and mercury.

Sulfide plays an interesting role in binding metals, as very stable complexes are formed with numerous metals [8]. Although sulfide is mostly expected to be present under reducing conditions, evidence for the presence of trace amounts of sulfide under oxic conditions, which bind metals in very stable complexes, has been recently put forward [9–12]. The presence of small multinuclear sulfide clusters such as Cu_3S_3 and Cd_4S_6, which are very stable, has been shown in oxic river waters [10]. Traces of sulfide in the nanomolar range in oxic waters may be sufficient to influence the speciation of many metals. The stability of inorganic complexes is generally relatively weak, with the exception of some of the sulfide complexes.

Natural organic matter plays a predominant role in complexing metals in natural waters. Natural organic matter may include a wide range of low molecular and high molecular mass compounds which nearly all possess some complexing properties for metals. Riverine organic matter has been shown to include a variety of organic compounds in all size ranges [13]. Simple organic molecules such as small carboxylic acids (oxalate, acetate, malonate, and citrate), amino acids, phenols, or catechols are all ligands for metals which may be released upon decomposition from organic matter or excreted by organisms. Complexes of moderate stability are formed between metals and small carboxylic acids and amino acids. Although these compounds are readily biodegradable, small steady-state concentrations may affect the speciation of trace metals. These complexes may contribute to a pool of labile metal species.

A very important part of natural organic matter with regard to interactions with metals consists of the humic (HA) and fulvic (FA) acids, which are polymeric compounds with complex structures and molecular masses of typically $500-2000$ g mol^{-1} for FA and 2000 to >5000 g mol^{-1} for HA, and which exhibit heterogeneous binding sites [14]. Complexation of metals by HA and FA has been extensively examined with a variety of methods and of model approaches, e.g., [14–19]. The heterogeneity of binding sites and the polyelectrolyte nature of metal ion binding have to be taken into account in modeling complexation by HA and FA. Phenolic and carboxylic groups bound to various aliphatic and aromatic structures are the most abundant complexing functional groups in HA and FA. However, small amounts of N- and S-containing functional groups may also play an important role for metal complexation, as shown, for example, for mercury [20].

Several models have been developed that predict the extent of complexation by HA and FA under specified conditions [15–17,19,21], taking into account the effects of pH, ionic strength, and competition between trace metals and major ions. The stability of HA and FA complexes depends on the metal to ligand ratio, with higher stability at low metal to complex ratios, because the stronger complexing sites are first occupied at low metal concentrations.

Figure 1 illustrates binding of Cu and Cd by HA and FA under typical freshwater conditions (HA and FA as dissolved organic carbon (DOC) = 2.8 mg L^{-1}; pH 8). Cu is strongly bound and therefore, free Cu^{2+} is in the range of 10^{-4} to 10^{-12} M for total dissolved Cu = 10^{-8} to 10^{-7} M in the presence of HA or FA. Cd is somewhat less strongly bound by HA and FA and free Cd^{2+} is in the range of $10-30\%$ of total Cd (Fig. 2). For Cd, complexation by FA and HA is also shown as calculated from the Windermere humic aqueous model (WHAM) model, a widely used model for binding of metals by HA and FA [21]. The affinity of HA and FA for metals generally corresponds to the Irving–Williams series: Cu(II) > Ni(II) > Zn(II) > Co(II) > Cd(II) > Ca(II) > Mg(II). With respect to the bioavailability of metals, it is important to keep in mind that metals complexed by HA and FA are bound to various functional groups with a range of thermodynamic stability and of kinetic lability and to material of molecular mass ranging from ~500 to >5000 g mol^{-1}.

There is some evidence from the experimental determination of metal speciation that stronger ligands are present in freshwater in comparison to HA and FA [22]. A comparison of the binding of Cu and Cd by HA and FA and the binding by ligands in lake waters, as determined by ligand/exchange voltammetric methods, is illustrated in Figs. 1 and 2. This comparison indicates that stronger binding ligands than HA and FA are present in the lake waters at low concentrations, as the free Cu^{2+} and Cd^{2+} determined in unfractionated lake waters appear to be lower than in the presence of HA and FA under similar conditions. The occurrence

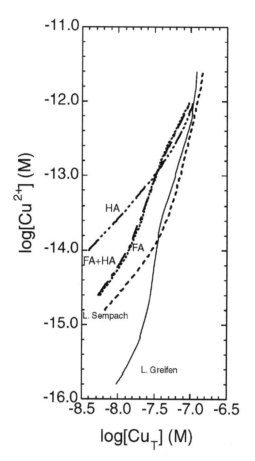

Figure 1 Calculated concentration of free Cu^{2+} in the presence of FA and HA in comparison to measured free Cu^{2+} in lake water (Lake Greifen and Lake Sempach, Switzerland) in dependence on total Cu present (Cu_T), as determined by the same method under similar conditions of DOC (2.8 mg L^{-1}) and pH 8. In both cases, Cu is mostly bound to organic ligands, but free Cu^{2+} is lower in the case of the lake waters, indicating the presence of strong organic ligands. Reprinted with permission of Kluwer Academic Publishers from Fig. 4(a) in Ref. [22].

of strong binding ligands has been first put forward in seawater, in which their role for several trace elements has been discussed [23–25].

A high specificity for some metals, especially for Cu, may indicate binding by S- or N-containing groups. A strong preference for binding of Cu over binding of Zn has been observed for ligands in lake water [26]. These ligands may be released by organisms during growth or upon decomposition of organic matter. Biomolecules with thiol functional groups, such as phytochelatins, glutathione, and metallothioneins possibly contribute to a pool of strong ligands. Some

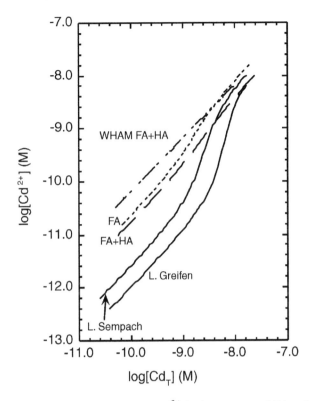

Figure 2 Calculated concentration of free Cd^{2+} in the presence of FA and HA in comparison to measured Cd^{2+} (Cd_T = total Cd) in lake water (Lake Greifen and Lake Sempach, Switzerland) in dependence on total Cd present (Cd_T), as determined by the same method under similar conditions of DOC (2.8 mg L^{-1}) and pH 8, and as calculated by the WHAM model. Reprinted with permission of Kluwer Academic Publishers from Fig. 6 in Ref. [22].

specific determinations have indicated the presence of thiols in various natural waters, even under oxic conditions [27–30] (Table 1). Some attempts to isolate specific Cu binding ligands from natural waters have been performed, using affinity chromatography methods [31–35]. The issue of the role of specific ligands in comparison to HA and FA is controversially discussed in the literature [36]. The occurrence of strong binding sites of HA and FA at low concentrations has been proposed as an alternative explanation to the occurrence of specific strong binding ligands [36,37]. The nature and characteristics of strong and possibly specific binding ligands remain to be further elucidated.

To conclude, trace metals in freshwater such as copper, zinc, and cadmium can be expected to be distributed in solution between these various ligands, including some weaker complexes with inorganic ligands and with small organic ligands, more stable complexes with more specific organic ligands, and a range of complexes with heterogeneous ligands of the FA and HA types.

2.2. Colloidal Particles

If considering the distribution of metals from the point of view of size distribution, one may distinguish between the dissolved range (defined as molecules smaller than a certain molecular mass, typically <1 or <10 kDa), the colloidal size range (usually operationally defined as the size range <0.45 or <0.2 μm and >1 or 10 kDa), and the particulate size range (>0.2 or >0.45 μm). The colloidal particles play a special role in an intermediate size range; that is, they are easily transported with the water phase, but have different properties from truly dissolved smaller molecules, with respect, for example, to diffusion and reactivity. Colloidal particles may consist of organic or inorganic components. Organic colloids may be of humic nature, consisting of large macromolecules with molecular masses of >1 or >10 kDa. Inorganic colloids may consist of oxides, e.g., iron and manganese oxides have often been detected in the colloidal size range [38–41]. Clay minerals may also occur in this size range.

Metals may be bound to these various types of colloids. Complexation by humics, as described above, may as well occur in the size range of colloidal macromolecules. Binding to inorganic particles may occur by adsorption, for which colloidal particles offer large specific areas and thus numerous binding sites. Co-precipitation of trace metals with solid phases is another possible process for metal ion binding in colloidal particles.

The distribution of metals over the colloidal size range has been evaluated in freshwater environments in several studies [42–50]. These studies indicate that substantial parts of the total concentrations of copper, zinc, cadmium, nickel, manganese, and iron are bound in colloidal particles. This fraction may however greatly vary depending on time and location in various streams and rivers. Colloidal particles thus play a significant role in the speciation and transport of metals in freshwaters. Their role in determining the bioavailability of trace metals needs to be carefully evaluated.

2.3. Particulate Matter

The solubility of trace metals under freshwater conditions is limited by solid phases that may be formed, in particular hydroxides and oxides, carbonates, and under anoxic conditions sulfides. The solubilities of the hydroxides and carbonates of copper, zinc, cadmium, and lead are rather high, with dissolved concentrations of these metals at equilibrium with the solid phases in the range of 10^{-6} to 10^{-8} M for typical freshwater conditions [51]. Solubility of these solid phases therefore rarely becomes limiting under conditions of the water column, where these elements are present in the nanomolar and subnanomolar range. In the porewater of sediments some of these solubility limits may be reached, as shown, for example, in metal-polluted sediments [52].

Important solid phases limiting the concentrations of trace metals are the sulfides under anoxic conditions. Sulfides present in the sediment porewaters may be solubility limiting for the trace metals in the sediments [11]. Supersaturation of Cu and Pb sulfides has been shown, for example, in the anoxic part of a lake [53]. Co-precipitation of trace metals with iron sulfides is also an important process leading to binding of trace metals in sediments [54]. The role of sulfides in limiting the bioavailability of trace metals from sediments will be discussed later.

The bulk particulate matter in freshwater consists of oxides and hydroxides of iron, manganese, aluminum, of silicates, of clay minerals, of calcium carbonate in waters with carbonate geological background, and of organic matter (algae, bacteria, organic debris). Trace metals may be bound by adsorption to the surface functional groups on all these phases. The distribution of metals over solid phases depends on the available adsorption sites and on the binding strength to various types of sites. Estimates of the available adsorption sites for a typical composition of suspended matter in rivers and lakes have indicated the importance of iron oxides, manganese oxides, and of organic matter for metal adsorption under natural freshwater conditions [55].

The low trace metal concentrations in freshwater are therefore often limited by adsorption reactions on the bulk particulate matter rather than by precipitation of solid phases.

2.4. Case Studies of Metal Speciation in Freshwater

Metal speciation in freshwater has been extensively studied in our laboratory, including in particular Cu, Zn, Cd, Ni, and Co [22,26,56–62]. The methods used aimed mostly at determining the free ion activity of these elements and the complexation parameters for binding by organic ligands. The methods used were based on competitive ligand exchange coupled to voltammetric determination [63]. Using these methods, the free metal ion concentrations are calculated based on complexation equilibria with added known ligands (Table 2). These studies have mostly been carried out in hardwater lakes and rivers.

2.4.1. Copper

Extensive data on Cu speciation in various environments (lakes, rivers, groundwater) have been obtained. The Cu speciation is characterized by strong complexation, presumably by natural organic ligands, in nearly all the environments studied (Table 2).Very strong complexation of copper has been observed in eutrophic lakes, where the ratios of free Cu^{2+} to total dissolved Cu were in the order of magnitude $[Cu^{2+}]/Cu(tot) = 10^{-7}$ to 10^{-6}, as measured in the lakes Greifen and Sempach (Switzerland) [26,56,58]. The conditional stability constants (at $pH \approx 8$) of the strongest complexes were in the range of $\log K_1 = 14.5 - 15.5$. The strongest complexation was observed in the epilimnion of these lakes, in particular in the euphotic zone during the productive season.

Table 2   Complexation of Cu, Zn, Cd, and Ni Under Freshwater Conditions (Range in Hardwater Lakes and Rivers, pH 7.5–8.5), as Determined by Ligand-Exchange/Voltammetry[a]

	Dissolved concentrations (nM)	$-\log[M^{2+}]_{free}$ (M)	$\log K_1$ conditional[b]	References
Cu	5–30	13.5–16	13.5–16	[56,58,64]
Zn	10–40	8.6–10	7.8–9.5	[26,57,64]
Cd	0.03–0.2	10.4–12.5	9.5–10.5	[59]
Ni	4–30	13.0–14.5	12–14	[60]

[a]Modified from Ref. [63].
[b]Conditional stability constants of the stronger complexes, pH 7.5–8.5.

In contrast, the Cu complexation in an oligotrophic lake (Lake Lucerne, Switzerland) was weaker, with $[Cu^{2+}]/Cu(tot) = 1–3 \times 10^{-6}$ and conditional stability constants $\log K_1 = 13.0–14.6$ (at pH ≈ 7.8).

The earlier observations led to the hypothesis that strong ligands originate from phytoplankton and are thus more abundant in eutrophic lakes. Much weaker Cu complexation was observed in some alpine acidic lakes, with $[Cu^{2+}]/Cu(tot) = 0.03–0.1$ at pH 5.5–6.0. Strong Cu complexation was also observed in river waters at circumneutral pH, where in a number of Swiss river waters the ratios were in the range of $[Cu^{2+}]/Cu(tot) = 5 \times 10^{-8}$ to 5×10^{-6}. In a small stream (Furtbach) with large variations in copper concentration during rain events, the ratios of $[Cu^{2+}]/Cu(tot)$ were highest for the concentration maximum of copper in solution, indicating inputs of copper in weakly complexed form [64]. A study on the size distribution of metals in a river indicated the distribution of the strong ligands binding Cu over the size ranges >10 and <10 kDa and a fraction of $\sim11\%$ of Cu in the colloidal size range [48].

Strong complexation of Cu has also been observed in an English lake, where low Cu^{2+} was in particular observed in the productive surface layer of the lake [65]. In a study of Cu speciation in New England rivers [66], strong complexation of Cu was attributed to binding by natural organic ligands and by sulfide. The determined free Cu^{2+} was in the range of 10^{-9} to 10^{-11}, in the pH range 5.2–7.9. In a similar way, binding of Cu by ligands with a range of $\log K = 7.8–8.1$ was observed in Connecticut rivers [67]. Complexation of Cu by FAs was observed in several streams with relatively high Cu and low pH (around 6) [68]. The free Cu^{2+} was in these cases in the range of several percentages of the total Cu. By using diffusion gradients in thin films (DGT), it was also shown that a large part of Cu in some softwater freshwaters at circumneutral pH was bound in organic complexes, whereas in a lake with acidic pH the extent of complexation was much lower [69,70].

From these studies on Cu complexation in freshwaters, it appears that Cu is mostly strongly bound in stable complexes with organic ligands under freshwater conditions. The stability of the complexes appears to decrease with decreasing pH, resulting in higher free Cu^{2+} at lower pH. Differences between sites with predominantly FA and HA and sites with more specific ligands may exist, but remain to be more clearly demonstrated.

2.4.2. Zinc

Zn appears to be much less strongly complexed than Cu (Table 2). Ratios of $[Zn^{2+}]/Zn(tot)$ in eutrophic lake water were observed in the range of 0.01–0.1. About 40–80% of zinc was present in strong complexes in eutrophic lake waters, whereas a substantial part of ~20–60% was bound to weaker ligands, as defined by the electrochemical method used [26,57]. Competition between zinc and copper for strong ligands in lake water was demonstrated [26]. The average conditional stability constants for the stronger complexes were $\log K_1 = 7.8–9.6$. In a metal-polluted lake with low pH more than 90% of Zn was present as Zn^{2+} [71].

In a small stream, where variations of total Zn were observed during rain events, a higher fraction of Zn (10–20%) was in the form of free Zn^{2+} at the highest concentrations observed [64]. DGT-measurements of Zn in this stream and in another one indicated a large fraction of Zn in DGT-labile forms, indicating weak complexes [72]. Only a small Zn fraction was found in the colloidal fraction in a river [48]. The typical speciation of Zn in freshwater as observed in these studies includes thus a substantial fraction present as Zn^{2+}, usually 1–10% of total Zn dissolved, some Zn bound to weaker ligands and a further Zn fraction bound to stronger ligands.

In a study of Zn speciation in a selection of European river waters, Zn was found by an electrochemical method to be mostly bound in labile complexes, and free Zn^{2+} ranged from 12% to 45% in rivers in the pH range of 6.8–8.3 [73].

2.4.3. Cadmium

Cadmium is also less strongly complexed than copper (Table 2). The ratios of $[Cd^{2+}]/Cd(tot)$ in water from eutrophic lakes were in the range of 0.01–0.03, and up to 0.09 in a river [59]. The conditional stability constants for the stronger complexes were in the range $\log K_1 = 9.5–10.3$. Stronger complexation of Cd was also observed in the eutrophic lakes in comparison to oligotrophic and acidic lakes. The comparison with HA and FA also indicated stronger complexation of Cd by natural ligands in the lake [22]. The speciation of Cd in these studies thus indicated a small fraction present as Cd^{2+}, a rather large fraction being in moderately strong complexes and probably some weaker complexes.

Cd was observed to be only weakly complexed in Connecticut rivers [67]. On the other hand, the speciation of Cd in river waters indicated the presence of rather strong complexes with high molecular weight organic matter [74], in which kinetically distinguishable complexes were identified.

2.4.4. Nickel

Complexation reactions of nickel are characterized by slow exchange reactions, and this is also observed for exchange reactions with natural organic ligands [60]. Nevertheless, ligand exchange reactions indicated the presence of strong organic complexes in lake and river waters (Table 2). Ratios of $[Ni^{2+}]/Ni(tot)$ were in the range of 2×10^{-7} to 1.8×10^{-5}. They are thus of a similar order of magnitude as those for copper and indicate that Ni is mostly bound in strong organic complexes. Ni in these strong complexes may be exchanged very slowly. In a study on metal distribution in a river only a small fraction of Ni ($\sim 5\%$) was found in the colloidal size range [48].

In an English lake, a substantial fraction of Ni was observed to be bound in strong organic complexes [65]. In a study of kinetic dissociation of Ni complexes in nickel-polluted river waters several kinetic components were distinguished [75] and they indicated that complexes with similar dissociation rate constants as nickel aqua complexes were present.

There are only a few studies available on nickel speciation in different types of freshwaters. Additional data are needed to obtain a more complete picture of Ni speciation in freshwaters.

2.4.5. Cobalt

The complexation of Co(II) was examined at several lake and river sites in Switzerland [61]. Moderately strong complexation was observed, with ratios of $[Co^{2+}]/Co(tot)$ in the range of $0.03-0.1$, with some values up to 0.3. Co(II) was thus to a large extent bound to rather strong organic complexes which represent $\sim 80-96\%$ of total dissolved Co. Conditional stability constants (pH 8.0 ± 0.1) of these complexes were in the range of $\log K = 9.5-11.6$. In a study on metal distribution in a river, only a small fraction of Co ($< 5\%$) was found in the colloidal size range.

Data on metal speciation in freshwaters are thus still rather scarce. A complete picture of the role of natural organic ligands for the speciation of trace metals in freshwaters has still to emerge, as more data become available. The results on complexation have only in a few cases directly been related to the availability to organisms.

3. BIOAVAILABILITY OF TRACE METALS AS A FUNCTION OF AQUEOUS SPECIATION

3.1. Free Ion Activity Model and Biotic Ligand Model

Interactions of trace elements with algae have been extensively studied under laboratory conditions with well-defined culture media ([5] and references therein;

[4,76–91]). This body of work has established some key features of interactions of metal ions with algae. In particular, the preponderant role of speciation in solution for uptake and effects of metals has been clearly demonstrated. On the basis of many laboratory experiments, the free ion activity model (FIAM) has been established, which relates biological uptake and effects to the activity of free aqua metal ions in equilibrium with ligands ([5] and references therein).

The FIAM states that the interactions of metals with an organism depend on the free ion activity in solution, which is regulated by the interactions with the various ligands available. The metal ions are assumed to interact with the organism by binding to biological metal carriers. Binding to the biological carrier ligands is thus in competition with the ligands in solution, and in the case of chemical equilibrium between these various ligands it is dependent on the free ion activity. The FIAM relies upon the assumption that the equilibrium between free metal ions and complexed species is sufficiently rapidly established so that equilibrium in solution is not disturbed by the interactions with the biological ligands. Furthermore, the FIAM relies on the assumptions that the transport of metals by membrane transporters is limited by the internalization step over the cell membrane and that the diffusion of metal to the cell membrane is not rate limiting.

The FIAM has been very successful in describing the role of speciation of metals, such as Cu and Zn, for their effects both as trace nutrients and as toxic elements for algae [79–81]. More recent work has focused on the limitations of this model and on exceptions to the FIAM [84–86,89–92]. Exceptions to the FIAM may be due to several factors that include uptake of specific organic or inorganic complexes, e.g., citrate complexes or silver thiosulfate [84,86], or uptake by passive diffusion of lipophilic complexes [93–95]. Uptake of metals clearly depends in some cases on other species than the free metal ions because of additional uptake pathways. Some specific complexes may be transported into the cells by anion transporters (e.g., citrate or thiosulfate). Lipophilic complexes may be directly transported over the cell membranes by diffusion into the hydrophobic part of the membrane [96].

The biotic ligand model (BLM) represents a further development of the FIAM by considering the interactions of a metal ion of interest with a biotic ligand and the competition by various other ions that may also bind to this biotic ligand [97–100]. The BLM allows the inclusion of the effects of parameters such as the concentrations of major cations (Ca^{2+}, Na^+) and anions and of pH on metal uptake by considering the interactions of these ions at a biotic ligand site, as well as the consideration of solution complexation. These interactions may be defined by equilibrium constants between cations and the biotic ligands, assuming that the interactions may be described by competing equilibria. This model has been used and calibrated in particular for interactions of metals with fish gills [98,101], for which competition between silver or copper, and major ions have been described [99–101]. Using the BLM, the effects of various factors of importance in natural freshwaters, such as hardness, chloride concentration, DOC, can be taken into account as well as the influence of toxic metal ions. The use of the BLM for regulatory purposes for metals in the aquatic environment is presently in discussion [97].

Both the FIAM and the BLM do not directly consider kinetic effects and are actually equilibrium-based models. These models assume that the rate determining step for metal uptake is the transport across the cell membrane and that the mass transfer of metals to the membrane is not limiting. A more detailed approach to metal uptake by organisms considers the flux of metals to the cell membrane by diffusion and the uptake flux [6,7,102,103]. Consideration of these fluxes evidences several cases, depending on the relative magnitude of uptake and diffusion fluxes. This approach considers the mass transfer of the metals to the biological interface, the adsorption step to biological carrier ligands, and the internalization step [102]. It demonstrates the importance of the relative rates of these different steps in a detailed description of metal uptake. Depending on the exposure conditions and on the organism character-istics (such as internalization rate of a metal, size of the organism), mass transfer of metals to the biological interface may become rate-limiting, rather than internalization. If the diffusion step of the free metal to the biological interface becomes limiting, the dissociation of labile complexes is relevant for uptake. Conditions under which the FIAM does or does not apply have been derived from this theoretical approach [6,102,103]. Some examples of limitation of uptake by the diffusion flux of a metal to the cell membrane have been shown experimentally in algae cultures [85,92].

With regard to the conditions in freshwater, these model approaches imply that the concentration of the free metal ions is the key parameter for metal bioavailability under conditions where the FIAM is valid. Under conditions of limited diffusion to the biological interface, the concentration of labile complexes becomes important. The lability needs to be defined as a function of the size and the uptake fluxes of the organisms of interest. Some specific complexes may also play a role.

3.2. Interactions of Various Types of Metal Complexes with Organisms

On the basis of the above discussion of uptake models, the role of various types of complexes observed in natural waters for metal uptake may be assessed in more detail. The free aqua metal ions clearly play a key role in all cases where metal uptake is governed by binding of the metal ion to a biotic ligand in equilibrium with the ligands in solution. The concentration of the free aqua metal ions is then a key parameter indicating the overall complexation under given conditions. The free aqua ion concentration is regulated by the complex assemblage of ligands present in a natural water, as described earlier.

In cases where the diffusion of metal ions to the biological membrane becomes limiting, in particular if the uptake rate by the organism is high, the supply of metal ions from labile complexes becomes important. In such cases, the concentration of metals bound in weak complexes, including inorganic complexes and weak organic complexes, such as those with simple organic ligands, determines uptake of the metal. The lability of the complexes with regard to dissociation kinetics is important in these cases [6].

More specific interactions are possible for some small organic complexes, as demonstrated for citrate complexes [84,89]. It is important to realize that direct uptake of a metal-organic complex, or also of a metal-inorganic complex, requires a specialized uptake path, such as a specialized carrier for the ligand that may eventually transport a metal–ligand complex.

Organic complexes of higher molecular mass and with a more hydrophilic character, such as those with FA and HA can hardly be taken up as complexes. However, FA and HA may interact with biological membranes and cell surfaces in various ways. Sorption of FA and HA at cell surfaces may affect the interactions of metal ions with cellular ligands and the characteristics of cell membranes [91,104–106].

Hydrophobic complexes represent a special category with regard to uptake, because they can be taken up by passive diffusion over the cell membrane. Uptake of hydrophobic complexes by algae has been shown, for example, for dithiocarbamate, and for oxine complexes [93–95]. The presence of hydrophobic complexes in natural waters is possible, both due to the presence of anthropogenic ligands and to the presence of natural ligands with hydrophobic properties. Uptake of hydrophobic complexes is thus an additional pathway, the importance of which needs to be further elucidated.

With regard to colloidal species with typical molecular masses of >1 kDa, it is unlikely that they can cross cell membranes. Furthermore, their diffusion is much slower than that of free aqua ions or of small complexes. The role of colloidal species in providing metals by dissociation of the complexes is yet unclear. Their importance in providing metals needs to be assessed especially with regard to higher organisms that may take up metals by ingestion, or by filter-feeding. The role of colloidal species for metal uptake by organisms has been examined experimentally only in few cases [107,108], which evidenced a complex behavior of colloids depending on the metal considered and on the uptake behavior of the organisms. Further research is necessary to understand the role of colloids as a reservoir of metals in natural freshwaters and the pathways by which the colloid-bound metals may become available.

3.3. Biological Factors Influencing Metal Bioavailability

Though the extent to which metals are bioavailable to algae is largely influenced by metal speciation in water, the metal accumulated under given exposure conditions will be also influenced by several biological factors. In each particular species, the accumulated metal will depend on the mechanism of uptake and accumulation and is interrelated to the growth rate and nutritional state of the cells [109]. The accumulation of metals reflects the cellular metabolism of metals in which several types of ligands and interactions are involved. Depending on the available metal concentrations, metals are taken up by constitutive or induced membrane-bound transport proteins that differ in number and metal affinity.

Kinetic studies of metal uptake by algae have indicated that metal-transport is under negative feedback regulation. At least two transport systems have been identified for copper [88], zinc [79,87,110], iron [111], and manganese [112]. Following uptake, the metal interacts with metallochaperones that act as metal-trafficking proteins to deliver the metal to metal-requiring proteins in the cytosol and to membrane transporters located in various organelles of the cell [113].

Cells are also provided with mechanisms of metal detoxification that involve compartmentalization of excess metal in vacuoles [114], metal exclusion, and metal chelation by metal-induced proteins such as phytochelatins, metal-lothioneins, and small organic acids [115]. Considering the network of cellular mechanisms and the time scales involved in metal metabolism, it is thus not surprising that interspecific differences in accumulation and tolerance do occur. The understanding of the biological mechanisms that serve to control the intra-cellular availability of metals has increased in recent years and will undoubtedly contribute to a better interpretation of metal accumulation in relation to both metal bioavailability and metal toxicity in freshwaters.

3.4. Case Studies of Metal Accumulation in Periphyton Under Freshwater Conditions

Only a few studies have so far directly demonstrated the relationships between metal concentrations and speciation in freshwater and accumulation or effects in organisms. Especially, speciation has only been considered experimentally in a few cases. Some examples from our own work will be summarized here.

The relationship between metal concentrations in water and in periphyton (algae growing on solid substrates in the river bed) has been examined under different conditions of several rivers and streams [64,116,117]. In a study concerning two rivers in Switzerland, Birs and Thur, the concentrations of copper and zinc were measured at several sites with different metal inputs [116]. Large gradients of copper and zinc in water were observed at several sites in the Birs river, due to several industrial sources of these metals. Higher concentrations of copper and zinc in water were reflected in higher concentrations of these metals in periphyton. Speciation of copper and zinc was not examined in this case. The clear accumulation in periphyton indicated that the high concentrations in water were to some extent bioavailable.

In the Thur river, a background site was compared to a more polluted site, due to inputs from agriculture and from sewage. The total dissolved concentrations of copper and zinc were clearly more elevated at the polluted site (by a factor of about 3). However, the accumulated Cu and Zn were not clearly different in periphyton from these two sites. Limited observations of the Cu speciation led to the hypothesis that Cu was strongly complexed at both sites and that the free Cu ion concentrations did not differ much between the two sites, and that thus available Cu was similar at the two sites [48]. The difference in metal bioaccumulation in the Birs and Thur is likely to be due to differences in the speciation and thus in the bioavailability of Cu and Zn at these various sites.

In a subsequent study, a detailed evaluation of the speciation of Cu and Zn and of their accumulation in periphyton was performed in a small stream in which the concentrations of these metals increased during rain events [64]. Accumulation of Cu and Zn in periphyton was observed when the concentrations of dissolved Cu and Zn increased during rain events. The concentrations of dissolved Cu and Zn, of the free metal ions and of labile forms of Cu and Zn, varied with time in different ways (Figs. 3 and 4). A detailed examination of the speciation and of the accumulation response led to the conclusion that for Zn the free aqua ion concentration, as measured indirectly by ligand-exchange/voltammetry, was the determining parameter for Zn accumulation in periphyton (Fig. 4). In contrast, the Cu accumulation rather followed the concentration of weak labile complexes (Fig. 3).

This relationship could be explained by the very low free Cu ion concentrations and high degree of complexation, which under these conditions may be limiting for uptake, so that supply of copper from the labile complexes becomes determining. These conclusions were corroborated by a study in microcosms, in which periphyton was exposed to natural river water, which was modified by the addition of metals or of ligands to obtain a range of different conditions [117]. In this case, relationships between zinc accumulation in periphyton and zinc free aqua ions, on the one hand, and copper accumulation in periphyton and weakly complexed Cu, on the other hand, were also observed.

These examples illustrate that deeper insights into the speciation are essential to understanding the bioavailability of metals in natural freshwaters and that several different parameters may be relevant, depending on the particular conditions.

3.5. Case Studies of Metal Uptake by Aquatic Insects

For benthic organisms, aquatic insects or worms, several uptake routes of metals need to be considered, namely uptake from water, uptake from food, and uptake from particles and sediments taken up in filter feeding. The relative importance of these different sources may differ among organisms and among metals [118]. Detailed investigations have been carried out on the cadmium sources for the aquatic insect *Chaoborus* [119–121].

These studies have shown that food, consisting of zooplankton, was the predominant source of cadmium for *Chaoborus*, rather than water [121]. Nevertheless, relationships between Cd in *Chaoborus* and Cd in water, and free Cd^{2+} have been observed in lakes [119]. Competition by protons in acidic lakes and

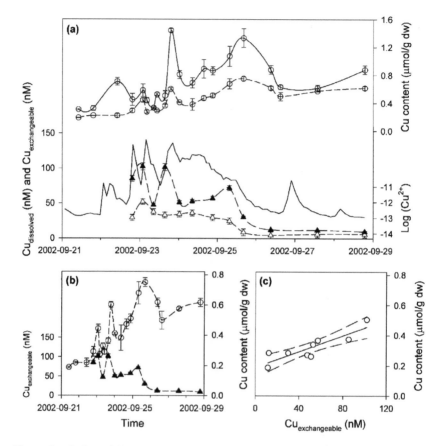

Figure 3 Cu in periphyton as related to speciation parameters of this metal during a rain event in a small stream. (a) Time course of Cu content in periphyton: (——— and ○) total and (– – – and ○) intracellular, (———) Cu dissolved, (weakly complexed, ▲) Cu exchangeable, and (△) Cu^{2+} during a rain event; (b) (○) enlargement of intracellular Cu content in periphyton and (▲) Cu exchangeable; and (c) (———) relationship between Cu content in periphyton and Cu exchangeable, regression line with 95% confidence interval. These relationships indicate that Cu in periphyton follows the weakly complexed Cu. Reprinted with permission of the American Chemical Society from Fig. 4 in Ref. [64].

complexation of Cd by natural organic matter had to be included in the model for Cd uptake (Fig. 5). The relationship between accumulated metals in *Chaoborus* and metals in water must thus be indirect, by relationships between accumulated metals in zooplankton and in algae, and free metal ions. These studies indicate that metals may be transferred between organisms by food over various trophic levels and that indirect relationships to speciation of metals in water may exist. Examination of several species of benthic animals with respect to Cd

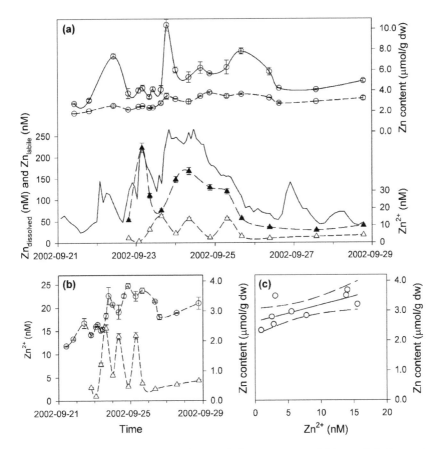

Figure 4 Zn in periphyton as related to speciation parameters of this metal during a rain event in a small stream. (a) Time course of Zn content in periphyton: (——— and ○) total, and (– – – and ○) intracellular Zn, (———) Zn dissolved, (weakly complexed, ▲), Zn labile, and (△) Zn^{2+} during a rain event; (b) (○) enlargement of intracellular Zn content in periphyton and (△) Zn^{2+}; and (c) (———) relationship between Zn content in periphyton and Zn^{2+}, regression line with 95% confidence interval. Zn in periphyton appears to depend on Zn^{2+}. Reprinted with permission of the American Chemical Society from Fig. 5 in Ref. [64].

accumulation indicated that either water or sediment may be the main source of Cd for different species, depending on their feeding and behavior [118].

3.6. Metal Bioavailability to Fish

Interactions of metals at the fish gills have been studied, which appear to represent the most important exposure pathway for metals and to be determining for metal toxicity to fish [98,101,122,123]. For these interactions the free aqua

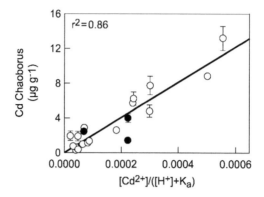

Figure 5 Cd in *Chaoborus* larvae as a function of free Cd^{2+} in Canadian lakes, normalized for competition between Cd and H^+ ions and for interactions of Cd with dissolved organic matter. (●) pH < 5.0; (○) pH > 5.0. Reprinted with permission of Macmillan Magazines Ltd. from Fig. 2 in Ref. [119].

ions may also be determining because uptake occurs by binding of metals to biotic ligands at the gills [98]. As mentioned in Section 3.1, the BLM model has been developed from experiments with fish and from modeling the metal–gill interactions as binding of metals to biotic ligands [98,101,122,123]. This model takes into account the metal speciation in water and the interactions between cations, anions, and organic ligands for binding to biotic ligands at the gills. Dissolved organic matter mostly plays the role of a competing ligand in solution for metals that decreases binding to the gill ligands [98,101,124,125]. For example, uptake of Cd by carp was shown to follow the FIAM in presence of HA [124] (Fig. 6). Natural organic matter was observed to decrease the toxicity effects of Cu and Cd on rainbow trout [125].

Limitations of the FIAM, as indicated in Section 3.1, also apply in the case of metal uptake by fish and have been observed in some studies, for example specific effects of some small organic ligands [126]. Using the BLM for fish, only the direct interactions of metals from water are considered, whereas other exposure pathways are not included. The role of dietary uptake of metals, and uptake of metals from sediments, vs. direct interactions between metal in water and biotic ligands remains to be further elucidated [127].

4. BIOAVAILABILITY OF METALS FROM SEDIMENTS

In metal-polluted waters, metals are usually accumulated in sediments, which provide a historical record of metal pollution. Sediments with elevated metal concentrations are often encountered in freshwater, due to present or past metal pollution. Several pathways have to be considered for the availability of metals from sediments to aquatic organisms. Metals in sediments are in contact with sediment

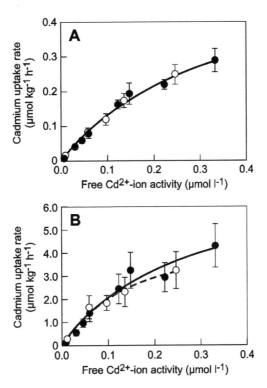

Figure 6 Cd uptake rates in carps (top) and carp gills (bottom) as a function of free Cd^{2+} activity in the presence (O) and absence (●) of humic acids. Reprinted with permission of Elsevier Science from Fig. 3 in Ref. [124].

porewaters, and the metal concentrations in porewaters are regulated by solubility equilibria of metal solid phases and by adsorption reactions, as discussed in Section 2.3. Exchange between porewater and water column may lead to fluxes of metals from sediments. Sediments may also be directly ingested by benthic organisms [128], depending on their feeding behavior. Resuspension of sediments in the water column and reequilibration of metals between particulate and dissolved phase, as well as release of metals upon reoxidation of sulfidic sediments, may also occur.

Sulfides are very important solid phases in many sediments under anoxic conditions. It has been postulated that the solubility of sulfides is a determining factor for the availability and the toxicity of metals in sediments [129–131]. According to the acid volatile sulfide (AVS) model, the porewater concentration of metals such as Cd, Ni, Pb, Zn, and Cu is determined by the solubility of their sulfides, which in turn depends on the availability of sulfide in the sediments [130]. AVS refers to sulfide that is reactive to acid extraction and consists mostly of $FeS(s)$. This sulfide may react with the trace metals to form solid sulfides with low solubility. If the trace metal content in sediments exceeds the available sulfide, higher metal concentrations occur in the porewater, which therefore become available for organisms.

The AVS model of metal toxicity in sediments relies upon the assumption that the available metals in porewater are determining for the toxicity for benthic organisms. This model has been demonstrated to be valid in a number of cases in laboratory and field experiments, in particular regarding the toxicity of sediments to benthic organisms [129–132]. However, examination of metal accumulation in various types of invertebrates has indicated that feeding behavior and dietary uptake was controlling uptake of several metals and that AVS did not in all cases provide reliable predictions of metal uptake [118,128,133]. Consideration of exposure to the metals bound in the solid phase of sediments may therefore also be necessary to understand the availability of metals from sediments, depending on the organisms.

5. CONCLUSIONS AND OUTLOOK

Metal speciation in freshwater environments strongly depends on the complex interactions of metals with natural organic ligands. Further work on the experimental determination of speciation needs to be done in order to systematically relate the complexation of trace elements with other characteristics of freshwaters. Direct relationships between metal speciation and metal bioavailability under freshwater field conditions have only been demonstrated in a few cases. Further work relating metal speciation and bioavailable concentrations is therefore needed to better understand these relationships.

On the basis of the simple models of FIAM and BLM, it appears that methods yielding the free ion concentrations should be used to define the bioavailable concentrations under field conditions. However, these widely used models for interactions of metals with organisms appear to be limited, as increasingly more sophisticated models including kinetic considerations become available [6,102]. These new developments influence the parameters that should be determined for a prediction of the bioavailability. Relationships between experimentally determined parameters by various methods and bioavailable concentrations for various organisms need to be further examined.

ACKNOWLEDGMENTS

We thank Hanbin Xue and Sébastien Meylan for their important contributions to the topic of this chapter.

ABBREVIATIONS

| AVS | acid volatile sulfide |
| BLM | biotic ligand model |

DGT   diffusion gradients in thin films
DOC   dissolved organic carbon
DOM   dissolved organic matter
FA   fulvic acid
FIAM   free ion activity model
HA   humic acid
WHAM   Windermere humic aqueous model

REFERENCES

1. Nriagu JO, Pacyna JM. Nature 1988; 333:134–139.
2. Tessier A, Turner DR, eds. Metal Speciation and Bioavailability in Aquatic Systems, Vol. 3. Chichester: John Wiley & Sons, 1995:679.
3. Meyer JS. Mar Environ Res 2002; 53:417–423.
4. Morel FMM, Hudson RJM, Price NM. Limnol Oceanogr 1991; 36:1742–1755.
5. Campbell PGC. Metal Speciation and Bioavailability in Aquatic Systems. In: Tessier A., Turner DR, eds. IUPAC Series on Analytical and Physical Chemistry of Environmental Systems. Chichester: John Wiley & Sons, 1995:45–102.
6. Van Leeuwen HP. Environ Sci Technol 1999; 33:3743–3748.
7. Hudson RJM. Sci Total Environ 1998; 219:95–115.
8. Dyrssen D. Mar Chem 1988; 24:143–153.
9. Rozan TF, Benoit G, Luther GW. Environ Sci Technol 1999; 33:3021–3026.
10. Rozan TF, Lassman ME, Ridge DP, Luther GWI. Nature 2000; 406:879–882.
11. Morse JW, Luther GWI. Geochim Cosmochim Acta 1999; 63:3373–3378.
12. Mylon SE, Benoit G. Environ Sci Technol 2001; 35:4544–4548.
13. Kaiser E, Ph.D., Swiss Federal Institute of Technology (ETH). Zürich, Switzerland, 2002.
14. Tipping E. Cation Binding by Humic Substances. Cambridge Environmental Chemistry Series. Cambridge: Cambridge University Press, 2002:434.
15. Benedetti MF, Milne CJ, Kinniburgh DG, Van Riemsdijk WH, Koopal LK. Environ Sci Technol 1995; 29:446–457.
16. Benedetti MF, Van Riemsdijk WH, Koopal LK, Kinniburgh DG, Gooddy DC, Milne CJ. Geochim Cosmochim Acta 1996; 60:2503–2513.
17. Kinniburgh DG, Milne CJ, Benedetti MF, Pinheiro JP, Filius J, Koopal LK, Van Riemsdijk WH. Environ Sci Technol 1996; 30:1687–1698.
18. Tipping E. Colloids Surf A: Physicochem Eng Aspects 1993; 73:117–131.
19. Tipping E. Aquat Geochem 1998; 4:3–48.
20. Haitzer M, Aiken GR, Ryan JN. Environ Sci Technol 2003; 37:2436–2441.
21. Tipping E. Comput Geosci 1994; 20:973–1023.
22. Xue H, Sigg L. Aquat Geochem 1999; 5:313–335.
23. Coale KH, Bruland KW. Limnol Oceanogr 1988; 33:1084–1101.
24. Bruland KW. Limnol Oceanogr 1989; 34:267–283.
25. Bruland KW. Limnol Oceanogr 1992; 37:1008–1017.
26. Xue HB, Kistler D, Sigg L. Limnol Oceanogr 1995; 40:1142–1152.
27. Le Gall AC, Van den Berg CMG. Deep-Sea Res 1998; 45:1903–1918.
28. Laglera LM, Van den Berg CMG. Mar Chem 2003; 82:71–89.
29. Al-Farawati R, Van den Berg CMG. Environ Sci Technol 2001; 35:1902–1911.

30. Tang D, Hung C-C, Warnken KW, Santschi PH. Limnol Oceanogr 2000; 45:1289–1297.
31. Gordon AS, Dyer BJ, Kango RA, Donat JR. Mar Chem 1996; 53:163–172.
32. Gordon AS, Donat JR, Kango RA, Dyer BJ, Stuart LM. Mar Chem 2000; 70:149–160.
33. Midorikawa T, Tanoue E. Mar Chem 1996; 52:157–171.
34. Midorikawa T, Tanoue E. Fresenius J Anal Chem 1999; 363:584–586.
35. Vachet RW, Callaway MB. Mar Chem 2003; 82:31–45.
36. Town RM, Filella M. Limnol Oceanogr 2000; 45:1341–1357.
37. Kogut MB, Voelker BM. Environ Sci Technol 2001; 35:1149–1156.
38. Perret D, DeVitre RR, Leppard G, Buffle J. In: Tilzer MM, Serruya C, eds. Large Lakes: Ecological Structure and Function. Heidelberg: Springer, 1990:224–244.
39. Buffle J, DeVitre RR, Perret D, Leppard GG. Geochim Cosmochim Acta 1989; 53:399–408.
40. Perret D, Newman ME, Nègre J-C, Chen Y, Buffle J. Water Res 1994; 28:91–106.
41. Davison W, DeVitre R. In: Buffle J, Van Leeuwen H, eds. Environmental Particles I. Vol. 1, Boca Raton, FL: Lewis Publishers, 1992:315–355.
42. Eyrolle F, Benedetti MF, Benaïm JY, Février D. Geochim Cosmochim Acta 1996; 60:3643–3656.
43. Babiarz CL, Hurley JP, Hoffmann SR, Andren AW, Shafer MM, Armstrong DE. Environ Sci Technol 2001; 35:4773–4782.
44. Benoit G, Oktay-Marshall SD, Cantu AI, Hood EM, Coleman CH, Corapcioglu MO, Santschi PH. Mar Chem 1994; 45:307–336.
45. Muller FLL. Mar Chem 1996; 52:245–268.
46. Ross JM, Sherrell RM. Limnol Oceanogr 1999; 44:1019–1034.
47. Sañudo-Wilhelmy SA, Rivera-Duarte I, Flegal AR. Geochim Cosmochim Acta 1996; 60:4933–4944.
48. Sigg L, Xue H, Kistler D, Schönenberger R. Aquat Geochem 2000; 6:413–434.
49. Wen L-S, Santschi P, Gill G, Paternostro C. Mar Chem 1999; 63:185–212.
50. Wells ML, Smith GJ, Bruland KW. Mar Chem 2000; 71:143–163.
51. Stumm W, Morgan JJ. Aquatic Chemistry. Chemical Equilibria and Rates in Natural Waters, 3d ed. New York: John Wiley & Sons, 1996:1022.
52. Carroll SA, O'Day PA, Piechowski M. Environ Sci Technol 1998; 32:956–965.
53. Balistrieri LS, Murray JW, Paul B. Geochim Cosmochim Acta 1994; 58:3993–4008.
54. Morse JW. Mar Chem 1994; 46:1–6.
55. Sigg L. In: Allen HE, Garrison AW, Luther GWI, eds. Metals in Surface Waters. Chelsea, MI: Ann Arbor Press, 1998:262.
56. Xue HB, Sigg L. Limnol Oceanogr 1993; 38:1200–1213.
57. Xue HB, Sigg L. Anal Chim Acta 1994; 284:505–515.
58. Xue H, Oestreich A, Kistler D, Sigg L. Aquat Sci 1996; 58:69–87.
59. Xue HB, Sigg L. Anal Chim Acta 1998; 363:249–259.
60. Xue H, Jansen S, Prasch A, Sigg L. Environ Sci Technol 2001; 35:539–546.
61. Qian J, Xue HB, Sigg L, Albrecht A. Environ Sci Technol 1998; 32:2043–2050.
62. Odzak N, Kistler D, Xue H, Sigg L. Aquat Sci 2002; 64:292–299.
63. Xue H, Sigg L. In: Rozan TF, Taillefert M, eds. Environmental Electrochemistry: Analyses of Trace Element Biogeochemistry. Vol. 811, Washington: Symposium Series, ACS, 2002:336–370.
64. Meylan S, Behra R, Sigg L. Environ Sci Technol 2003; 37:5204–5212.

65. Achterberg EP, Van den Berg CMG, Boussemart M, Davison W. Geochim Cosmochim Acta 1997; 61:5233–5253.
66. Rozan TF, Benoit G. Geochim Cosmochim Acta 1999; 63:3311–3319.
67. Mylon SE, Twining BS, Fisher NS, Benoit G. Environ Sci Technol 2003; 37:1261–1267.
68. Breault RF, Colman JA, Aiken GR, McKnight D. Environ Sci Technol 1996; 30:3477–3486.
69. Zhang H, Davison W. Anal Chem 2000; 72:4447–4457.
70. Gimpel J, Zhang H, Davison W, Edwards A. Environ Sci Technol 2003; 37:138–146.
71. Knauer K, Ahner B, Xue H, Sigg L. Environ Toxicol Chem 1998; 17:2444–2452.
72. Meylan S, Odzak N, Behra R, Sigg L. Anal Chim Acta 2004; 510:91–100.
73. Jansen RAG, Van Leeuwen HP, Cleven RFMJ, Van den Hoop MAGT. Environ Sci Technol 1998; 32:3882–3886.
74. Lam MT, Murimboh J, Hassan NM, Chakrabarti CL. Electroanalysis 2001; 13:94–99.
75. Mandal R, Hassan NM, Murimboh J, Chakrabarti CL, Back MH, Rahayu U, Lean DRS. Environ Sci Technol 2002; 36:1477–1484.
76. Anderson MA, Morel FMM, Guillard RRL. Nature 1978; 276:70–71.
77. Sunda W. Biol Oceanogr 1988/89; 6:411–442.
78. Sunda W, Guillard RRL. J Mar Res 1976; 34:511–529.
79. Sunda WG, Huntsman SA. Limnol Oceanogr 1992; 37:25–40.
80. Sunda WG, Huntsman SA. Limnol Oceanogr 1995; 40:132–137.
81. Sunda WG, Huntsman SA. Limnol Oceanogr 1996; 41:373–387.
82. Sunda WG, Huntsman SA. Limnol Oceanogr 1998; 43:1467–1475.
83. Sunda WG, Huntsman SA. Limnol Oceanogr 2000; 45:1501–1516.
84. Errecalde O, Seidl M, Campbell PGC. Water Res 1998; 32:419–429.
85. Fortin C, Campbell PGC. Environ Toxicol Chem 2000; 19:2769–2778.
86. Fortin C, Campbell PGC. Environ Sci Technol 2001; 35:2214–2218.
87. Knauer K, Behra R, Sigg L. Environ Toxicol Chem 1997; 16:220–229.
88. Knauer K, Behra R, Sigg L. J Phycol 1997; 33:596–601.
89. Errecalde O, Campbell PGC. J Phycol 2000; 36:473–483.
90. Slaveykova VI, Wilkinson KJ. Environ Sci Technol 2002; 36:969–975.
91. Slaveykova VI, Wilkinson KJ, Ceresa A, Pretsch E. Environ Sci Technol 2003; 37:1114–1121.
92. Hassler CS, Wilkinson KJ. Environ Toxicol Chem 2003; 22:620–626.
93. Croot PL, Karlson B, Van Elteren JT, Kroon JJ. Environ Sci Technol 1999; 33:3615–3621.
94. Phinney JT, Bruland KW. Estuaries 1997; 20:66–76.
95. Phinney JT, Bruland KW. Environ Toxicol Chem 1997; 16:2046–2053.
96. Escher BI, Sigg L. In: Van Leeuwen HP, Köster W, eds. Kinetics and Transport at Biointerfaces. Vol. 9, IUPAC Series on Analytical and Physical Chemistry of Environmental Systems, John Wiley & Sons, 2004:205–269.
97. Paquin PR, Gorsuch JW, Apte SA, Batley GE, Bowles KC, Campbell PGC, Delos CG, Di Toro DM, Dwyer RL, Glavez F, Gensenmer RW, Goss GG, Hogstrand C, Janssen CR, McGeer JC, Naddy RB, Playle RC, Santore RC, Schneier U, Stubblefield W, Wood CM, Wu KB. Comput Biochem Phys Part C 2002; 133:3–35.

98. Playle RC. Sci Total Environ 1998; 219:147–163.
99. Santore RC, DiToro DM, Paquin PR, Allen HE, Meyer JS. Environ Toxicol Chem 2001; 20:2397–2402.
100. DiToro DM, Allen HE, Bergman HL, Meyer JS, Paquin PR, Santore RC. Environ Toxicol Chem 2001; 20:2383–2396.
101. McGeer JC, Playle RC, Wood CM, Galvez F. Environ Sci Technol 2000; 34:4199–4207.
102. Galceran J, Van Leeuwen HP. In: Van Leeuwen HP, Köster W, eds. Kinetics and Transport at Biointerfaces. Vol. 9, IUPAC Series on Analytical and Physical Chemistry of Environmental Systems, John Wiley & Sons, 2004:147–203.
103. Wilkinson KJ, Buffle J. In: Van Leeuwen HP, Köster W, eds. Kinetics and Transport at Biointerfaces. Vol. 9, IUPAC Series on Analytical and Physical Chemistry of Environmental Systems, John Wiley & Sons, 2004:445–533.
104. Vigneault B, Percot A, Lafleur M, Campbell PGC. Environ Sci Technol 2000; 34:3907–3913.
105. Parent L, Twiss MR, Campbell PGC. Environ Sci Technol 1996; 30:1713–1720.
106. Campbell PGC, Twiss MR, Wilkinson KJ. Can J Fish Aquat Sci 1997; 54:2543–2554.
107. Wang W-X, Guo L. Environ Sci Technol 2000; 34:4571–4576.
108. Carvalho RA, Benfield MC, Santschi PH. Limnol Oceanogr 1999; 44:403–414.
109. Sunda WG, Huntsman SA. Sci Total Environ 1998; 219:165–181.
110. Sunda WG, Huntsman SA. Limnol Oceanogr 1998; 43:1055–1064.
111. Harrison GI, Morel FMM. Limnol Oceanogr, 1986; 31:989–997.
112. Sunda WG, Huntsman SA. Limnol Oceanogr 1985; 30:71–80.
113. Finney LA, O'Halloran TV. Science 2003; 300:931–936.
114. Clemens S. Planta 2001; 212:475–486.
115. Mason AZ, Jenkins KD. In: Tessier A, Turner DR, eds. Metal Speciation and Bioavailability in Aquatic Systems. Vol. 3, Analytical and Physical Chemistry of Environmental Systems, Chichester: John Wiley & Sons, 1995:479–608.
116. Behra R, Landwehrjohann R, Vogel K, Wagner B, Sigg L. Aquat Sci 2002; 64:300–306.
117. Meylan S, Behra R, Sigg L. Environ Sci Technol 2004; 38:3104–3111.
118. Warren LA, Tessier A, Hare L. Limnol Oceanogr 1998; 43:1441–1454.
119. Hare L, Tessier A. Nature 1996; 380:430–432.
120. Hare L, Tessier A. Limnol Oceanogr 1998; 43:1850–1859.
121. Munger C, Hare L, Tessier A. Limnol Oceanogr 1999; 44:1763–1771.
122. Richards JG, Curtis PJ, Burnison BK, Playle RC. Environ Toxicol Chem 2001; 20:1159–1166.
123. Playle RC, Dixon DG, Burnison K. Can J Fish Aquat Sci 1993; 50:2678–2687.
124. Van Ginneken L, Bervoets L, Blust R. Aquat Toxicol 2001; 52:13–27.
125. Richards JG, Burnison BK, Playle RC. Can J Fish Aquat Sci 1999; 56:407–418.
126. Van Ginneken L, Chowdury MJ, Blust R. Environ Toxicol Chem 1999; 18:2295–2304.
127. Bervoets L, Blust R, Verheyen R. Ecotoxical Environ Saf 2001; 48:117–127.
128. Lee BG, Griscom SB, Lee J-S, Choi HJ, Koh C-H, Luoma SN, Fisher NS. Science 2000; 287:282–284.
129. DiToro DM, Mahony JD, Hansen DJ, Scott KJ, Hicks MB, Mayr SM, Redmond MS. Environ Toxicol Chem 1990; 9:1487–1502.

130. DiToro DM, Mahony JD, Hansen DJ, Scott KJ, Carlson AR, Ankley GT. Environ Sci Technol 1992; 26:96–101.
131. Ankley GT. Environ Toxicol Chem 1996; 15:2138–2146.
132. Sibley PK, Ankley GT, Cotter AM, Leonard EN. Environ Toxicol Chem 1996; 15:2102–2112.
133. Lee B-G, Lee J-S, Luoma SN, Choi HJ, Koh C-H. Environ Sci Technol 2000; 34:4517–4523.

4

Bioavailability and Biogeochemistry of Metals in the Terrestrial Environment

Kerstin Michel and Bernard Ludwig

Department of Environmental Chemistry, University of Kassel,
Nordbahnhofstrasse 1a, D-37213 Witzenhausen, Germany

1. INTRODUCTION

Metals comprise about two-thirds of the known elements. They have low electro-negativity values ranging from 0.7 to 1.8 (Pauling scale) and tend to lose electrons and thus form positive ions. Metals are divided into light and heavy metals. The term heavy metal refers to any metallic chemical element with a density exceeding a certain threshold value which varies in the literature between 3.5 and 6 g/cm^3 [1,2]. Heavy metals are the largest subgroup of metals. Chemically, the metals differ from the non-metals in that they form basic oxides and hydroxides. Other forms in which metals occur in soils are other sparingly soluble salts (e.g., primary and secondary silicates, carbonates, sulfates), solid solutions, surface complexes, and exchangeable cations in the solid–liquid interphase. Important components of soil solutions are free ions (aqua complexes) and other inorganic complexes (e.g., oxo, hydroxo, chloro, sulfato complexes) and complexes with organic ligands.

Many metals are essential for life. They are required for completion of the life cycle and cannot be replaced by any other element [3]. Metals that are essential for plants, animals, and humans include Fe, Zn, Mn, Cu, and Co, whereas Pb, Cd, and Hg have no known physiological function. However, essential metals also may become toxic when exposures to biota become excessive. In general, two factors are decisive for the effects of both essential and potentially toxic metals on an organism: the exposure of an organism to a given metal, i.e., contact with the metal, and the bioavailability of the metal [4,5]. However, unfortunately there is no general agreement on the definition of bioavailability and the approaches used for its determination.

The objective of this chapter is to critically review methods and approaches to determine bioavailable metals. Furthermore, potentials and limitations of existing concepts are discussed and data from various bioavailability studies on selected metals are presented. The biogeochemistry of selected metals is also considered. Soil plays a key role in terrestrial ecosystem functioning. Therefore, the focus of this chapter will be on soils, soil-dwelling organisms, and plants.

2. THE CONCEPT OF BIOAVAILABILITY

Bioavailability is a qualitative term conveying the concept that exposure of organisms to a substance or element can often not directly be derived from total concentrations present in the environment. The concept of bioavailability was first proposed by Pavlou et al. [6] in 1975. Since then the term bioavailability increasingly appeared in the literature—often ill-defined. It has been used to describe very different fractions [7].

The following (and other) definitions of the term "bioavailability" can be found in literature:

- The fraction of an ingested chemical that is absorbed and available for metabolism, storage, or excretion [8].
- The degree to which a contaminant in a potential source is free for uptake (movement into or onto an organism) [9].

- A metal is considered to be in bioavailable form when it is taken up by organisms and is subsequently accessible to the physiological processes upon which it exerts an effect [10].
- The extent to which a toxic metal is available for uptake [11].
- Metal availability for uptake into roots [12].
- The fraction of total metal that is available to exert action and effect within the receptor organism [4].

2.1. Approaches to Determine Bioavailability

Different approaches have been developed to define bioavailability. As a consequence thereof, a great variety of methods exists today that were developed for measuring bioavailable metal fractions in soils.

2.1.1. Chemical Approach

From a chemical point of view, bioavailability is defined in terms of the chemical form in which an element occurs in a given system at a given time. There is evidence that the uptake of metals is highly dependent on the chemical speciation of the metals in solution [13]. In aquatic systems and for higher plants, the best correlations were often found between total uptake and the activity of free, uncomplexed metals ions in solution. This relationship is normally described as the free ion activity model (FIAM). It is therefore assumed that, with the exception of Hg in aqueous systems, the free metal ion is the chemical species which is most bioavailable and therefore predominantly taken up by biota [14,15].

2.1.1.1. Metal speciation in solution: Soil solution, or interstitial pore water, has been defined as the aqueous liquid phase of the soil and its solutes [16]. Soil solution and pore water can be obtained in both field and laboratory experiments. In the field soil solution is commonly collected in different depths of the soil profile by tensiometers or lysimeters. A relatively simple technique to obtain soil solution in the laboratory is centrifugation [17]. Compared to lysimeters, centrifugation tends to extract higher concentrations of most solutes in soil solution [18,19]. Displacement techniques or saturation extracts are also used to obtain soil solution in the laboratory [14,19].

Various techniques exist to measure free metal activity in solution. Free metal species can be determined by potentiometric techniques, for example, ion-selective electrodes (ISEs). The determination of free metal ions by ISEs is mainly restricted to Cu^{2+} [20]. ISEs also exist for other metals such as Cd and Pb. However, their use is limited by a lack of sensitivity and poor detection limits. Other possibilities are voltammetric methods, the use of ion-exchange resins or dialysis techniques. Voltammetric techniques such as anodic stripping voltammetry measure the electrochemically labile species of metal in solution. They are not species-specific [21]. Besides the free metal ions, metal ions paired with inorganic ions or weak organic ligands are also determined. Ion-exchange resins are employed in column-equilibration procedures to separate the free metal ions from their complexes [22,23]. In principle, the exchange resin only sorbs the free metal ions. These are finally eluted from the resin and measured by atomic absorption spectroscopy. According to Nolan et al. [24], the Donnan equilibrium method developed by Cox et al. [25] is the most accurate technique for the determination of free metal ions that is available today. This method employs a dialysis separation which is done by a semi-permeable membrane that can be crossed only by certain species. A species can be excluded on the basis of its charge or its size.

It is also possible to calculate metal speciation in soil solution. Chemical equilibrium models such as PHREEQC [26,27], MINEQL+ [28], or WHAM [29] are used for this purpose. The calculations are based on known solution composition and on assumptions that are made with regard to the interaction of metals with organic ligands.

2.1.1.2. Extractions and digestions: Single extractant procedures are widely used to determine metals in soils that are available to plants. Various extractants, for example, $CaCl_2$, diethylenetriaminepentaacetate (DTPA), ethylenediaminetetraacetate (EDTA), mineral acids like HNO_3 or HCl, and Mehlich 1 or 3, have been used for this purpose [30,31]. The released quantities often vary widely. A comparison of the extraction of several heavy metals by different extractants showed that Mehlich 3, a mixture of diluted acid, salt, and EDTA, removed on average 32% of the total metal content in soils. In contrast, DTPA extracted only 11% [32]. DTPA is also used to predict metal concentrations available to soil animals such as earthworms (e.g., [33–35]).

Sequential extractions were developed to study metal speciation and behavior in soils and sediments. The general principle underlying such extraction procedures is the subsequent use of different extractants with increasing strength. It is supposed that each extractant dissolves a specific adsorbent. Sequential extraction procedures are often used to differentiate between mobile, i.e., actually bioavailable, potentially bioavailable, and inert metal fractions (e.g., [36,37]). The suitability of such extractions to predict or reflect bioavailable metal fractions can be validated by comparing the various extracted metal concentrations with plant uptake, i.e., tissue concentrations (e.g., [38]).

A variety of sequential extraction procedures exists [39]. Two of the most frequently used are the ones developed by Tessier et al. [40] and Zeien and Brümmer [41], they are compared in Table 1. Table 1 shows that the fractions identified by sequential extractions are operationally defined and that they differ between the various procedures.

Regulatory guidelines concerning soil quality criteria or risk assessments often refer to the total metal concentration in soil which is determined by digestion procedures. Usually, strong acids like nitric acid (HNO_3), sulfuric acid (H_2SO_4), hydrofluoric acid (HF), or mixtures of these (e.g., aqua regia) are involved [42] leading to a partial breakdown of the lattices of silicates. Thus, metals belonging to the inert fraction which consists of insoluble stable forms

Table 1 Comparison of Two Sequential Extraction Procedures Commonly Used for the Determination of the Main Heavy Metal Bonding Forms in Soils

	Tessier et al. [40]		Zeien and Brümmer [41]
Fraction/bonding form	Extractant	Fraction/bonding form	Extractant
(1) Exchangeable	1 M $MgCl_2$	(1) Mobile	1 M NH_4NO_3
(2) Bound to carbonates	1 M NaOAc (pH 5)	(2) Easily mobilizable	1 M NH_4OAc (pH 6)
(3) In Mn–Fe oxides occluded	0.04 M NH_2OH–HCl in HOAc	(3) In Mn oxides occluded	0.1 M NH_2OH–HCl + 1 M NH_4OAc
(4) Organically bound	H_2O_2 (30%) + HNO_3	(4) Organically bound	0.025 M NH_4EDTA
		(5) In poorly crystalline Fe oxides occluded	0.2 M NH_4 oxalate buffer (pH 3.25)
		(6) In well crystalline Fe oxides occluded	0.1 ascorbic acid in 0.2 M NH_4 oxalate buffer
(5) Residual	Conc. $HClO_4$/HF	(7) Residual	Conc. $HClO_4$ + conc. HNO_3

and stable crystalline metal forms are also determined. Digestions are normally processed under elevated temperature (using hotplates, sand baths, aluminum blocks, or pressure digestion bombs) [43–46]. In recent years, microwave-assisted methods with both open and pressurized systems were developed (e.g., [47,48]). Microwave digestion often produces higher precision and increases elemental recovery [49–51].

2.1.1.3. Diffusive gradients in thin films: Zhang and Davison [52] proposed the so-called diffusive gradients technique in thin films (DGT) to separate free metals from organically complexed metals in aqueous solutions. Several metals, e.g., Al, Fe, Zn, and Cu, have been determined in soil solutions using DGT [53,54]. In principle, the hydrated metal cations and soluble inorganic complexes are accumulated on an ion-exchange resin after diffusing through a thin layer of gel. Like plants, DGT locally lowers metal concentrations in the soil solution. DGT responds to metals re-supplied via dissociation of labile metals in solution and via desorption from the labile metal pool in the solid phase.

2.1.2. Biological Approach

Biologists determine bioavailability based on the proportion of a metal that could pass into an organism. According to the biological approach, the chemical form is relevant only to the presence of a biological receptor. This approach includes acute tests considering germination, plant growth, and effects on microbial (respiration, nitrification, colony-forming units, etc.) and faunal parameters (e.g., reproduction, mortality) [55]. These tests are often carried out without chemical analyses. Commonly used methods for the biological approach are bioassays, bioaccumulators, biomarkers, and biosensors (Section 2.1.2.1) and isotopic dilution techniques (IDT; Section 2.1.2.2). In other studies, the tissue concentrations of plants or animals obtained by digestion (Section 2.1.1.2) were considered as bioavailable fraction [56,57].

2.1.2.1. Bioassays, bioaccumulators, biomarkers, and biosensors: A bioassay is a way to test for the presence of a chemical contaminant by using living organisms as an indicator. A substance can thus be quantitatively determined by measuring its direct effect on a (suitable) microorganism, plant, or animal. Bioassays are carried out under controlled conditions [16]. A number of bioassays have been introduced that measure mortality, growth, reproduction, or behavior of organisms. Frequently determined parameters are, for example, LD_{50} and LC_{50}. LD_{50} is the amount of an element or compound which causes the death of 50% of a group of test organisms in one dose. It is a measure for short-term poisoning potential or acute toxicity. The concentration of an element or compound that kills 50% of the test organisms in a given time (usually 4 h) is the LC_{50} value.

To measure the amount of bioavailable metals, sentinels or bioaccumulators are also used. They are defined as biological monitors that accumulate a pollutant in their tissues without significant adverse effects [58]. Especially, soft-bodied animals such as earthworms or snails have been used as sentinels (e.g., [56,59]). On the basis of their habits, for example, their feeding behavior, conclusions can be drawn for other species concerning the bioavailability of a pollutant from the same source.

The biomarker approach uses test organisms as environmental condition sensors. The molecular mechanisms which these organisms activate to cope with adverse environmental conditions are indicative of the stresses they are encountering. Biomarkers are thus biological (physiological, histological, cellular, or biochemical) manifestations of pollutant stress [60]. Numerous biomarkers have been proposed and examined. The activity of δ-aminolevulinic acid dehydratase (ALAD), a metalloprotein that plays a role in heme synthesis, has been used as a biomarker for Pb exposure in humans and multicellular aquatic eukaryotes (e.g., [61–63]). Recently, bacterial ALAD was proposed as a biomarker for Pb bioavailability in contaminated environments [10]. The ALAD activity of certain bacterial isolates (e.g., *Pseudomonas putida*) can be used as a biomarker for biologically available Pb, since Pb can replace the metallic component, normally Zn or Mg, which is required for enzyme activity and thus inhibits ALAD. Metallothioneins are also considered to be potential biomarkers in terrestrial ecosystems. They are metal-binding, low molecular-weight proteins that are characterized by high cysteine content and occur in animals, plants, and eukaryotic microorganisms. Their synthesis can be induced by a wide variety of metal ions, for example, Cd, Cu, Co, and Zn. The biomarker function of metallothionein, especially for Cd, has been investigated for terrestrial invertebrates such as soil insects [*Orchesella cincta* (Collembola)] [64], gastropods like the Roman snail (*Helix pomatia*) [65,66], or lumbricides, e.g., the common brandling worm (*Eisenia fetida*) [66]. Other potential biomarkers that have been measured in earthworms include DNA alteration, energy reserve responses, or lysosomal membrane stability (neutral-red retention) [67–69]. For plants, the induction and amounts of phytochelatins, the fatty acid composition of leaves and peroxidase activity were proposed as biomarkers for heavy metal stress [70–72].

Recently, the use of metal-specific luminometry based assays has received much attention. Such assays belong to the group of so-called biosensors. A biosensor is an analytical device incorporating a biological material or a biomimic (microorganisms, organelles, cell receptors, enzymes, antibodies, etc.) that is intimately associated with or integrated within a physicochemical transducer (e.g., optical, electrochemical, magnetic). For example, Tibazarwa et al. [73] constructed a bacterial luminometry based biosensor to predict bioavailable Ni in soil with high selectivity for Ni and a detection limit of 0.1 μM for Ni^{2+}. For this purpose, *Ralstonia eutropha* was used that is characterized by Co and Ni resistance. The regulatory genes of the Co and Ni resistance were cloned upstream of the bioluminescent reporter system leading to inducible light production controlled by the resistance regulatory genes. The Ni levels determined with this biosensor corresponded well to Ni accumulated in different parts of maize and potato tubers [73].

  2.1.2.2. IDT: IDT were initially used to characterize the phytoavailable pool of macronutrients, especially of P [74,75]. Recently, the availability of some metallic trace elements (e.g., Cd, Zn) to plants and soil-dwelling animals has been evaluated by using this technique [76,77]. IDT are combined solution- and solid-phase fractionation techniques that allow to distinguish between the isotopically non-exchangeable (fixed) metal pool and the isotopically exchangeable, i.e., labile, metal pool in soils. The latter one corresponds to the solid phase pool of an element in soil that may supply the soil solution during a given period of time, whereas the metals in the fixed pool are strongly bound to the solid phase. The introduction of a radioactive isotope in a soil solution leads to its rapid distribution within the labile metal pool without modifying the equilibrium between the different metal pools. From the isotopic composition of the labile pool, i.e., from the ratio of the radioisotope to the stable element, the isotopically exchangeable fraction can be calculated. The isotopic composition of the fixed pool is zero, since the radioisotope does not mix with the non-exchangeable metal pool. Organisms that are able to mobilize non-labile forms of metals are characterized by a lower isotopic composition than that of the labile pool. In contrast, there is no difference in the isotopic composition if metal uptake by an organism occurs only from the labile pool.

2.1.3. Environmental Approach

Another approach is the use of empirical relationships to predict the bioavailability of metals in different environments. Tissue concentrations or accumulation characteristics such as bioconcentration factors or uptake-rate constants are associated with soil characteristics [pH, cation exchange capacity (CEC), dissolved organic matter (DOM), organic matter content, etc.]. The influence of the different soil characteristics on metal concentrations or accumulation is usually determined by multiple regression analyses. This approach was, for example, used to investigate the plant availability of heavy metals in sewage sludge amended soils [78] and to assess the importance of soil properties for metal accumulation in springtails [79].

2.2. Problems and Limitations

The chemical approach was successfully applied to aquatic ecosystems. However, in soils links between metal speciation and uptake or accumulation by organisms are scarce. The quantification of metal speciation in soils is difficult. Neither single or sequential extraction procedures nor pore water analyses are suited to determine metal speciation. Measurements of metal speciation in natural systems disturb the solution equilibrium, i.e., the ratio of the various forms of a metal present, and therefore lead to changes in metal speciation. Some methods such as ISEs are not applicable due to poor detection limits. Extraction procedures only identify operationally defined fractions without any significant chemical meaning. None can be consistently related to metal availability. For example, DTPA is often used to determine metal fractions available to biota in soils. However, in several studies no correlation between DTPA-extractable concentrations and uptake by plants or soil-dwelling organisms were found (e.g., [80,81]). Especially for Ni DTPA-extractable soil contents did not reflect plant concentrations well [78,80,82]. Hooda et al. [78], who compared the ability of different extractants (EDTA, DTPA, $CaCl_2$, NH_4NO_3) to predict plant-available metals, concluded that the quality of the results depends on crop species, metal, and extractant used.

It has to be borne in mind that there are various abiotic parameters that affect the bioavailable fraction of metals in soil (Table 2). The most important factor is pH, since it governs metal solubility, ion exchange, biological activity, etc. The fact that chemical availability does often not mimic bioavailability may at least partly be explained by altered soil chemistry, especially pH or ionic strength, due to extraction procedures.

Additional problems derive from the heterogeneous nature of the soil system. Exposure of organisms to metals occurs not only through the liquid, but also through the solid and gaseous phase. This is of special importance for soil-dwelling organisms, since different uptake routes are possible. Therefore, the metal fraction that can actually be taken up may vary so that it may not be sufficient to consider only the activity of the free metal ion. Plants and microorganisms are primarily exposed through the aqueous phase [83]. Predominant pore-water uptake is also likely for soft-bodied animals such as earthworms. Soft-bodied animals are characterized by a water-permeable integument [84]

Table 2 Selected Abiotic Parameters Influencing Processes that Determine Bioavailable Metal Fractions (According to Ref. [4])

Parameter	Process affected
pH, organic matter content, clay, inorganic oxides	Sorption, leaching, uptake by biota
Ionic strength	Sorption, uptake by biota
Cation exchange capacity	Sorption, leaching
Dissolved organic carbon, inorganic ligands	Complexation in the aqueous phase, leaching, uptake
Temperature	Exchange and equilibration processes, activity of biota
Redox conditions	All processes

enabling uptake of soluble metals through the skin. In the case of semi-soft bodied invertebrates such as springtails pore water concentrations may be of minor importance, although it was hypothesized that the ventral tube, a special organ of springtails that is important for moisture uptake, may play a role in metal uptake [85]. For example, Vijver et al. [79] reported that the accumulation of Cd and Pb in the springtail *Folsomia candida* correlated best with solid-phase properties. For hard-bodied soil organisms (arthropods), especially surface-active species, dietary uptake and thus uptake through the intestine is considered to be the major pathway. In particular, this path may be of special importance for metals that are strongly bound to organic matter [85]. Van Straalen et al. [81] also concluded that metal concentrations in invertebrates are presumable mainly determined by feeding mechanisms.

The universal validity of the FIAM for metal uptake by plants has also been challenged. There is evidence that the free metal is not the only species that can be taken up by plants. In fact, some inorganic complexes, such as chloro and sulfato complexes, may also be available [57,86].

According to the biological approach, metal concentrations determined in plants and soil-dwelling organisms are true measures of bioavailability. Body residues are often directly related to effects on the organ level [87]. Conder et al. [88], however, stated that metal concentrations in earthworms can be over- or under-estimated due to ingested soil that can account for up to 50% of the apparent body weight. The determination of the bioavailable portion of metals, therefore, requires the removal of soil particles adhering on the surface or being present in the intestine of an organism. It has also to be taken into account that metals taken up by biota may be partitioned into biologically available, biologically unavailable, and storage fractions [89]. Another limitation of the biological approach is that the results obtained by investigations of soil-dwelling organisms or plants depend on numerous organism-specific factors such as species, age, growth stage, physiological state, exploited soil volume, or interactions with other biota. Moreover, it has to be borne in mind that some organisms possess mechanisms to regulate the uptake of contaminants (e.g., [87]). Others are able to alter their environment and thus metal availability. Examples are acidification by soil microorganisms and plant roots, exudation of metal-binding compounds (metallothionein, siderophore, etc.) by various biota, and symbiotic associations of plant roots with mycorrhizal fungi. Thus, generalizations are difficult.

The main limitation of the empirical approach is the fact that it does not provide any insight into underlying mechanisms. In general, the fundamentals of the processes that determine the behavior of metals in the environment are only poorly understood. Multiple regression analyses are therefore primarily used to identify parameters that control or reduce metal uptake by soil organisms or plants (e.g., [90]). In some studies, however, no relationship between soil characteristics and parameters that presumably reflect bioavailability could be found (e.g., [91]).

A great variety of questions concerning the "bioavailability" of organic and inorganic compounds in different environments exists. This is reflected by the various definitions (Section 2). In general, the definition of this term used in any scientific study will always depend on the background and perspective of the researcher. Therefore, it is difficult to find a definition that is the best choice under all conditions. A promising approach for metals in terrestrial environments was presented by Peijnenburg et al. [85], who recommended handling bioavailability as a kinetic process. The process consists of several distinct phases: a physicochemically driven desorption process ("environmental availability"), a species-specific physiologically driven process, the so-called "environmental bioavailability" and the internal concentration of an organism ("toxicological bioavailability"). The latter one is related to organ-effect levels and, therefore, decisive for the actual bioavailability. However, it is not possible to distinguish between these three phases, if metal accumulation in biota is relatively fast [92].

Maybe the best way to overcome the problems related to bioavailability is the use of bioassays, especially biomarkers and biosensors. They allow the combined determination of exposure, accumulation, and toxic effects at the receptor level and are, therefore, well suited to monitor actual bioaccessibility and bioavailability.

3. BIOGEOCHEMISTRY AND BIOAVAILABILITY OF METALS

In this section, the results of different studies concerning bioavailable metals in terrestrial ecosystems are presented. As mentioned earlier, these results have to be considered with some caution.

3.1. Aluminum

Aluminum is the third most abundant element in the Earth's crust (after silica and oxygen) and the most abundant metal of the world ($\sim$8.1% by weight). It has only one isotope (^{27}Al). In rocks, Al primarily exists as alumino-silicates such as feldspar [e.g., $Na(AlSi_3O_8)$ or $Ca(Al_2Si_2O_8)$] or clay minerals. Weathering of Al-containing silicates leads to the formation of gibbsite [$Al(OH)_3$], the predominant crystalline Al oxide in soils. Besides oxides, silicates and hydroxides are the dominant forms of Al in soils [93]. In aqueous solutions, Al exists in a large variety of species. Mechanisms controlling Al speciation and solubility are highly pH-dependent [94]. In acid solutions and soils (pH $<$ 5), Al^{3+} dominates, whereas $Al(OH)_4^-$ is the most important species in alkaline solutions. In soils with pH $>$ 6.5, Al is mainly present as hydroxo complexes. Polymeric species, for example, $Al_{13}(O_4)(OH)_{24}^{7+}$, rarely occur in streamwaters and soil solutions [95]. The highest Al solubility is observed between pH 4.5 and 4.0 [96].

Aluminum is not essential for biological functions. In contrast, toxic effects have been reported in numerous investigations [97–100]. The most toxic form is Al^{3+}, although phytotoxicity has also been shown for $Al(OH)_4^-$ and $Al_{13}(O_4)(OH)_{24}^{7+}$ [100,101]. The bioavailability of Al in soils and water is generally low. This is due to its adsorption to mineral surfaces and to interactions with organic matter. In soils polluted with metallurgical dusts, total Al concentrations ranged between 1.4 and 114 g kg^{-1}. Sequential extractions according to Zeien and Brümmer [41] (cf. Table 1) revealed that the major part of Al (78.5%) was bound in silicate lattices (fraction 7/"residual"). Potentially available and mobilizable Al (fractions 1 and 2) accounted together with fractions 3 and 4 for <4% of the total Al present [102]. In cherry orchard soils, available Al concentrations increased with decreasing pH below 5.5 from 0.1 to 2.4 milliequivalents per 100 g. Aluminum content determined in the root system of sweet cherry trees (*Prunus avium*) was proportional to its availability in the soil [103].

3.2. Cadmium

Cadmium is a highly toxic, non-essential metal (for its biogeochemistry, see Chapter 8 of Volume 43). Numerous investigations have been carried out to assess its bioavailability. In various studies, it was shown that Cd uptake by biota, especially by plants, increased with decreasing pH and/or increasing Cd content of the soil [87,104–106]. McBride [107] concluded that total Cd concentrations and soil pH are the two most important factors controlling Cd bioavailability. However, in the study of Huang et al. [108] total Cd concentrations were not correlated to the Cd content of lettuce (*Lactuca sativa* L.). In 20 Dutch field soils and one OECD (Organization for Economic Cooperation and Development) artificial soil, no Cd was taken up by the compost worm *Eisenia andrei* if total Cd concentration of the soil was below 0.01 mmol kg^{-1} [92]. According to Lock and Janssen [109], the bioavailability of Cd and Zn to the pot worm *Enchytraeus albidus* was primarily controlled by pH and CEC. Clay and organic matter content are two other factors that may affect Cd bioavailability [107]. However, close relationships between these parameters and Cd uptake by biota are often lacking (e.g., [104,106,109]).

It is suggested that Cd is almost exclusively taken up as free Cd^{2+} from labile pools [110]. In contrast to this assumption, on the basis of isotopic dilution experiments it was shown that snails (*Helix aspersa*) are able to take up Cd from the non-labile pool, i.e., from the solid phase, which is in general suggested to be not bioavailable [77]. The non-labile pool contributed 16% to the total Cd concentration in snail tissue (on average 26 mg Cd kg$_{dry\,mass}^{-1}$). In contrast, plants do not seem to take up Cd from the non-labile pool even if they modify the physicochemical conditions in the rhizosphere to enhance metal mobility (e.g., [111,112]). However, the uptake of Cd in Swiss chard (*Beta vulgaris* L.) was increased by chloride. It was hypothesized that the increased uptake of Cd in the presence of chloride may be due to bioavailable Cd–chloro complexes [57,86]. In contrast, Peakall and Burger [113] stated that chloro complexes are not taken up by organisms.

3.3. Chromium

In the order of the most abundant elements in the Earth's crust, chromium is in the 20th place (0.019%). The most abundant isotope is ^{52}Cr (83.79%) followed by ^{53}Cr (9.50%), ^{50}Cr (4.35%), and ^{54}Cr (2.36%). Naturally, Cr occurs as chromite ($FeCr_2O_4$) and to a lesser extent as krokoite ($PbCrO_4$).

Chromium occurs in various oxidation states ranging from Cr^{2-} to Cr^{6+} with trivalent Cr as the most stable oxidation state. Trivalent Cr is essential to animals and humans, whereas Cr^{6+} is highly toxic and carcinogen. The Cr content of German soils usually ranges from 5 to 100 mg Cr kg^{-1}. Soils that have developed on Cr-rich bedrock often contain up to 3000 mg Cr kg^{-1} [114]. Plants accumulate Cr primarily in the roots. The translocation to the shoots is extremely limited [115]. Zayed et al. [115] reported that with X-ray absorption spectroscopy no Cr^{4+}, Cr^{5+}, or Cr^{6+} were detectable in the tissue of various plant species, even if plants were treated with CrO_4^{2-}. There is evidence that the conversion of CrO_4^{2-} to Cr^{3+} occurred in the root tissue and not in the rhizosphere or in solution [115,116].

3.4. Cobalt

Cobalt is a rare element. It is in the 30th place on the index of the most commonly occurring elements in the Earth's crust (0.004%). There is only one isotope (^{59}Co). Cobalt is often associated with Ni, Ag, Pb, Cu, and Fe ores. Important Co minerals are smaltite ($CoAs_{2-3}$), cobaltine (CoAsS), and linnéite (Co_3S_4). Common oxidation states are Co(II) and Co(III).

Cobalt is essential for humans and all higher animals, since it is a component of vitamin B_{12}. It is still not known whether it is essential for plants or not. Its content in soil varies in dependence on bedrock and ranges between 1 and 40 mg Co kg^{-1}. At pH values higher than 6, Co in soils is strongly bound to oxides, especially Mn oxides, and complexed by DOM. Adsorption and thus availability of Co increase with decreasing pH below 6 [114]. In general, Co availability is rather low. Normally, the fraction of available Co in soil accounts for ~6% [117]. However, in agricultural soils irrigated by lake water only 1% of the total Co content was actually bioavailable as determined by extraction with EDTA [118]. Uptake of Co by plants is species-specific. Cobalt concentrations in different vegetables grown on the same soil (surface concentration 10.9 ppm) in Saudi Arabia varied by a factor of 5.5 with watercress showing the highest (3.68 ppm) and lettuce the lowest concentrations (0.67 ppm). These corresponded to a removal of Co from soil solution in the range of 8–44% [119].

3.5. Copper

Copper is the 22nd most abundant element constituting ~0.01% of the Earth's crust. It consists of the isotopes ^{63}Cu (69.17%) or ^{65}Cu (30.83%). Copper mainly occurs in sulfidic copper ores. The principal Cu minerals are chalcopyrite ($CuFeS_2$), covellite (CuS), chalcosite (Cu_2S), and bornite (Cu_5FeS_4).

Copper is an essential element for all biota, but in excess it is highly toxic for plants, some animals, especially ruminants, and microorganisms. The Cu content of unpolluted soils amounts to $2-40$ mg Cu kg^{-1} [114]. It is concentrated in the clay and organic matter fraction of soils and is one of the least mobile metals in the soil environment. Biological available forms are Cu^+ and Cu^{2+} in organic complexes and in inorganic salts. In acid soils (pH < 7), Cu^{2+} dominates in soil solution. It is also the most mobile species, although a variety of ionic species may occur in soil. Examples are $CuOH^+$, $Cu(OH)_2$ (dominant species above pH 7), $Cu(CO_3)_2^{2-}$, or CuO_2^{2-} [96]. In general, Cu is easily complexed by organic and inorganic ligands. It is suggested that its mobility in soil is especially affected by interactions with DOM. Complexation of Cu by DOM significantly reduces its sorption in soils (e.g., [120]). The internal level of Cu in plants and animals is obviously regulated by physiological mechanisms. In several studies, relatively constant tissue concentrations were reported independently of the characteristics and the Cu content of the soils under investigation (e.g., [79,121]). Chaignon et al. [122] also reported that Cu concentrations in the shoots of tomato plants cultivated on 24 calcareous and 5 acidic Cu-contaminated top soils were not related to soil Cu content, extractable Cu, or soil properties. However, strong positive correlations between Cu concentrations in the roots, extractable Cu (by EDTA or $K_4O_7P_2$) and organic C content were found when only calcareous soils were taken into account. Thus, root concentrations were better suited as indicator for bioavailable Cu in calcareous soils than shoot concentrations. There was no similar relationship for acidic soils.

3.6. Iron

Iron is in the fourth place on the index of the most commonly occurring elements in the Earth's crust (4.7%). After aluminum, it is the second most abundant metal. Common Fe minerals are magnetite (Fe_3O_4), hematite (Fe_2O_3), limonite (FeOOH), and pyrite (FeS_2). The naturally occurring isotopic mixture mainly consists of ^{56}Fe (91.7%) together with three other isotopes [^{54}Fe (5.8%), ^{57}Fe (2.2%), ^{58}Fe (0.3%)].

Iron is essential for almost all biota. Phytotoxicity can be observed under anaerobic conditions, for example, in paddy soils. The Fe content of unpolluted soils usually ranges between 0.2% and 5% [114]. In aqueous solutions, Fe may occur as Fe(II), Fe(III), organic ferrous (Fe^{2+}), or organic ferric (Fe^{3+}) complexes. In biological environments, Fe(II) which is relatively soluble is rapidly oxidized by atmospheric oxygen to Fe(III). The solubility of Fe(III) strongly depends on pH: Fe(III) is precipitated under neutral and alkaline (and oxidizing) conditions, its solubility, thus, decreases with increasing pH. At neutral pH, Fe in aerobic soils mainly occurs as insoluble Fe(III) oxides and hydroxides. The Fe^{3+} concentration only amounts to 10^{-18} mol L^{-1}, whereas plants require between

10^{-4} and $10^{-8}\,\text{mol L}^{-1}$ for optimal growth [123,124]. In extremely acidic soils (pH < 3.5), Fe^{3+} may occur in concentrations up to $10^{-6}\,\text{mol L}^{-1}$ in soil solution [125].

Plants have developed different strategies to make Fe more available for uptake. These include acidification of the root environment by secretion of H^+, changes in root morphology, reduction of Fe^{3+} to Fe^{2+} which is more soluble by a specific root reductase and secretion of iron-chelating molecules, so-called siderophores [124,126]. Siderophores are also produced by microorganisms [123,127,128]. As for other essential elements such as Cu or Zn, homeostatic mechanisms regulating the internal Fe concentration exist: in plants, the Fe transport across various membrane structures is controlled including uptake from the environment and distribution to various organs and tissues [126].

3.7. Lead

Lead is a toxic, non-essential metal with a biological half-lifetime of 5–20 years (for its biogeochemistry, see Chapter 10 of Volume 43). Its solubility and thus availability increase with increasing acidity below pH 4.5 [114]. However, the influence of soil reaction on Pb uptake by plants is less pronounced than for other metals [129]. The results on bioavailable Pb fractions in soil systems are not consistent. Basta and Gradwohl [130] could not relate tissue concentrations of lettuce to the "available" soil Pb fraction which was determined by extraction with $Ca(NO_3)_2$. Vijver et al. [84] reported that Pb body concentrations of larvae of *Tenebrio molitor* (mealworms) were primarily correlated with the total Pb content of the soil. In contrast, the results of a bioluminescent microbial assay supported the FIAM that mainly Pb^{2+} is bioavailable [11].

Uptake and translocation of Pb in plants is obviously species-related. Mohamed et al. [119] reported that various vegetables accumulated between 0 (leek, cabbages) and 46.2 ppm (salq, *Beta vulgaris*) Pb in their tissue corresponding to transfer coefficients from soil solution to the plants of 0% or 7837%, respectively. In both hydroponic cultures and Pb-contaminated soils, Pb concentrations in maize shoots were significantly higher than in ragweed. In contrast to shoot concentrations, Pb uptake by ragweed roots and thus root concentrations were threefold higher compared to maize [131].

3.8. Nickel

Nickel constitutes ~0.01% of the Earth's crust and is the 23rd most abundant element. It occurs in iron meteorites and is often associated with Co, As, and Sb. Examples for Ni-containing minerals are pentlandite [$(Fe, Ni)_9S_8$] and pyrrhotite ($Fe_{1-x}S$). The most important isotope is ^{58}Ni accounting for 68.27% of the Ni present. Other isotopes are ^{60}Ni (26.10%), ^{61}Ni (1.13%), ^{62}Ni (3.59 %), and ^{64}Ni (0.91%). Nickel is closely related to Co in both its chemical and its biochemical properties.

Nickel is essential for biological functions in animals and some micro-organisms, but probably not in higher plants. In unpolluted German soils, Ni contents between 5 and 50 mg kg^{-1} were determined. The solubility of Ni increases with decreasing pH, especially below 5.5 [114]. Karczeswka [102] reported that in soils contaminated with metallurgical dusts the main portion of Ni was occluded in Fe oxides (56.9%) or bound in silicates (30.1%) (determined by sequential extraction). Thus, <8% of the total Ni present were potentially available and mobilizable. Total Ni concentrations ranged between 1.18 and 13.8 mg kg^{-1}.

Plant uptake of Ni varies greatly among species. In the study of Mohamed et al. [119], the tissue concentration of leek was 0.1 ppm, whereas watercress accumulated 42.6 ppm. The Ni concentration in watercress was almost three times higher than that in soil (14.6 ppm in 20 cm depth). These findings are in contrast to Peijnenburg et al. [121] who reported that no significant uptake by lettuce was found for Ni. It was hypothesized that this result may be due to low bioavailable concentrations in the investigated soils as a result of low solubility. Using a Ni-specific biosensor, Tibazarwa et al. [73] found that the bioavailability of Ni was greatly reduced by beringite—an alkaline alumino-silicate—and steel shots, additives that are used for *in situ* mobilization of metals. The bioavailable fraction detected in Ni-enriched soils by a biosensor reflected Ni accumulation by different plants (maize, potato tubers). Compared to Ca(NO$_3$)$_2$-exchangeable Ni concentrations determined for the same soils, the Ni-levels measured by a biosensor were 40–50% lower [73].

3.9. Zinc

Zinc is in the 26th place on the index of the most commonly occurring elements constituting to ~0.01% the Earth's crust. The key source of Zn is sphalerite (ZnS). Other important minerals are smithonite (ZnCO$_3$), willemite (Zn$_2$SiO$_4$), and zincite (ZnO). Natural Zn consists of an isotopic mixture of ^{64}Zn (48.6%), ^{66}Zn (27.9%), ^{67}Zn (4.1%), ^{68}Zn (18.8%), and ^{70}Zn (0.6%).

Zinc is essential for microorganisms, plants, animals, and humans. Solution concentrations >1 or 2 mg Zn L^{-1} lead to toxic effects on microorganisms or plants, respectively. Zinc has been considered to be non-toxic for soil animals. Recently, this assumption was refuted by different workers (e.g. [132,133]). The Zn content of unpolluted soils ranges from 10 to 80 mg kg^{-1}. Solution concentrations vary between 0.001 and 0.8 mg L^{-1} in agricultural soils, in extremely acidic forest soils concentrations up to 4 mg L^{-1} can be reached [114].

The mobility of Zn is controlled by factors similar to those determining Cu mobility in soils [96]. In various studies, it was demonstrated that Zn partitioning and uptake are mainly affected by pH with Zn availability being highest in acidic soils [133]. Lock and Janssen [133] reported that Zn availability to red clover (*Trifolium pratense*) was closely related to pore water concentration and the CaCl$_2$-extractable fraction. The results of a bioluminescent microbial assay also supported the FIAM [11]. In contrast, acute lethal toxicity tests (LC_{50}) with the earthworm *E. fetida* showed that the $Ca(NO_3)_2$-extractable Zn fraction in spiked artificial soil was suitable to measure Zn bioavailability [132]. The suitability of $Ca(NO_3)_2$-extractable concentrations to predict Zn uptake was also confirmed for plants [130]. However, in numerous investigations uptake or tissue concentrations of Zn were not related to soil characteristics or Zn concentrations in pore water or bulk soil (e.g., [79,121,134]). This was usually explained by the fact that Zn as well as Cu and Fe is an essential element and organisms have developed regulation mechanisms allowing them to maintain homeostasis under various environmental conditions.

4. CONCLUDING REMARKS

A reliable assessment of environmental risks is only possible on the basis of bio-available concentrations. In terrestrial environments, especially in soils, the determination of bioavailable metals is limited by several reasons: no definition of the term "bioavailability" exists that is generally applicable, the bioavailable fraction determined in numerous studies differed in dependence on the metal and the organisms under investigation (cf. Section 3) and metal availability in soils is influenced by various abiotic parameters. It is therefore necessary to clearly define the term "bioavailability" and to further develop simple techniques that allow the determination of available metal fractions *in situ*.

The metal of greatest concern in research is still Cd due to its high toxicity and its ability to bioaccumulate. Plant tissue concentrations of Cd below phytotoxic values might pose a risk to human or animal health. Other non-essential metals, especially Pb, may be less important, since their concentrations in the environment currently decline in many countries. However, there is still a great need to enhance the knowledge about the bioavailability of essential metals such as Zn which are required for plant growth and animal health.

ABBREVIATIONS

ALAD	δ-aminolevulinic dehydratase
CEC	cation exchange capacity
DGT	diffusive gradients in thin films
DOM	dissolved organic matter
DTPA	dieethylenetriaminepentaacetate
EDTA	ethylenediaminetetraacetate
FIAM	free ion activity model
IDT	isotopic dilution techniques
ISE	ion-selective electrode
OECD	Organization for Economic Cooperation and Development

REFERENCES

1. Falbe J, Regitz M, eds. Roempp Chemie-Lexikon. 10th ed. Version 2.0, Stuttgart: Thieme, 1999 (electronic version).
2. Phipps DA. In: Lepp NW, ed. Effects of Heavy Metal Pollution in Plants. London: Applied Science Publishers, 1981:1–54.
3. Brown PH, Welch RM, Cary EE. Plant Physiol 1987; 58:801–803.
4. Baker S, Herrchen M, Hund-Rinke K, Klein W, Kördel W, Peijnenburg W, Rensing C. Ecotoxicol Environ Saf 2003; 56:6–19.
5. Caussy D, Gochfeld M, Gurzau E, Neagu C, Ruedel H. Ecotoxicol Environ Saf 2003; 56:45–51.
6. Pavlou SP, Dexer RN, Mayer FL, Fisher C, Hague RH. In: Neuhold JM, Ruggiero LF, eds. Ecosystem Processes and Organic Contaminants. Washington, D.C.: U.S. Government Print Office, 1977:21–26.
7. Landrum PF, Hayton, WL, Lee H, McCarty LS, Mackay D, McKim JM. In: Hamelink JL, Landrum PF, Bergman HL, Benson WH, eds. Bioavailability. Boca Raton: Lewis Publishers, 1994:203–219.
8. Fries GF, Marrow GS. Chemosphere 1992; 3:109–113.
9. Newman MC, Jagoe CH. In: Hamelink JL, Landrum PF, Bergman HL, Benson WH, eds. Bioavailability. Boca Raton: Lewis Publishers, 1994:39–62.
10. Ogunseitan OA, Yang S, Ericson J. Soil Biol Biochem 2000; 32:1899–1906.
11. Ritchie JM, Cresser M, Cotter-Howells J. Environ Pollut 2001; 114:129–136.
12. Lasat MM. J Environ Qual 2002; 31:109–120.
13. Parker DR, Chaney RL, Norvell WA. In: Loeppert R, Schwab AP, Goldberg S, eds. Chemical Equilibrium and Reaction Models. Madison: Soil Science Society of America, 1995:163–200.
14. Wolt JD. Soil Solution Chemistry. New York: John Wiley, 1994:1–345.
15. Checkai RT, Corey RB, Helmke PA. Plant Soil 1987; 99:335–345.
16. Internet Glossary of Soil Science Terms (http://soils.org/sssagloss/index.html). Madison: Soil Science Society of America, 2003. Verified 13 October 2003.
17. Davis BE, Davis RJ. Nature 1963; 198:216–217.
18. Giesler R, Lundström US, Grip H. Eur J Soil Sci 1996; 47:395–405.
19. Ludwig B, Meiwes KJ, Khanna PK, Gehlen R, Fortmann H, Hildebrand EE. J Plant Nutr Soil Sci 1999; 162:343–351.
20. Sauvé S, McBride M, Hendershot WH. Arch Environ Contam Toxicol 1995; 29:373–379.
21. Mota A, Correira Dos Santos M. In: Tessier A, Turner DR, eds. Metal Speciation and Bioavailability in Aquatic Systems. New York: John Wiley, 1995:205–257.
22. Chow PYT, Cantwell FF. Anal Chem 1988; 60:1596–1573.
23. Persaud G, Cantwell FF. Anal Chem 1992; 64:89–94.
24. Nolan AL, Lombi E, McLaughlin MJ. Aust J Chem 2003; 56:77–91.
25. Cox JA, Slownawska K, Gatchell DK, Hiebert AG. Anal Chem 1984; 56: 650–653.
26. Parkhurst DL, Appelo CAJ. User's Guide to PHREEQC2 – A hydrogeochemical transport model. Water Res Invest Rep Colorado, 1999:1–312.
27. Ludwig B, Khanna PK, Hölscher D, Anurugsa B. Eur J Soil Sci 1999; 50:717–726.
28. Schecher W, McAvoy D. Comput Environ Urban Syst 1992; 16:65–76.
29. Tipping E. Aquat Geochem 1998; 4:3–47.

30. Adriano DC. Trace Elements in the Terrestrial Environment. Heidelberg: Springer, 1986:1–533.
31. McBride MB, Nibarger EA, Richards BK, Steenhuis T. Soil Sci 2003; 168:29–38.
32. Sims JT, Igo E, Skeans Y. Commun Soil Sci Plant Anal 1991; 22:1031–1045.
33. Gonzalez MJ, Ramos I, Hernandez LM. Int J Environ Anal Chem 1994; 57:135–150.
34. Lock K, Janssens F, Janssen CR. Eur J Soil Biol 2003; 39:25–29.
35. Dai J, Becquer T, Rouiller JH, Reversat G, Bernhard-Reversat F, Nahmani J, Lavelle P. Soil Biol Biochem 2004; 36:91–98.
36. Düring R-A, Hoss T, Gäth S. Soil Till Res 2002; 66:183–195.
37. Olajire AA, Ayodele ET, Oyediran GO, Oluyemi EA. Environ Mon Assess 2003; 85:135–155.
38. Ma H, Wang X, Zhang C. Chem Spec Bioavail 2003; 15:15–22.
39. Filgueiras AV, Lavilla I, Bendicho C. J Environ Monit 2002; 4:823–857.
40. Tessier A, Campbell PG, Bisson M. Anal Chem 1979; 51:844–851.
41. Zeien H, Brümmer GW. Mitteilgn Dtsch Bodenkundl Gesellsch 1989; 59:505–510.
42. Ure AM. In: Alloway BJ, ed. Schwermetalle in Böden. Berlin: Springer, 1999: 63–110.
43. Nieuwenhuize J, Poley-Vos CH, van den Akker AH, van Delft W. Analyst 1991; 116:347–351.
44. McGrath D. Sci Total Environ 1995; 164:125–133.
45. Sánchez J, Marino N, Vaquero MC, Ansorena J, Legórburu I. Water Air Soil Pollut 1998; 107:303–319.
46. Tam NFY, Yao MW. Bull Environ Contam Toxicol 1999; 62:708–716.
47. Bettinelli M, Beone GM, Spezia S, Baffi C. Anal Chim Acta 2000; 424:289–296.
48. Sastre J, Sahuquillo A, Vidal M, Rauret G. Anal Chim Acta 2002; 462:59–72.
49. Chen M, Ma LQ. Soil Sci Soc Am J 2001; 65:491–499.
50. Zhang S, Lu A, Shan X-Q, Wang Z, Wang S. Anal Bioanal Chem 2002; 374: 942–947.
51. Fuh C-B, Lin H-I, Tsai H. J Food Drug Anal 2003; 11:39–45.
52. Zhang H, Davison W. Anal Chem 1995; 67:3391–3400.
53. Jansen B, Kotte MC, van Wijk AJ, Vestraten JM. Sci Total Environ 2001; 277: 45–55.
54. Zhang H, Zhao F-J, Sun B, Davison W, McGrath SP. Environ Sci Technol 2001; 35:3602–2607.
55. Hund-Rinke K, Kördel W. Ecotoxicol Environ Saf 2003; 56:52–62.
56. Gomot de Vaufleury A, Pihan F. Chemosphere 2000; 40:275–284.
57. Smolders E, McLaughlin MJ. Soil Sci Soc Am J 1996; 60:1443–1447.
58. Beeby A. Environ Pollut 2001; 112:285–298.
59. Maenpaa KA, Kukkonen JVK, Lydy MJ. Arch Environ Contam Toxicol 2002; 43:389–298.
60. Shugart L, Bickman J, Jackim E, McMahon G, Ridley W, Stein J, Steinert SA. In: Hugget RJ, Kimerle RA, Mehrle PM Jr, Bergman HL, eds. Biomarkers: Biochemical, Physiological, and Histological Markers of Anthropogenic Stress. Chelsea: Lewis Publishers, 1992:125–153.
61. Overman SR, Krajicek JJ. Environ Toxicol Chem 1995; 14:689–695.
62. Burden VM, Sandheinrich MB, Caldwell CA. Environ Pollut 1998; 101:285–289.
63. Fleming DEB, Chettle DR, Wetmur JG, Desnick RJ, Robin J-P, Boulay D, Richard NS, Gordon CL, Webber CE. Environ Res 1998; 77:49–61.

64. Hensbergen PJ, van Velzen MJM, Nugroho RA, Donker MH, van Straalen NM. Comput Biochem Physiol [C] 2000; 118:267–270.
65. Berger B, Dallinger R, Thomaser A. Environ Toxicol Chem 1995; 14:781–791.
66. Dallinger R, Berger B, Gruber C, Hunziker P, Stürzenbaum S, Bordin G. Cell Mol Biol 2000; 46:331–346.
67. Kille P, Stürzenbaum SR, Galay M, Winters C, Morgan AJ. Pedobiologia 1999; 43:602–607.
68. Spurgeon DJ, Svendsen C, Rimmer VR, Hopkin SP, Weeks JM. Environ Toxicol Chem 2000; 19:1800–1808.
69. Scott-Fordsmand JJ, Weeks JM. Rev Environ Contam Toxicol 2000; 165:117–159.
70. Keltjens WG, van Beusichem ML. Plant Soil 1998; 203:119–126.
71. Tahlil N, Rada A, Baaziz M, Morel JL, El Meray M, El Aatmani M. Biol Plant 1999; 42:75–80.
72. Verdoni N, Mench M, Cassagne C, Bessoule JJ. Environ Toxicol Chem 2001; 20:382–388.
73. Tibazarwa C, Corbisier P, Mench M, Bossus A, Solda P, Mergeay M, Wyns L, van der Lelie D. Environ Pollut 2001; 113:19–26.
74. Frossard E, Fardeau JC, Brossard M, Morel JL. Soil Sci Soc Am J 1994; 58: 846–851.
75. Sinaj S, Frossard E, Fardeau JC. Soil Sci Soc Am J 1997; 61:1413–1417.
76. Sinaj S, Mächler F, Frossard E. Soil Sci Soc Am J 1999; 63:1618–1625.
77. Scheifler R, Schwartz C, Echevarria G, de Vaufleury A, Badot P-M, Morel J-L, Environ Sci Technol 2003; 37:81–86.
78. Hooda PS, McNulty D, Alloway BJ, Aitken MN. J Sci Food Agric 1997; 73: 446–454.
79. Vijver M, Jager T, Posthuma L, Peijnenburg W. Environ Toxicol Chem 2001; 20:712–720.
80. Mellum HK, Arnesen AKM, Singh BR. Commun Soil Sci Plant Anal 1998; 29:1183–1198.
81. van Straalen NM, Butovsky RO, Pokarzhevskii AD, Zaitsev AS, Verhoef SC. Pedobiologia 2001: 45:451–466.
82. Qian J, Shan X-Q, Wang Z-J, Tu Q. Sci Total Environ 1996; 187:131–141.
83. Plette AC, Nederlof MM, Temminghoff EHM, van Riemsdijk WH. Environ Toxicol Chem 1999; 18:1882–1890.
84. Vijver M, Jager T, Posthuma L, Peijnenburg W. Ecotoxicol Environ Saf 2003; 54:277–289.
85. Peijnenburg WJGM, Posthuma L, Eijsackers HJP, Allen HE. Ecotoxicol Environ Saf 1997; 37:163–172.
86. McLaughlin MJ, Andrew SJ, Smart MK, Smolders E. Plant Soil 1998; 202: 211–216.
87. Peijnenburg WJGM, Posthuma L, Zweers PGPC, Baerselmann R, de Groot AC, Van Veen RPM, Jager T. Ecotoxicol Environ Saf 1999; 43:170–186.
88. Conder JM, Seals LD, Lanno RP. Chemosphere 2002; 49:1–7.
89. Lanno R, LeBlanc S, Knight B, Tymowski R, Fitzgerald D. In: Sheppard S, Bembridge J, Holmstrup M, Posthuma L, eds. Advances in Earthworm Ecotoxicology. Pensacola: SETAC Press, 1998:41–53.
90. Marr K, Fyles H, Hendershot W. Can J Soil Sci 1999; 79:385–387.
91. Zhang S, Shan X-Q. Chem Spec Bioavail 2000; 12:117–123.

92. Peijnenburg WJGM, Baerselman R, de Groot AC, Jager T, Posthuma L, Van Veen RPM. Ecotoxicol Environ Saf 1999; 44:294–310.
93. Savory J, Wills MR. In: Merian E, ed. Metals and Their Compounds in the Environment. Weinheim: VCH, 1991:715–741.
94. Lundström US, Giesler R. Ecol Bull 1995; 44:114–122.
95. Gérard F, Boudot J-P, Ranger J. Appl Geochem 2001; 16:513–529.
96. Kabata-Pendias A, Pendias H. Trace Elements in Soils and Plants. 2nd ed. Boca Raton: CRC Press, 1992:1–365.
97. Gardiduenas-Pina R, Cervantes C. BioMetals 1996; 9:311–316.
98. van Gestel CAM, Hoogerwerf G. Pedobiologia 2001; 45:385–395.
99. Malathi N, Sarethy IP, Paliwal K. J Plant Biol 2002; 29:29–32.
100. Ma G, Rengasamy P, Rathjen AJ. Aust J Exp Agric 2003; 43:497–501.
101. Shann JR, Bertsch PM. Soil Sci Soc Am J 1993; 57:116–120.
102. Karczeswka A. Appl Geochem 1996; 11:35–42.
103. Melakeberhan H, Jones AL, Hanson E, Bird GW. Plant Dis 1995; 79:886–892.
104. Lehoczky E, Marth P, Szabados I, Palkovics M, Lukacs P. Commun Soil Plant Anal 2000; 31:2425–2431.
105. Lehoczky E, Kiss Z, Kalra YP. Commun Soil Plant Anal 2002; 33:3177–3187.
106. Nan Z, Zhao C, Li J, Chen F, Sun W. Water Air Soil Pollut 2002; 133:205–213.
107. McBride MB. Soil Sci 2002; 167:62–67.
108. Huang B, Kuo S, Bembenek R. Water Air Soil Pollut 2003; 147:109–127.
109. Lock K, Janssen CR. Chemosphere 2001; 44:1669–1672.
110. Sauvé S, Norvell WA, McBride M, Hendershot W. Environ Sci Technol 2000; 34:291–296.
111. Grinsted MJ, Hedley MJ, White RE, Nye PH. New Phytol 1982; 91:19–29.
112. Gérard E, Echevarria G, Sterckeman T, Morel J-L. Environ Qual 2000; 29: 1117–1123.
113. Peakall D, Burger J. Ecotoxicol Environ Saf 2003; 56:110–121.
114. Blume H-P, Brümmer GW, Schwertmann U, Horn R, Kögel-Knabner I, Stahr K, Auerswald K, Beyer L, Hartmann A, Litz N, Scheinost A, Stanjek H, Welp G, Wilke BM. Lehrbuch der Bodenkunde. 15th ed. Heidelberg: Spektrum, 2002:1–543.
115. Zayed A, Lytle CM, Qian J-H, Terry N. Planta 1998; 206:293–299.
116. McGrath SP. New Phytol 1982; 92:381–390.
117. Schrauzer GN. In: Merian E, ed. Metals and Their Compounds in the Environment. Weinheim: VCH, 1991:879–892.
118. Stalikas CD, Pilidis GA, Tzouwara-Karayanni SM. Sci Total Environ 1999; 236: 7–18.
119. Mohamed AE, Rashed MN, Mofty A. Ecotoxicol Environ Saf 2003; 55:251–260.
120. Zhou LX, Wong JWC. J Environ Qual 2001; 30:878–883.
121. Peijnenburg W, Baerselmann R, de Groot A, Jager T, Leenders D, Posthuma L, Van Veen R. Arch Environ Contam Toxicol 2000; 39:420–430.
122. Chaignon V, Sanchez-Neira I, Herrmann P, Jaillard B, Hinsinger P. Environ Pollut 2003; 123:229–238.
123. Schlegel H. Allgemeine Mikrobiologie. 7th ed. Stuttgart: Thieme, 1992:1–634.
124. Hell R, Stephan UW. Planta 2003; 216:541–551.
125. Neilands JB, Konopka K, Schwyn B, Coy M, Francis T, Paw BH, Bagg A. In: Winkelmann G, van der Helm D, Neilands JB, eds. Iron Transport in Microbes, Plants and Animals. Weinheim: VCH, 1987:3–34.

126. Curie C, Briat J-F. Annu Rev Plant Biol 2003; 54:183–206.
127. Renshaw JC, Roboson GD, Trinci APJ, Wiebe MG, Livens FR, Collison D, Taylor RJ. Mycol Res 2002; 106:1123–1142.
128. Terano H, Nomoto K, Takase S. Biosci Biotechnol Biochem 2002; 66:2471–2473.
129. Davies, BE. In: Alloway B, ed. Schwermetalle in Böden. Berlin: Springer, 1999:131–149.
130. Basta N, Gradwohl R. J Soil Contam 2000; 9:149–164.
131. Huang JW, Cunningham SD. New Phytol 1996; 134:75–84.
132. Conder JM, Lanno RP. Chemosphere 2000; 41:1659–1668.
133. Lock L, Janssen CR. Environ Pollut 2004; 126:371–374.
134. Terrés C, Navarro M, Martín-Lagos F, Giménez R, Olalla M, López H, López MC. Bull Environ Contam Toxicol 2002; 68:224–229.

5

Heavy Metal Uptake by Plants and Cyanobacteria

Hendrik Küpper and Peter M. H. Kroneck

*Universität Konstanz, Mathematisch-Naturwissenschaftliche Sektion,
Fachbereich Biologie, D-78457 Konstanz, Germany*

1.   INTRODUCTION: HEAVY METALS AND PLANTS— A COMPLEX RELATIONSHIP

Many heavy metals such as Cu, Ni, and Zn are well-known as essential trace elements for plants, and even Cd has been found to be a micronutrient for marine algae [1]. Deficiency in the supply of certain heavy metals, in particular Zn, is a serious problem for agriculture in many parts of the world [2]. Furthermore, productivity of phytoplankton in the oceans is often limited by the availability of heavy metals, not just Fe as some people tend to think, but also Cd, Cu, Ni, and Zn [3]. In the first part of this chapter, we will focus on beneficial aspects of these transition metals in plant metabolism, with emphasis on their structural and functional roles in various metalloenzymes. In view of the complexity of the topic and the enormous number of publications accumulating in the field, we recommend as primary references the *Handbook on Metalloproteins* [4], the *Handbook of Metalloproteins* [5], and the special issue of *Chemical Reviews* on Biomimetic Chemistry [6]. Furthermore, we will give a brief summary of the latest developments concerning the uptake of Cd, Cu, Ni, and Zn into the plant and the effects of their deficiency on plant metabolism. The uptake, sequestration, and export of Zn, Cu, and Co has been summarized most recently for cyanobacteria by Cavet et al. [7] and for plants by Hall and Williams [8] and Cobbett et al. [9].

However, all of these metals can also be toxic to plants and cyanobacteria, and in many cases, the beneficial range is extremely narrow. In the case of Cu, for example, even the picomolar concentrations that occur in the oceans seem to be toxic to some cyanobacteria [10]. Above the heavy metal threshold leading to growth inhibition, a variety of toxic effects have been observed in all plant species, the relative importance of which depends on the nature and concentration of the metal, the plant species, and different environmental conditions including irradiance and soil conditions. The most comprehensive recent review on this subject is Ref. [11]. We will focus in the second part of our review on (a) the inhibition of photosynthesis and (b) the most recent findings concerning other sites of heavy metal toxicity in plants.

Plants have developed a number of strategies to resist the toxicity of heavy metals, such as efflux pumps [12], complexation of heavy metals inside the cell by strong ligands such as phytochelatins (PCs) [13] or histidine [14], and several other mechanisms [11,15]. Plants that actively prevent metal accumulation inside the cells are called excluders, these represent the majority of metal-resistant plants [16]. Other resistant plants deal with potentially toxic metals in just the opposite way, i.e., they actively take up metals and accumulate them. These plants, which have been named "hyperaccumulators" [17], are able to accumulate several percent metals in the dry weight of their above-ground parts. The active accumulation in the above-ground parts of hyperaccumulator plants provides a promising approach for both cleaning anthropogenically contaminated soils (phytoremediation) and for commercial extraction (phytomining) of metals from naturally metal-rich (serpentine) soils [18,19]. We will summarize recent findings concerning the function of the metal hyperaccumulation trait in plants, the biochemical, physiological, and molecular biological mechanisms of metal hyperaccumulation. Finally, we will conclude with an overview of past and present attempts to use metal hyperaccumulators for cleaning up soils and mining metals.

2. POTENTIALLY TOXIC HEAVY METALS AS PLANT MICRONUTRIENTS

Figure 1 shows typical pathways of micronutrients in plants, including factors that may contribute to efficiency in comparison to inefficiency of microelement utilization. Table 1 summarizes the most important enzymes with one or more atoms of the discussed heavy metals in the active center.

2.1. Cadmium

For a long time, cadmium has been known as a highly toxic metal that represents a major hazard to the environment. Only recently, results from oceanographic research have shed a new light on the environmental role of Cd. Initially, Bruland found that the concentrations of Cd in the oceans follows a pattern that is generally characteristic of micronutrients and not of toxic substances [20]. The close correlation between the concentrations of phosphate and Cd has been used in palaeontology [21]. It was then found in a phytoplankton species belonging to the diatoms, *Thalassiosira weissflogii*, that addition of Cd to Zn-limited cultures enhanced growth and led to the expression of a carbonic

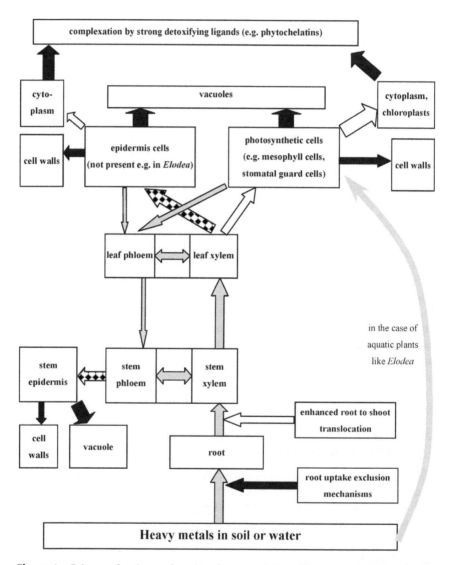

Figure 1 Scheme of pathways for microelement nutrition. Black arrows show pathways contributing to inefficiency in the use of micronutrients, white arrows show mechanisms of increased micronutrient efficiency. Pathways that may have both effects are labeled with a black/white pattern and those that are most likely neutral to efficiency are labeled in gray.

anhydrase with properties different from the normal enzyme of Zn-replete cultures [22]. Cd migrated with the band of a carbonic anhydrase in gel electrophoresis [22,23], but it still remained unclear whether Cd was merely substituting for Zn in an enzyme that would normally contain Zn. Lane and Morel [1] finally

Table 1 Representative Examples of Metalloproteins (Ni, Cu, Zn, Cd) in Plants and Cyanobacteria

Metalloprotein	Metal site	X-ray structure	Selected references
Urease	2 Ni	+	[72–74,78–80]
Plastocyanin	1 type-1 Cu	+	[36–39]
Laccase[a]	4 Cu (1 type-1 Cu; 1 type-2 Cu; 1 type-3 Cu pair)	+	[43,44,346]
Cytochrome *c* oxidase (COX)	Dinuclear CuA; CuB	+[b]	[45,347]
Multi-copper oxidase (FET3)		Not determined[c]	[47,48,348,349]
Superoxide dismutase (Cu,Zn SOD)	1 type-2 Cu; 1 Zn	+	[35,40–42,93]
Ethylene receptor (ETR1)	1 Cu	Structural model	[46]
Zn-carbonic anhydrase	1 Zn	+	[94]
Cd-carbonic anhydrase	1 Cd	Not determined	[1,22,23,24]
Purple acid phosphatase (PAP)	1 Zn; 1 Fe(III)	+	[96]
Zn finger protein/C2H2 domain	1 Zn	+	[95]

[a]Type-2 Cu and type-3 Cu pair form a trinuclear center [346].
[b]Determined for bacterial and mitochondrial enzyme [347].
[c]Multi-copper oxidase; the FET3 protein of yeast exhibits similarity to the human multi-copper protein ceruloplasmin [348]. Ceruloplasmin X-ray structure: 6 Cu, 3 type-1 Cu, and one trinuclear cluster like in laccase [349].

succeeded in proving that the Cd-containing carbonic anhydrase of *T. weissflogii* is indeed a different protein than the normal Zn-carbonic anhydrase, by showing that the Cd-carbonic anhydrase is much larger (43 kDa vs. 27 kDa) and is not recognized by antibodies for the Zn-carbonic anhydrase of the same organism, TWCA1. And, even more important, the Cd-carbonic anhydrase did not bind Zn, demonstrating that it is indeed specifically synthesized as a Cd enzyme. Nevertheless, it can most likely be regarded as a replacement of the Zn enzyme in the evolutionary sense.

In the oceans, most organisms suffer from Zn deficiency, in particular in the surface layer because of its high cell density [3]. While concentrations of dissolved Zn are ~ 10 nM in the deep sea, they are in the picomolar range at the surface [20]. It seems that some organisms were able to reduce their need for Zn by using Cd for part of the functions that normally require Zn; this would have given them an evolutionary advantage, since there was no competition for Cd [3]. In support of this theory, the expression of Cd-carbonic anhydrase in *T. weissflogii* was shown to be inversely correlated to the availability of Zn [24]. Cadmium deficiency in this species, and most likely a range of other planktonic algae with Cd-carbonic anhydrases, seems to specifically cause a reduced activity of the carbon-concentrating mechanism and, in consequence, reduced growth [1].

It remains to be seen how many other beneficial or even essential roles this metal may play in cyanobacterial and plant metabolism. Note that there are already a few studies that suggest such roles in fungi [25,26], cyanobacteria [27], various species of marine phytoplankton [28], and even higher plants [29]. One known beneficial role of Cd in higher plants is found in the Cd hyperaccumulator plant *Thlaspi caerulescens* (*Ganges ecotype*), which can accumulate Cd in concentrations of $>1\%$ of its dry mass [30]. Hyperaccumulation of toxic metals has been shown to deter herbivores and pathogens (see below), and this has also been observed in this particular case [H. Küpper, unpublished results].

The uptake of Cd into the plant seems to occur via various channels for the transport of other divalent cations, in particular Zn [8]. A channel which took Cd but not Zn was detected by Lombi et al. [31] in the Ganges ecotype of *T. caerulescens*, but later results indicated that this is the iron transporter IRT1 [32], which had previously been shown to transport Cd as well [33]. The only Cd-specific transport proteins known so far are involved in Cd detoxification and will be therefore discussed in Section 3.

2.2. Copper

As mentioned in Section 1, the beneficial range for Cu concentrations is very narrow. Copper is often in limiting supply in the oceans ultimately leading to N_2O release into the atmosphere [3], but even there it is still toxic to some cyanobacteria [10,34], so that *Prochlorococcus* has to avoid zones with highest Cu levels.

Copper is needed as an active center in many proteins and enzymes (Table 1). The best studied and most important of these proteins are the following:

(a) *Plastocyanin.* This type-1 (blue) copper protein transfers electrons between the Rieske protein and PS I in all Chlorophyta (higher plants and green algae), Embryophyta, and most cyanobacteria [35]. In other groups of photosynthetic organisms, it is either absent (e.g., Rhodophyta, Phaeophyta, Euglenophyta), or its presence has not been shown unambiguously (e.g., Cryptophyta, Chlorarachnio-phyta, Dinophyta). In all organisms, where plastocyanin is absent, it is replaced by cytochrome c_6, an iron-dependent electron carrier. In those organisms that can express both PC and cytochrome c_6, the two soluble proteins can replace each other depending on the availability of iron and copper [35].

The crystal structures of plastocyanin from different organisms have been solved. The structure of poplar Cu(II)–plastocyanin (from *Populus nigra* var. *italica*) was first determined by Freeman and colleagues at 2.7 Å resolution, and was further refined to 1.6 Å [36,37]. At the molecular level, its function in photosynthesis is to transfer electrons from the membrane-bound cytochrome b_6f complex in photosystem II to the membrane-bound reaction center of photosystem I [38]. The plastocyanin molecule is an eight-stranded β-sandwich (cupredoxin fold). The Cu site is located near one end, conventionally described as the "northern" end of the barrel. Cu is coordinated by the side chains of His37, Cys84, His87, and Met92. Neither the Cu atom nor its ligands are accessible from the solvent, with the exception of the imidazole of the northern histidine (His87). One edge of the imidazole ring is exposed to the solvent and probably forms the electron transfer pathway to the photo-oxidized chlorophyll dimer $P700^+$. There is another potential binding site for redox partners, the so-called acidic patch that is located around the conserved Tyr83 residue, 19 Å away from the Cu atom vs. 6 Å for the imidazole ring of His87. The overall structure of poplar plastocyanin and the coordination sphere of the Cu site apply with only minor qualifications to the known structures of plas-tocyanin from plants, green algae, and cyanobacteria. For a compre-hensive review of the structural, spectroscopic, and functional properties of plastocyanin refer to Ref. [39].

(b) *Superoxide dismutase* (Cu,Zn SOD). A type-2 (non-blue) Cu center constitutes the redox-active site of this important metalloenzyme. Superoxide dismutases are ubiquitously distributed in all biological systems including prokaryotic, eukaryotic, and plant cells. In addition to the Cu,Zn enzyme, iron-, manganese-, and nickel-dependent SODs have been detected since the early discovery of SOD [40]. Cu,Zn SOD comes as a homodimer of $\sim 2 \times 16$ kDa, in eukaryotes it is localized in the cytosol, in the intermembrane space, and in lysosomes, while a distinct Cu,Zn SOD is found as an extracellular form. In sequences derived from plants, two distinct forms have been detected, localized in the cytoplasm and in the chloroplast (Table 1). Each subunit hosts a Cu,Zn pair responsible for the catalytic dismutation of the superoxide anion, O_2^- [41]. As discussed in detail (including energetic aspects) by Raven et al. [35], Cu,Zn SOD detoxifies O_2^-, generating H_2O_2 that is further processed by the iron-dependent enzyme catalase, or diffuses out of the cell. Under turnover conditions, the copper ion at the active site cycles between the Cu(II) and Cu(I) redox states, following successive interactions with the superoxide anion. The electron transfer between substrate and Cu is extremely efficient, the second-order catalytic rate for the enzymatic reaction being limited only by substrate diffusion [42]. In the first part of the reaction, Cu(II) becomes reduced by O_2^-, dioxygen is then released and the Cu(I)-bridging histidine bond broken with concomitant protonation of this histidine. The reduced Cu(I) protein then reacts with a second superoxide radical to form H_2O_2 and the reoxidized enzyme.

(c) The first plant Cu enzyme that was isolated is the *laccase* from the Japanese lacqer tree. This multi-copper oxidase was isolated in 1938 by Keilin and Mann [43]; but only in 1958, finally, the redox role of Cu in this enzyme was demonstrated by Malmström et al. [44] by using EPR.

(d) The first plant enzyme in which Cu was shown to be the active center is the *cytochrome oxidase* (COX) of the respiratory electron chain [45a]. No crystal structure has been determined for any cyanobacterial or plant COX, but a cyanobacterial COX has been characterized biochemically and spectroscopically [45b–d].

(e) The *sensor for* the plant hormone *ethylene* (ETR1) is a homodimeric protein that binds one Cu and is most likely located in the plasma membrane [46].

(f) A *multi-copper oxidase II* seems to be involved in iron uptake of green algae by reoxidation of Fe^{2+} during its uptake into the cell [47,48]. As discussed in these reports, reoxidation of the Fe^{2+} released from chelating agents may be necessary to prevent the generation of hydroxyl radicals via the Fenton reaction.

The pathway of Cu uptake into and trafficking inside plants and cyanobacteria is still only partially understood; the current knowledge about this was recently reviewed [7–9,49]. Early on it was demonstrated that Cu uptake is strictly regulated. This regulation involves changes in the levels of the transport proteins resulting in changes of V_{max} of Cu uptake (e.g., [50]). Long-range transport in the xylem may be mediated by Cu-complexing ligands, such as nicotianamine and histidine; both coordinate Cu with high affinity and were found in the xylem sap of chicory and tomato in sufficient concentrations to bind all Cu [51]. The transport into shoot cells was proposed to be mediated by COPT1, a transporter that was shown to repair yeast strains defective in Cu uptake [52]. This transporter consists of 169 amino acids, which build three potential transmembrane domains and an N-terminal methionine- and serine-rich putative Cu binding site. It seems to form a small family together with a few other putative metal transporters [53]. The mechanism of metal transport, however, is still unknown. Transport between organelles seems to occur via CPx-type ATPases (P_{1B} subfamily). RAN1 (=AtHMA7) seems to deliver Cu(I) across post-Golgi membranes for ethylene receptors [54,55] and PAA1 (=AtHMA6) seems to transport Cu(I) across the chloroplast envelope into the chloroplast stroma [56]. Altogether, seven P_{1B} CPx-ATPase genes have been detected in the *Arabidopsis thaliana* genome [57]. In the cyanobacterium *Synechocystis*, the CPx-ATPase CtaA seems to transport Cu into the cell [58]. In addition, transporters of the ZIP family (ZIP = "ZRT, IRT-like proteins") seem to be involved in Cu uptake into the cell, in particular under Cu-deficiency stress [59]. Trafficking inside the organelles to the assembly of Cu-requiring proteins seems to be mediated by Cu binding chaperones [7,60,61]. A comparison of their predicted structures has been computed [62]. Two plant Cu chaperones have been characterized in more detail. Pufahl et al. [63] have characterized ATX1 and suggested that it delivers Cu to a Cu ATPase, Ccc2, which transports the Cu into post-Golgi vesicle [64]. Himelblau et al. [65] suggested that CCH transports Cu within the cytosol to the ATPase RAN1 (see above). In addition, CCH seems to be involved in Cu recycling during senescence [66]. Also other (putative) transition metal transporters may play a role in Cu uptake and homeostasis, like AtMGT1 that has been proposed to transport Cu(II) in addition to Mg(II) [67].

Symptoms of Cu deficiency are usually chlorosis (i.e., loss of chlorophyll), which in severe cases later on leads to necrosis of affected tissues; the overall biomass of affected plants is subsequently greatly reduced. These symptoms are known for a long time; a recent study describing them and the thresholds of their appearance is in Ref. [68]. The Cu deficiency symptoms can be explained in view of the normal functions of Cu in plant metabolism. For example, lack of the Cu proteins plastocyanin and superoxide dismutase will cause malfunction of photosynthesis and thus lead to oxidative stress. Because of its involvement in the iron assimilation pathway, Cu deficiency stress increases the effects of iron deficiency stress [69]. But in contrast to animal systems, in some plants, e.g., *Chlamydomonas*, Cu deficiency combined with higher iron concentrations seems not to induce iron deficiency stress [48]. This indicates the existence of a Cu-independent iron uptake pathway in plants.

Metallothioneins (MT) seem to have a role in recycling metals, in particular Cu, from catabolized metalloproteins in senescing plant tissues (originally found by Buchanan-Wollaston [70] and reviewed in Ref. [13]). Recently, Guo et al. [71] localized different MTs that bind Cu. Their expression was upregulated in response to high Cu levels in the plant as well as during senescence. MT1a and MT2b were found in the phloem of all plant organs; MT2a and MT3 were expressed in mesophyll cells of older leaves, but at high Cu levels they were also expressed in young leaves and root tips. All MTs except MT4 were highly expressed in leaf trichomes; MT4 was found only in seeds.

2.3. Nickel

Nickel was one of the last metals found to be essential for the growth of plants. In 1975, Dixon et al. [72] and Fishbein et al. [73] found it to be the active center of urease [72,73], an enzyme that converts urea to ammonia and carbon dioxide, both of which can be re-used in plant metabolism. Urease itself and its function has been known for a very long time; it was the very first enzyme to be purified and crystallized in 1926 [74]. James Sumner, at Cornell, isolated and crystallized urease from jack bean over a period of 9 years and he received the Nobel prize in Chemistry in 1946 for this work. However, it took half a century before Ni was found to be its active center, and even now only a few studies have shown that Ni is an essential plant micronutrient *sensu strictu* [75–77]. The first crystal structure of urease became available in 1995 using recombinant enzyme from *Klebsiella aerogenes* [78]. The protein is arranged as a tightly associated symmetrical trimer of a basic $\alpha\beta\gamma$-unit. The coordination of the dinuclear Ni center involves four histidines, one aspartate, one lysine carbamate, one bridging oxo ligand (water, hydroxide, or oxo dianion), and two terminal waters (or hydroxides), with a Ni–Ni distance of ≈ 3.5 Å. In the original catalytic mechanism for urease [79], a two Ni site was proposed, with the metals within a distance of 6 Å. Hereby, the oxygen of the urea carbonyl coordinates to one Ni with additional activation provided by a carboxylate. This mechanism could be extensively refined on the basis of several crystal structures, while retaining many of its basic features [80].

In most plant species, nickel deficiency is rarely observed, because only very minute amounts of this metal are needed for normal metabolism, and the adequate range between limiting and toxic concentrations is exceptionally large compared with other heavy metals [81]. For this reason, it took long before Ni was identified as essential for plant growth (see above), and Ni has been termed an "ultra-micronutrient" [82]. Symptoms of Ni deficiency are a consequence of the lack of urease, which leads both to an accumulation of urea to toxic levels and to a lack of usable nitrogen sources. Visible symptoms of urea toxicity are leaf tip necroses [75,76], while symptoms of nitrogen deficiency are interveinal necrosis and patchy necrosis of younger leaves [83]. In addition, it was reported that the disturbance of nitrogen metabolism, even when nitrogen is supplied as nitrate, leads to strongly reduced levels of alanine and thereby impaired protein synthesis [84]. Interestingly, urease activity seems not to be regulated apart from the limitation by Ni availability; as long as Ni is available, urease is expressed constitutively [85].

In contrast to all "normal" plants referred to so far, plants that hyperaccumulate Ni can easily suffer from Ni deficiency. As recently observed in a study of three Ni hyperaccumulating Brassicaceae [86], these plants show severe symptoms of Ni deficiency on normal soil, and only an addition of several hundred parts per million Ni(II) leads to normal growth. In addition to its function in urease, in hyperaccumulators Ni is used for deterring pathogens and herbivores for which it is predominantly stored in epidermal vacuoles (see below). Most likely the mechanism for Ni sequestration into these vacuoles is so efficient that it easily leaves all other parts of the plants Ni-deficient.

Another function of Ni apart from urease activity is found in N_2-fixing legumes. The symbiotic N_2-fixing bacteria of these plants require Ni as the active center of hydrogenase, an enzyme that recycles H_2 produced by a side reaction of nitrogenase [87]. For this reason, legumes [81] and some diazotrophic cyanobacteria [88] have been reported to have an elevated Ni requirement.

Hardly anything is known about the transport of Ni into the plant and translocation within the plant. The only Ni(II)-specific transporter found so far (TgMTP1t2) seems to be involved in the hyperaccumulation of Ni in *Thlaspi goesingense* [89]. It was shown to be an alternative splicing product of an mRNA that can also produce a transporter (TgMTP1t1) for the divalent ions of Cd, Zn, and Co, but the protein itself has not been characterized. Some activity for Ni uptake was also shown for the Mg transporter AtMGT1 [67].

Like most metals, Ni is in short supply in the oceans; Ni deficiency has been shown to limit the use of urea by phytoplankton [90]. Probably, for this reason, some green algae have an alternative urea-degrading system, the ATP-dependent urea amidohydrolase–allophanate pathway [91].

2.4. Zinc

Zinc has been known for a long time to be a trace element that is essential for the growth of plants; it was first discovered to be essential in the fungus *Aspergillus niger* (at that time fungi were regarded as plants) in 1869 by Raulin [92]. Zinc is indispensable to all forms of life and is required as an essential constituent of numerous proteins and enzymes (Table 1). High-resolution structures of Zn binding sites have now been identified in all six classes of the IUB nomenclature. Early on, three types of binding sites have been recognized: catalytic, co-catalytic, and structural. More recently, a new type of site at protein interfaces has become apparent [93]. Here, we discuss a few prominent examples of Zn proteins.

(a) *Carbonic anhydrase.* Most aquatic plants use the enzyme carbonic anhydrase for converting bicarbonate to carbon dioxide. This was the very first enzyme with Zn(II) in the active center that was discovered

(Keilin and Mann [94a]). It is part of the so-called "carbon concentrating mechanism", which is important because in water only a small fraction of CO_2 is available as such; the equilibrium is much in favor of bicarbonate. The enzyme belongs to the class of lyases and carries a catalytic Zn coordinated by three imidazole residues from histidines. The fourth ligand of catalytic Zn is always a water molecule, which is activated for ionization, polarization, or displacement by the identity and arrangement of ligands coordinated to Zn [94b]. In carbonic anhydrase, the three histidine ligands readily allow formation of a $Zn(II)-OH$ species, which then can add OH^- to CO_2 to form HCO_3^-.

(b) *Cu,Zn superoxide dismutase.* This enzyme is highly expressed in all organisms performing oxygenic photosynthesis. It is involved in the defense against reactive oxygen species and has a dinuclear Cu,Zn center as discussed earlier. The two metals ($Cu-Zn$ distance 6.2 Å) are bridged by the imidazole ring of a histidine. Three other histidine residues bind to the catalytic Cu site, whereas two histidines and an aspartic acid are the Zn ligands [93]. The role of Zn is generally considered supportive to that of Cu. However, Zn might be important for substrate specificity.

(c) *Zinc fingers.* The term zinc finger defines a Zn binding motif that can fold autonomously into a separate functional unit or domain where the Zn is bound tetrahedrally by a combination of cysteines and histidines [95]. Originally, it was thought that zinc finger domains are DNA-binding units. However, with the identification of proteins that interacted with various zinc finger domains, it became clear that these domains might also mediate protein–protein interaction, or might interact with membranes or second messengers.

(d) *Phosphatases.* This class of Zn enzymes carries heteronuclear metal sites, such as a Zn,Mg site in alkaline phosphatase, or Fe(III),Zn in the purple acid phosphatase. The characteristic purple color of these enzymes, which function as non-specific phosphomonoesterases, comes from a phenolate–Fe(III) charge transfer transition around 550 nm. In the enzyme from kidney bean, the Zn is ligated by aspartate, asparagines, two histidines, and a water, the Fe(III) center carries a tyrosine, two aspartates, and one histidine. Hereby, both metals are bridged by an aspartate [96].

Zinc uptake into the plant is mediated by a large number of transporters belonging to several protein families. The largest number belongs to the ZIP family that was first discovered in *Arabidopsis thaliana* [97]. As reviewed in detail elsewhere [8,98], these are proteins of 309–476 amino acids (35–40 kDa) usually containing eight transmembrane domains, two of which contain histidine-rich regions as putative metal-binding sites. As for all other heavy metal transporters in plants, however, no crystal structures and no detailed data on the metal transport mechanism are available so far. ZIP transporter expression was found to be regulated by metals at the level of both transcript and protein accumulation [99]. While some ZIPs have a rather low specificity and therefore also transport, e.g., Cd (e.g., OsZIP1), some (e.g., OsZIP3) are very Zn-specific [100]. The ZIP transporters seem to be primarily expressed in roots [8], only a few (e.g., ZNT1 [101], AtZIP6 [102]) were found at similar levels in the shoots. While the ZIP transporters are involved in Zn uptake into the cell, the cation diffusion proteins (CDF [8]) seem to be involved in transporting it into intracellular organelles or out of the cell [103]. One Zn-specific CDF has been characterized in more detail by van der Zaal et al. [104]. ZAT was found in *Arabidopsis thaliana* (all plant parts), its expression was upregulated at high Zn, its overexpression led to both enhanced Zn tolerance and Zn accumulation, and it was therefore proposed to participate in the vacuolar sequestration (for storage and detoxification) of Zn [104]. A homologous transporter (ZTP1) was subsequently found to be highly expressed in leaves of the Zn hyperaccumulator *Thlaspi caerulescens* (see below [105]). The energy coupling of the CDF transporters is not yet understood; while transport against a concentration gradient certainly requires energy, these proteins do not have any nucleotide-binding domain [49]. Some of the CPx-type heavy metal ATPases (Section 2.2) may be involved in Zn transport as well, but so far this has only been proposed on the basis of sequence homologies and T-DNA insertion mutants [9]. Another transporter family that generally contains Zn-transporting members are the Nramps ("natural resistance associated macrophage proteins"), but so far none of the plant members of this family has been shown to be Zn-specific [8].

Zinc deficiency is a widespread problem for agriculture in many countries of this planet and for ~40% of the total land area [2,106]. Symptoms of Zn deficiency are chlorosis (in particular on young leaves), stunted growth, and in severe cases wilting. A more detailed description of Zn deficiency symptoms, including ultrastructural effects, was recently published by Kim and Wetzstein [107]. Many of these symptoms can be explained by oxidative stress that may be caused by deficiency in superoxide dismutase (see above). In the oceans, most organisms suffer from Zn deficiency, in particular in the surface layer because of the high cell density present [3]. While dissolved Zn concentrations are ~10 nM in the deep sea, they are in the picomolar range at the surface [20].

There are large differences between individual plant species and even among cultivars of the same species in their tolerance to suboptimal Zn nutrition; those which tolerate low Zn levels have been termed "zinc efficient" [108]. The mechanisms of Zn efficiency are not well understood, but in most plants analyzed so far the shoot seems to play a more important role than the root [2]. One exception was a correlation between higher Zn uptake with higher Zn efficiency as well as higher tolerance to Zn toxicity in the grass *Holcus lanatus* [109]. A recent study showed that Zn efficiency is correlated with elevated expression and activity of superoxide dismutase and carbonic anhydrase [110], but it remained unclear whether these elevated expression levels are the reason or the result of

Zn efficiency. A close correlation between Zn efficiency and tolerance to high levels of bicarbonate in the soil was reported [111]; the mechanisms, however, remained unclear. Other recent studies surprisingly indicated that Zn efficiency in barley and bean is controlled by a single gene [112,113]. If this is verified and valid also for other plants, Zn efficiency would be a much less complex phenomenon than currently believed.

Little is known so far about Zn metabolism in seeds during their formation and later germination, but one study has indicated that MTs are involved in supplying Zn to seeds during their formation [114].

3. HEAVY METALS AS INHIBITORS OF PLANT METABOLISM

Elevated concentrations of the metals mentioned earlier inhibit plant metabolism, leading to various effects depending on the metal applied, the type of affected plant, and the environmental conditions during the stress. An overview of pathways of toxic metals in plants is shown in Fig. 2. Important general reviews on this theme include the book of Prasad and Hagemeyer [11] concerning higher plants, and Wood and Wang [115] and Maeda and Sakaguchi [15] concerning algae and cyanobacteria. Reviews on specific damage or resistance mechanisms are mentioned in the specific sections later. Because of their sensitivity to heavy metal toxicity, plants can also be used for monitoring water quality, as reviewed recently [116].

The threshold concentration of a heavy metal that leads to toxicity strongly depends on the type of plant under investigation, because plants drastically differ in their ability to deal with metal toxicity. Since many detoxification mechanisms are highly metal-specific, also the order of toxicity thresholds varies between plant species. If largely non-resistant water plants are considered, where the metal toxicity is also not influenced by soil mobility of the metals, because the uptake occurs over the whole plant surface, the order of toxicity generally seems to be $Hg > Cu > Cd > Ni, Zn > Pb$. The most environmentally relevant toxicities are probably those of Cu, Cd, and Zn. Copper is widely used as a pesticide in agriculture, and field runoff may easily reach concentrations of several micromolar [117]. Cadmium and zinc have been released by the Zn processing industry leading to localized but high concentrations. Near gold mines, the mercury amalgamation method causes severe Hg pollution; mechanisms and relevance of Hg toxicity to plants have been reviewed [118]. The toxicity of heavy metals strongly depends on the bioavailability. Clearly, strongly complexed metals are unavailable for plant uptake, which made many authors think that toxicity is directly related to the free ion concentration [119–121]. This is, however, again simplifying the situation too much, as shown by experiments in which the uptake into the plant decreased much less upon chelation of the metal than one would have to expect from the concentration of free ions [122], or even an increased toxicity was found upon complexation [123]. Such results may indicate that the plant uptake can occur directly out of the ligand in contact to the transport

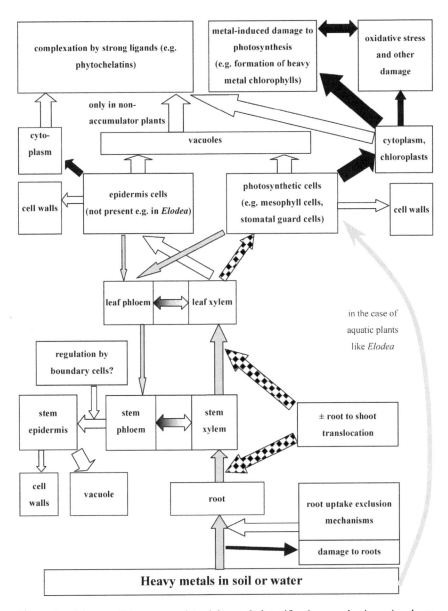

Figure 2 Scheme of heavy metal toxicity and detoxification mechanisms in plants. Black arrows show pathways primarily leading to toxicity, white arrow show detoxification mechanisms. Pathways that may have both effects are labeled with a black/white pattern and those that are most likely neutral to toxicity/detoxification are labeled in gray.

protein, and/or the ligand is taken up together with the metal. In both cases, the metal ion would not be released into water before being taken up, so that uptake is not only determined by the equilibrium [metal–ligand] $\rightleftharpoons$ metal + ligand, but also by the equilibrium [metal–ligand] + transporter $\rightleftharpoons$ [metal–transporter] + ligand. This has to be kept in mind when dealing with studies in which very high total metal concentrations were applied, but interpreted as if the plant would only "see" the free ions.

Among the mechanisms proposed to contribute to heavy metal damage are inhibition of enzymes [124], inhibition at various sites of photosystem II reaction centers (PS II RC) [125], enhancement of photoinhibition [126] and retarded recovery from it [127], oxidative stress [128–131], the impediment of plastocyanin function [132], changes in lipid metabolism [133], and disturbances in the uptake of essential microelements [134,135]. However, many of the effects on enzymes and electron transport were examined only *in vitro* (discussed, e.g., by Barón et al. [125] and Küpper et al. [136]), some of them either could not be observed to operate *in vivo* [135,137] or could not be confirmed by *in vitro* studies [138], and in many cases alternative explanations for the effects observed may be more likely [136]. In many studies, heavy metals were applied at extremely high concentrations, which are far from being ecologically relevant. This section will focus on studies which deal with mechanisms of metal-induced inhibition of plant metabolism that were observed under environmentally relevant conditions, preferably *in vivo*.

Plants have developed a number of strategies to resist the toxicity of heavy metals, such as efflux pumps [12], complexation of heavy metals inside the cell by strong ligands such as PCs [13] or histidine [14], and several other mechanisms [11,15]. As for mechanisms of damage, this section will focus on recent research and on those resistance mechanisms that are likely to be of importance *in vivo*. While metal tolerance is a constitutive feature in plants normally growing on metal-rich habitats such as hyperaccumulators (see below), in many other plants the defense mechanisms are only activated at an elevated level when the cells are actually stressed by metal toxicity; this has been observed both in algae and higher plants [139–142]. After the end of the stress, the resistance mechanisms seem to be soon downregulated again; however, so far this has only been investigated in the green alga *Scenedemus* [139,142].

3.1. Damage to the Photosynthetic Apparatus and Mechanisms Counteracting It

The photosynthetic apparatus, both its primary photochemical side and its biochemical carbon-fixing part, is one of the most important sites of inhibition by many heavy metals [143]. In all studies investigating this, a much stronger inhibition was found for photosystem II compared to photosystem I [144–147]. The relative importance of specific inhibition sites, however, strongly depends on the type and concentration of heavy metal, the irradiance conditions, and the organism under investigation [143,145]. Mechanisms of heavy metal-induced damage to photosynthesis that were observed under environmentally relevant conditions include the formation of heavy metal-substituted chlorophylls [34,136,142,147–154] and a delayed recovery from photoinhibition [127].

Several authors have shown that drastically different types of damage occur depending on the irradiance conditions, as discussed in detail later.

3.1.1. Low Irradiance with a Dark Phase

Already Cedeno-Maldonado et al. [155] found, using the unicellular green alga *Chlorella*, that damage induced by excess Cu^{2+} leads to stronger inhibition in high compared to low irradiance, and that the Cu bound to a component of the thylakoid membranes only in low irradiance. Küpper et al. [147] confirmed the latter finding, and furthermore showed that the binding is to chlorophyll (Chl) of the light harvesting antenna, in which the central ion Mg^{2+} was substituted by Cu and other toxic metals. Later on, it was shown that this type of reaction, which had been named "*shade reaction*", occurs in the chlorophyll molecules of the LHC II in Chlorophyta and the homologous Chl a/c proteins in brown algae [136]. Substitution of Mg^{2+} in Chl by heavy metals (Mg-substitution) results in an impairment of the correct function of the light harvesting antenna, because heavy metal substituted chlorophylls (hms-Chls) are not suitable for photosynthesis for many reasons as discussed in detail in a recent review [153]. The most important reasons are: (i) in contrast to Mg–Chl, they do not (e.g., Cu–Chl, Ni–Chl) or only weakly (e.g., Cd–Chl, Zn–Chl) bind axial ligands [156], which are necessary for correct folding of the pigment–protein complexes [157,158]. (ii) hms-Chls have a very unstable singlet excited state that relaxes thermally [159], so that in solution they have a low fluorescence quantum yield and *in vivo* they do not transfer captured excitons towards the reaction center but act as exciton quenchers [160].

In view of the latter fact, it is not surprising that only a few percent of the total Chl has to be converted to hms-Chls for complete inhibition of photosynthesis [136,142,147] and that their exciton quenching leads to a strong decrease of fluorescence during *shade reaction* [136,147,151,154,161–165]. The formation of non-fluorescent (e.g., Cu–Chl) or extremely weakly fluorescent (e.g., Cd–Chl) hms-Chls in the light harvesting antenna during the *shade reaction* reduces both F_0 and F_m proportionally, so that F_v/F_m remains unchanged or sometimes even increases [136,142,151,165,166] until a very late stage of damage. Only in samples in which almost all PS II fluorescence is already quenched, the remaining PS I fluorescence and the weak fluorescence of Cd–Chl may lead to a decrease of F_v/F_m. Formation of the fluorescent (~30% quantum yield compared to Mg–Chl [159]) but still non-functional Zn–Chl, in contrast, leads to a slightly stronger decrease of F_m compared to F_0 and thus a decrease of F_v/F_m (although still not as strong as during inhibition by the same metal in high irradiance, see below) [136]. In some plant species, the relative antenna size per reaction center as measured by Chl fluorescence kinetics remains unchanged during *shade reaction*. This is not surprising because heavy metal Chls act as quenchers of excitons [136,153], so that Mg-substitution in only a few percent of the total Chl knocks out a whole photosynthetic unit and only the undamaged units remain measurable [136].

In metal-sensitive plants and cyanobacteria, Mg-substitution occurs in the submicromolar concentration range, making it an environmentally relevant process [147,148,150]. With higher metal concentrations leading to high percentages of hms-Chls in the plants, the formation of Cu–Chl becomes easily visible in the UV/Vis absorbance spectra of plants and their extracts. Since Cu–Chl has a strongly blue-shifted red absorbance maximum, a blue shift of the absorbance maximum of the extracts [144,147,149,154] and whole chloroplasts or plants [136,151] can be observed. When stable hms-Chls are formed (e.g., Cu–Chl), such plants remain green even when they are already dead [151].

While in Chlorophyta the *shade reaction* is strictly connected to a low irradiance light period combined with a dark period (the length of which has to be increased with increasing irradiance in the light period), in brown algae (Phaeophyta) *shade reaction* occurs also in high irradiance [136,165]. This was correlated with differences in the supramolecular structure of the LHC II of Chlorophyta compared to the Chl a/c LHC of brown algae [136]. The *shade reaction* is not related to a direct inhibition of PS II protein synthesis; Yruela et al. [167] have shown that even very high Cu concentrations did not significantly affect the protein composition of PS II, and in particular the LHC II.

3.1.2. High Irradiance

In most studies on heavy metal toxicity published so far, heavy metals were applied to Chlorophyta (higher plants and green algae) in high irradiance or without a dark phase. Under such conditions, direct damage to the PS II RC occurs instead of the Mg-substitution in the antenna Chls. This reaction has been named "*sun reaction*" by Küpper et al. [147]. After inhibition of photosynthesis, under high irradiance the antenna pigments are degraded [136,147,151]. The bleaching ("chlorosis") is unrelated to inhibition of chlorophyll synthesis, as shown by experiments with Cd toxicity in barley [168], but may be related to chlorophyll oxidation by lipid peroxides [169]. Furthermore, bleaching during *sun reaction* is slower than the Mg-substitution during *shade reaction*, so that the very low ratio of Mg-substitution during *sun reaction* cannot be caused by competition with the bleaching process [136,147,151]. The latter study has further confirmed this by experiments with brown algae, in which the *shade reaction* was observed even under high irradiance [136]. This study also shed a new light on findings of other authors concerning the damage to the PS II RC that occurs during the *sun reaction* under high irradiance. Although most of the Chl is degraded under these conditions, small amounts of hms-Chls are formed during the *sun reaction*. The detection of Cu–Chl in Cu^{2+} treated red algae [136] demonstrated that hms-Chl formation can occur in complexes other than in LHC II. Since PS I resists attack of heavy metals even in high irradiance [144–146], the core of PS II is the most likely target. Many authors have come to the conclusion that the PS II RC is the target of heavy metal induced inhibition, but there is little agreement on which components are effected, and literally all components participating in the energy conversion within PS II RC have been proposed as targets [170,171]. Most probably different effects occur when diverse materials are treated under various conditions (cf. [170]); often high heavy metal concentrations were used, which may not be relevant in nature. Most authors locate the site of heavy metal inhibition very close to the primary photochemical event in PS II [155,161,170,172–176], either on its oxidizing or on its reducing side. The *in vivo* results of Küpper et al. [136], however, suggested that heavy metals inhibit charge separation in PS II. Li and Miles [177] and Hsu and Lee [178] came to the same conclusion based on convincing experiments with Cd and Cu treatments of isolated spinach chloroplasts. As discussed in detail by Küpper et al. [136], the inhibition of charge separation may occur via an insertion of heavy metals into the pheophytin of the PS II RC, which can also explain the results published by Yruela et al. [138,171,173,179].

Some authors, such as Pätsikkä et al. [126], interpreted the decrease of F_v/F_m during the sun reaction as an increase of photoinhibition. The data of Küpper et al. [136,151], however, rendered another explanation more likely. First, as mentioned earlier, no decrease of F_v/F_m was observed when brown algae were inhibited by Cu^{2+} in high irradiance; typical *shade reaction* (see above) occurred instead. Second, the *sun reaction* with its decrease in F_v/F_m was observed in red algae, which use phycobilisomes instead of the LHC II, even in very low irradiance [136,180]. These results indicate that the decrease in F_v/F_m during the *sun reaction* is not caused by photoinhibition, but by the insertion of heavy metals into the pheophytin of the PS II RC (see above). This reaction would render the basic fluorescence (F_0) little affected while decreasing the variable fluorescence, resulting in the decreased F_v/F_m. Furthermore, many authors observed an increased rate of Cu-induced inhibition of PS II activity in high compared to low irradiance [135,147,155] and usually attributed it to increased photoinhibition. However, this phenomenon may have a simpler explanation; the inaccessibility of the antenna chlorophylls for binding Cu under high irradiance makes the probability of a Cu^{2+} binding to the PS II RC statistically more likely. In addition, the recent study by Pätsikkä et al. [135] showed that Cu concentrations in the chloroplasts of Cu-inhibited plants are much lower than those required for causing direct photoinhibition in isolated chloroplasts [181] and thylakoids [182]. There may be, however, an indirect enhancement of photoinhibition under Cu-induced stress, as suggested by the results of Vavilin et al. [127]. These authors found, using submicromolar Cu concentrations and the unicellular green alga *Chlorella*, a slower recovery from photoinhibition under elevated Cu concentrations and concluded that this was due to a non-specific inhibition of the synthesis of the D1 protein of PS II.

Under some conditions, effects differing from those discussed earlier (*shade reaction*, *sun reaction*) have been found. Using a long light period and an intermediate irradiance during severe Cu toxicity stress (8 and 80 μM Cu^{2+}) in *Zea mays*, Ouzounidou et al. [183] observed a stronger Cu^{2+}-induced inhibition of the light-adapted steady-state photosynthesis compared to the dark-adapted initial photosynthetic performance. Similar results were obtained earlier by Krupa et al. [184,185] for Cd- and Ni-induced damage in *Phaseolus vulgaris* and by Xyländer et al. [186] for Hg-induced inhibition of the green alga *Haematococcus*. While Krupa et al. hypothesized that this effect indicates an inhibition of the Calvin cycle as the primary event, Xyländer et al. regarded it as non-reversible photoinhibition, and Ouzounidou et al. [183] concluded that the transition of the photosynthetic system itself from the dark-adapted to the light-adapted state is affected. Future investigations should show what is the true nature of these effects and under which conditions they generally occur.

In red algae, yet another effect on the photosynthetic apparatus was observed. In addition to the inhibition of the PS II RC characteristic of the *sun reaction* (see above), a strong decrease of the phycobilisome-specific peak in *in vivo* absorption spectra was observed [136]. The latter had also been found with Hg toxicity in the cyanobacterium *Spirulina* [187], but was not found after Cu treatment of the red alga *Gracilariopsis* [180].

Because of the sensitivity of the photosynthetic apparatus to heavy metal toxicity, plants have developed strategies that specifically protect this part of their metabolism, in addition to the general metal detoxification mechanisms described in the following text. In particular, in plants that accumulate large amounts of metals, the highest metal accumulation is usually found in cell types that are not photosynthetically active, as described in more detail later in Sections 3.5 and 4.2. Because of the presently very limited sensitivity of metal microanalysis, so far there is hardly any knowledge about the cellular compartmentation of heavy metals in non-accumulator plants. This is clearly an interesting subject for future research. For the same reason, also another defense reaction was so far investigated only in hyperaccumulators. Under conditions of heavy-metal toxicity, these plants seem to reduce the binding of heavy metals (Cd, Ni) to essential sites (such as metal enzymes or chlorophyll) by accumulating the essential metal that would be replaced by toxic metals [86,188,189] (for details see Section 4).

3.2. Damage to Root Function and Mechanisms Preventing It

In most terrestrial plants, the mobility of Cu is rather low, so that the highest concentrations of this metal are found in the roots [126,190]. In addition, some of the Cu entering the shoot is recycled to the roots via the phloem [190]. In such plants, the roots can also be a primary target for Cu-induced damage [191,192]. Based on a reduced Cu toxicity under elevated iron supply, Pätsikkä et al. [135] concluded that the inhibition they observed was caused by Cu-induced disturbances of iron uptake, but direct evidence for such an effect of Cu has not yet been published. In contrast, a recent study by Pietrini et al. [166] found no inhibition of iron uptake by cadmium toxicity; such an inhibition had been proposed by Siedlecka and Krupa [134].

Roots may defend themselves against metal-induced damage by expressing peroxidases, which can be used in lignification and thus reducing the uptake of further heavy metal as proposed by Cuypers et al. [193]. Another strategy for reducing damage to the roots, as well as preventing damage to other parts of the plant by reducing metal load, is actively pumping the excess metal out of the plant. Such ATP-dependent efflux pumps have been found by van Hoof et al. [12] for the detoxification of Cu in roots of *Silene vulgaris*. This mechanism may be even more important in aquatic microorganisms, in which case physical barriers may be less efficient [115].

3.3. Oxidative Stress and Changes in Lipid Metabolism Reducing Its Impact

Many authors reported an increase in oxidative stress under conditions of heavy metal toxicity, as reviewed in Refs. [11,131,194]. Although a large percentage of these studies was carried out under very high heavy metal concentrations that would not be environmentally relevant (up to the millimolar range [130]), some studies with lower (but still several micromolar) metal concentrations have shown the induction of oxidative stress as well. This was reported not only for redox-active metals such as Cu, but also for the redox-inert Cd and Zn. Sandalio et al. [195] found an increased lipid peroxidation coupled with a reduction in Zn/Cu-superoxide dismutase activity in the shoots, starting at $10 \, \mu M \, Cd^{2+}$. In this case, the decrease in SOD activity was most likely the primary effect, which then led to oxidative stress. The decrease of the Zn-dependent SOD (Section 2) was most likely caused by Cd-induced Zn deficiency, since Cd significantly reduced Zn translocation from roots to shoots [195].

The possibility of an involvement of Mg-substitution in the generation of oxidative stress was studied [152]. This was done by measuring the kinetics and efficiency of energy transfer between Chl derivatives and oxygen in solutions of Chl derivatives. It was found that heavy metal-substituted Chls are less efficient (for Cu–Chls) or equally efficient (for Zn–Chls) in the generation of singlet oxygen as Mg–Chls, but all the heavy metal substituted Chls had a reduced efficiency for quenching singlet oxygen, so that Mg-substitution indirectly contributes to oxidative stress.

Many authors reported that the oxidative stress caused by heavy metal-toxicity causes damage to cellular membranes by oxidation of lipids [128,133,169,191,195,196] and they concluded that this causes leakage of essential ions through the membranes out of the cell [191,196,197]. Demidchik et al. [198,199], however, have shown that the Cu toxicity-induced conductivity increase in the plasma membrane is not caused by lipid peroxidation, but by a disturbance of calcium fluxes, confirming the hypothesis of Stohs and Bagchi

[200]. Using reasonable heavy metal concentrations (starting at 1 μM in 1997 and 5 μM in 2001), Demidchik et al. [198,199] found that inhibitors of Ca^{2+} channels stopped the leakage, while oxidants and antioxidants did not influence it. In addition, an inhibition of the H^+-ATPase was observed.

One reason why the later studies found little oxidation-related damage to the plasma membrane may be that plants quickly suppress this type of damage. First, they reduce the oxidative stress by increasing the expression of enzymes that scavenge reactive oxygen species [131]. This mechanism does not seem to work, however, under Cd stress, which decreases Zn concentrations in the shoot and thereby inhibits the synthesis of superoxide dismutase (see above). Another mechanism counteracting heavy metal-induced oxidative stress is the induction of changes in the membrane composition such as a decrease of the ratios phosphatidylcholine/phosphatidyl-ethanolamine, unsaturated/saturated fatty acids, and lipid/protein. Such effects were consistently reported by many authors working with diverse heavy metals and plants [201–206], with variations mainly in the importance of individual changes. These changes make the membranes less vulnerable to reactive oxygen species.

Another strategy employed by plants for counteracting the metal-induced oxidative stress is the production of antioxidants such as glutathione. A recent detailed study on this pathway has been performed by Cuypers et al. [207].

3.4. Detoxification with Metal-Binding Ligands

In addition to the more specific detoxification mechanisms discussed so far, plants developed a number of general detoxification mechanisms that are active in almost all cells and protect plant metabolism as a whole. The best known of them is the complexation of heavy metals inside the cell by strong ligands such as PCs and MTs [13].

3.4.1. Phytochelatins

PCs were first discovered in 1985 by Grill et al. [208]. They are short polypeptides that are synthesized from glutathione by a Cd-activated specific γ-glutamylcysteine dipeptidyl transpeptidase commonly called PC synthase [209]. In contrast to the original suggestions by Grill et al., PC synthase is not activated by binding of the metal to the enzyme, but by the presence of glutathione-like peptides with blocked thiol groups [210]. PCs have been shown to complex metals in the cytoplasm, from which they are pumped into the vacuole [211]. In normal non-hyperaccumulating plants, the metals remain associated with the PCs in the vacuole, where they form high molecular weight complexes [13]. There have been doubts whether the protection from heavy metal toxicity by PCs is really relevant under natural conditions and thus whether there is another role for PCs. The study of Howden et al. [212], however, has clearly shown that even submicromolar Cd concentrations cause a severe reduction of growth in PC-deficient mutants, and to date no study has demonstrated any other role for PCs *in vivo* [13]. PCs detoxify Cd not only in plants, but also in yeast [213] and some animals [214]. The *in vivo* protection provided by PCs seems to be limited, however, to Cd, the metalloid arsenic [215], and possibly Hg; for other metals the affinity to PCs and/or the induction of PC synthesis is too low to be relevant under natural conditions [216,217]. This was confirmed by Schat et al. who showed that blocking PC synthesis does not affect the resistance of plants against Co, Cu, Ni, and Zn [218]. For this reason, plants use other ligands for scavenging other metals, in particular the highly toxic copper.

3.4.2. Metallothioneins

For Cu, MTs of type I and II seem to be important detoxification ligands, as reviewed by Cobbett and Goldsbrough [13]. The study of García-Hernández et al. [219] suggested that MT1 is important for scavenging Cu mainly in leaf veins; Cu-induced MT1 expression mainly in leaf veins and to a lesser extent in mesophyll cells, while hardly any induction of MT1 in response to Cu stress was observed in roots. In contrast to PCs, however, detoxification seems not to be the main role of MTs in plants; they may be more important in metal distribution during normal metabolism and during senescence [13].

3.4.3. Glutathione

In principle, also the thiol group of glutathione could act as a ligand for heavy metal sequestration and as an antioxidant counteracting the oxidative stress that may be caused by heavy metal toxicity (see above). The data published concerning this subject so far are, however, contradictory. A recent study of Cd and Cu toxicity in phytoplankton showed no upregulation of glutathione synthesis under metal stress conditions that caused a strong induction of PCs [220]. This suggested that glutathione in plants does not have a direct role in metal detoxification, but only via its integration into PCs. However, another study found, using Cd toxicity in the higher plant *Phragmites australis*, a 30-fold increase in the concentration of reduced glutathione in leaves [166]. One possible explanation for this discrepancy may be the different metal concentrations applied, as well as the different plant systems used. While Ahner et al. [220] used submicromolar metal concentrations ($0.03-3$ µM total Cd^{2+}; $0.12-120$ µM Cu^{2+}, pM–nM free metal after chelate buffering) and marine organisms that are adapted to extremely low metal concentrations (Section 2), Pietrini et al. [166] used micromolar (50 and 100 µM) Cd and a freshwater swamp plant.

3.4.4. Free Amino Acids

Plants seem to use monomeric amino acids for binding potentially toxic heavy metals. In hyperaccumulators, histidine has been shown to bind Ni [14], Zn [189,221], and possibly also Cd [189], as discussed in greater detail in Section 4.

The non-proteogenic amino acid nicotianamine was known for some time to be involved in Fe homeostasis in plants [222] and more recently was associated with Cu homeostasis [51]. Furthermore, Weber et al. [223] have now shown that it is produced at a much higher level by the Cd/Zn-tolerant hyperaccumulator *Arabidopsis halleri* than by the Cd/Zn sensitive non-accumulator *A. thaliana* [223], suggesting a role in the resistance and/or accumulation mechanism of *A. halleri*. Finally, an accumulation of free proline has been suggested to act as a mechanism of defense against heavy metal toxicity [224–226]. Many authors observed that heavy metal toxicity induces the accumulation of free proline in the plants (e.g., [224–226]) and metal-tolerant plants have higher proline levels than non-tolerant plants [227]. Schat et al. [227] found, however, that heavy metal-induced accumulation of proline in the terrestrial plant *Silene vulgaris* can almost totally be blocked by placing the plants in conditions preventing water loss, and concluded that the proline accumulation is related to a defense against water loss under metal stress rather than the metal stress as such. This is not a logical explanation, however, for the elevated proline levels in the freshwater alga *Chlorella* [225,226]. And it is also not logical in view of the fact observed by Schat et al. [227] that the metal-tolerant plants had higher proline levels than the non-tolerant ones. In view of the potential importance of proline accumulation as a mechanism against toxicity of many heavy metals (Cd, Cu, Cr, Ni, Zn), there is clearly a need for further research.

3.4.5. Other Metal Detoxifying Ligands

In addition to MTs, other proteins seem to be produced by plants specifically for binding and probably detoxifying metals. One such protein family was found in *Arabidopsis thaliana* [*Arabidopsis thaliana* farnesylated protein (ATFP)] [228]. A complexation of molybdenum by anthocyanins, combined with a direct correlation between anthocyanin content and molybdenum accumulation, has been discovered in the epidermis of *Brassica* species [229]. The cell walls can act as metal-binding ligands as well, in particular in algae and other aquatic plants where the complete cell wall surface is in direct contact with the metals [15]. In aquatic algae, the exudation of metal-binding thiol ligands of so far unknown nature has been observed as a response to Cu stress [230].

3.5. Detoxification by Compartmentation

Another very common mechanism of metal tolerance is metal exclusion, as mentioned previously in Section 3.2 about roots. This mechanism is, however, not restricted to roots; marine algae exhibit similar mechanisms as shown for the brown alga *Fucus serratus* by Nielsen et al. [165].

Many plants detoxify heavy metals by sequestering them, either as PC complexes or without specific ligands, in the vacuoles [231,232]. This plant-specific (animal and bacterial cells do not possess this organelle) metal detoxification strategy provides an efficient form of protection because the vacuole does not contain any sensitive enzymes. When vacuolar sequestration is the major detoxification mechanism, metal tolerance is often associated with elevated metal accumulation; an extreme form of this sequestration is found in hyperaccumulator plants (Section 4). In most heavy metal-tolerant plants, the vacuolar sequestration mainly occurs in non-photosynthetic cells of the epidermis, reducing toxicity to the heavy metal sensitive photosynthetic apparatus [86,233–237]. For a general review on mechanisms of such differential ion accumulations in leaves, see Ref. [238].

Both exclusion and vacuolar sequestration are metal transport processes against the concentration gradient, so that they require an active transport system [239]. So far, knowledge about such transport systems is limited, as described in Section 2. Protein families involved in vacuolar sequestration may be the Nramps, CDFs, and CAXs [8]. Three transporters for vacuolar sequestration have been characterized, all are homologous, almost identical in sequence, and belong to the CDF family. These are MTP1t1, ZAT, and ZTP transporting Zn (and possibly Cd and Co), and MTP1t2 transporting Ni (Section 2) [89,104,105]. Only one efflux carrier has been identified; this was the Cd^{2+}-detoxifying AtDTX1 (*Arabidopsis thaliana* detoxification1 [240]).

3.6. Other Damage Mechanisms and General Resistance Mechanisms

Substitution of essential by toxic metals most likely occurs not only in chlorophyll, but also other sites that bind the essential metal relatively loosely, a feature that applies to almost all Mg- and many Zn-binding sites in enzymes [241]. In most cases, however, the toxic effect of such a substitution would not be as severe as in the case of Chl. While Mg-substitution in one Chl molecule inactivates many others by quenching excitons (see above [136,153,154,160]), a substitution in an enzyme inactivates only one single enzyme molecule.

The metal binding resulting in metal transport by transport proteins is often not completely specific for the desired metal, so that toxic metals can compete with the uptake of essential metals [242]. Such an effect was proposed to cause iron deficiency under Cd toxicity [243], but could not be verified in a recent study using a different plant species [166]. This discrepancy indicates that the Cd-induced disturbance of iron uptake does not apply to all plants or all conditions. In diatoms and other silicate requiring algae, a heavy metal-induced irreversible inhibition of silicic acid uptake resulting in silicate deficiency was observed and attributed to a binding of Cu to a silicic acid transporter [244,245].

Respiration is affected by excess metals as well, but the concentrations required for its inhibition are usually much higher than those leading to inhibition of photosynthesis [136,147,165,180,246]. In principle, heavy metals can have a mutagenic effect on the genome, as it is well known for Cd and Ni in animals including humans, but so far there is no convincing evidence that such effects play a significant role in *in vivo* heavy metal toxicity in plants.

As reviewed in detail elsewhere [11,15,115] several more mechanisms may be involved in heavy metal detoxification, for example:

(a) Changes of the oxidation state of the metals, such as reduction of Hg^{2+} to Hg^0 or Cu^{2+} to Cu^+. In the former case, the reduced form of the metal is volatile so that it need not even be stored.

(b) Precipitation of insoluble metal complexes (e.g., sulfides) on the cell surface.

(c) Methylation of the element, e.g., many algae convert arsenic to dimethyl- and trimethylarsenic compounds.

4. HYPERACCUMULATION—HEAVY METAL UPTAKE AS A DEFENSE AGAINST HERBIVORES AND PATHOGENS

Plants developed a number of strategies to resist the toxicity of heavy metals, as discussed in the previous section. Plants that actively prevent metal accumulation inside their cells are called excluders; these represent the majority of metal-resistant plants [16]. Other resistant plants deal with potentially toxic metals in just the opposite way, i.e., they actively take up metals and accumulate them. This phenomenon was discovered as early as 1885 by Baumann [247] for *Viola calaminaria* and *Thlaspi calaminare*, later renamed [248] to *Thlaspi caerulescens* ssp. *calaminare*.

These plants, which have been named "hyperaccumulators" [249], are able to actively accumulate several percent metals in the dry weight of their above-ground parts. This ability provides a promising approach for both cleaning anthropogenically contaminated soils (phytoremediation) and for commercial extraction (phytomining) of metals from naturally metal-rich (serpentine) soils (Section 4.3). By now, more than 400 species of hyperaccumulators for diverse metals have been identified in $\sim$50 plant families in many parts of the world, but representing $<$0.2% of all plant species [250,251]. Of all hyperaccumulators identified, $\sim$75% are Ni hyperaccumulators. The best-known genus among these is *Alyssum*; a large number of species belonging to this genus accumulate Ni. Many other plant genera from various families hyperaccumulate Ni as well; two particularly noteworthy species among these are the tree *Sebertia acuminata* [252] that accumulates up to 26% Ni in the dry weight of its latex, and the high-biomass herbaceous plants *Alyssum bertolonii* and especially *Berkheya coddii* [253–255] that may become a profitable "phytomine" (Section 4.3). In contrast, although *Thlaspi caerulescens* was the first hyperaccumulator ever discovered (see above), only about 20 species of Zn hyperaccumulators have been described so far [250,251, and publications since then], the best known of which remains *T. caerulescens*. This species, and in particular the Ganges ecotype, is the first one that has been shown to be also a true Cd hyperaccumulator (referred to as

"French A" in Ref. [30]), in the sense of accumulating >10,000 ppm Cd in the dry weight of its above-ground parts. Because of these and other features (e.g., kinship with *Arabidopsis* and genetic characteristics), in recent years several authors proposed *T. caerulescens* as a model species for Zn/Cd/Ni hyperaccumulation [256,257].

Recently, Ma et al. [258] found that the fern *Pteris vittata* hyper-accumulates the highly toxic metalloid arsenic to several percent of its dry mass [258]; soon As-hyperaccumulation was reported for many closely related *Pteris* species [157]. Only one species analyzed so far, *Haumaniastrum katangense*, and only its population from the Shaban Copper Arc in Zaïre and adjacent parts of the Zambian Copperbelt, seems be a true Cu hyperaccumulator (up to 9200 ppm Cu [259]). However, no detailed physiological characterization has yet been reported. No Hg hyperaccumulators have been identified. It may be that the toxicity of Hg and possibly Cu is too severe, so that the mechanisms enabling plants to accumulate As, Cd, Ni, and Zn are not sufficient for accumu-lating, without toxicity to the plant itself, concentrations of Cu or Hg that would be effective as a defense against pathogens and herbivores.

4.1. Biological Importance of Hyperaccumulation

Many studies have shown that metal accumulation serves as a defense mechan-ism against herbivores and pathogens [260–263], as it was originally proposed by Boyd and Martens [260]. Herbivores that were given the choice between *T. caerulescens* plants grown on different Zn levels [264] or belonging to eco-types with different abilities of hyperaccumulation [265] were shown to choose those plants which accumulate the lowest amount of metal. The same was observed for the Ni hyperaccumulator *Senecio coronatus* [262]. This effect seems to be directly caused by the metal accumulation, and not related to other chemical defenses in hyperaccumulators [266]. On the contrary, Davis and Boyd have shown that hyperaccumulators have less need for other chemical defenses, so that they synthesize less allelochemicals than non-accumulator species of the same genus [266]. The chemical defense of hyperaccumulators does not only work against herbivores, but also fungal pathogens [263,267]. While it is clear from those many studies that hyperaccumulation does protect against a broad range of herbivores and pathogens, like any other defense strategy it has limitations. Zinc accumulation did not protect the Zn hyperaccumulator *Arabidopsis halleri* from attack by snails [268], and the Ni hyperaccumulator *Streptanthus polygaloides* was more susceptible to *Turnip mosaic* virus than the related non-accumulator *S. insignis*, in particular when Ni-fertilized plants were compared [269]. In this context, it should also be mentioned that hyperac-cumulated metals may be spread from hyperaccumulators back into the environ-ment via those herbivores that are not deterred by the high heavy metal content [270]. Since the amount of heavy metal content accumulated in the plant can easily be controlled by the concentration of the metal in the growth medium

[264], hyperaccumulators may be an ideal model for a systematic study of plant–pathogen/herbivore interactions, as discussed by Pollard [271].

In addition to the protection against herbivores and pathogens, Boyd and Jaffre recently proposed that hyperaccumulation may serve as "elemental allelopathy" [272]. This study suggested that hyperaccumulators increase the metal concentration in the surface soil next to them and thereby inhibit the growth of non-accumulators competing for space and nutrients. At the same time, the elevated metal concentrations would encourage growth of hyperaccumulator seedlings [86,234,273,274]. Elevated metal concentrations in the surface soil under hyperaccumulator plants were also measured by other authors [275,276], and Perronnet et al. [274] have shown that the hyperaccumulated metals in leaves of *T. caerulescens* easily become bioavailable again after incorporation of the leaves into the soil. These interesting ideas definitely deserve further investigations.

Another alternative hypothesis about the biological role of hyperaccumulation was the increase of osmotic pressure for increased tolerance to the drought stress that often characterizes the natural habitats of hyperaccumulators. This hypothesis, however, was falsified by a recent study of Whiting et al. with the Ni hyperaccumulator *Alyssum murale* and the Cd/Zn hyperaccumulator *T. caerulescens* [277].

The general value of the accumulated metals to the plants becomes obvious during leaf senescence. As it is generally known for metals that are essential plant nutrients [66], also hyperaccumulators recycle beneficial metals, which seem to include the hyperaccumulated heavy metals; Cd concentrations were found to be lower in senescent compared to mature and young leaves of *T. caerulescens* [189,278]. Furthermore, roots of *T. caerulescens* have been found to grow towards rather than away from heavy metals [279].

4.2. Mechanisms of Hyperaccumulation

The mechanisms by which hyperaccumulator plants accumulate the enormous amounts of heavy metals in their shoots, and prevent phytotoxicity of these metals, have been a subject of many studies, in particular during the past decade, but many of these mechanisms are still under debate [280]. In contrast to earlier belief, in *Arabidopsis halleri* it was found that metal tolerance and metal accumulation properties of hyperaccumulators are genetically independent characteristics [281,282]. While the former study suggested that most of the tolerance is controlled by a single gene, the latter study indicated the involvement of more than one major gene, and furthermore showed a partial co-segregation of the resistances against Zn and Cd toxicity, as well as a co-segregation of the accumulation ability for both metals.

Roots of *Thlaspi caerulescens* have been observed to grow towards rather than away from both Cd and Zn [279], but the enhanced metal uptake by hyperaccumulators compared to related non-accumulators was found not to be related to enhanced mobilization of the metal from the soil [283]. Instead, Lasat et al. [284,285] found an enhanced uptake of metals into the root symplasm in *T. caerulescens* compared to the non-accumulator *T. arvense* [284,285], and a reduced sequestration into the root vacuoles was associated with the higher root to shoot translocation efficiency of *T. caerulescens* compared to *T. arvense* and *T. ochroleucum* [273,285]. Recent work by Kerkeb and Krämer [286] on the Ni hyperaccumulator *Alyssum lesbiacum* has shown that the elevated histidine levels of some hyperaccumulators [14] play a role in xylem loading. Probably nicotianamine, which is synthesized at a higher level in *A. halleri* than in *A. thaliana* [223], plays a similar role. However, Lasat et al. [284] pointed out that the heavy metal uptake inside the shoot plays at least as important a role in hyperaccumulation as do the root mechanisms.

Clearly, the major ligands for both Cd and Zn in hyperaccumulator plants must be different from the PCs known from normal plants (Section 3.4.1 and Ref. [13]). Using X-ray absorption spectroscopy, Krämer et al. [14] did not find any sulfur ligands in the xylem sap of Ni hyperaccumulating *Alyssum* species, but found Ni binding by free histidine. Nickel in the hyperaccumulating tree *Sebertia acuminata* is bound mainly by citrate and not by PC [252]. Similar results were obtained for the metalloid arsenic in the As hyperaccumulator *Pteris vittata* [287,288] and even the classical PC inducing metal, Cd, is detoxified in another way in hyperaccumulators. Küpper et al. [188] reported that the total sulfur concentration in cells is lower than the Cd concentration. Further evidence to this end was obtained by Ebbs et al. [289], who demonstrated that PC levels in both shoots and roots of Cd-accumulating *T. caerulescens* are lower than in the non-accumulator *T. arvense* when challenged with Cd. Further, Küpper et al. [189] showed directly, using X-ray absorption spectroscopy of frozen-hydrated tissues, that most of the Cd in *T. caerulescens* is bound by weak oxygen ligands such as the organic acids that are always abundant in plant vacuoles. Sulfur ligands contributed less than one-third of the total Cd ligands in mature and senescent leaves. In *T. caerulescens*, Zn was never found to be associated with sulfur ligands in any part of the plant [188,189,221,234], but in stems 50% and in roots 70% of it was bound by histidine [189,221]. In this context, it should be noted that hyperaccumulators may not contain more organic acids than related non-accumulators [273], rather the metals bind to these ligands because the latter are anyway present. The same applies to the association of Zn with silicate in transport vesicles [290]. Despite the evidence discussed earlier, one could argue that PCs may still be involved in protecting the cytoplasm against heavy metal toxicity. Because of the low metal concentrations and the small volume of this compartment compared to the vacuole, such a binding would not contribute much to the average speciation measured in the earlier mentioned studies. Schat et al. [218], however, showed that blocking PC synthesis affects the Cd resistance of only non-accumulator plants, but not Cd hyperaccumulators, leading to the question of what substance contributes the sulfur ligands to the ligand environment of *Thlaspi caerulescens* [189].

A number of recent studies have indicated that metal compartmentation may be a key mechanism of metal tolerance in hyperaccumulator plants. Metals have been shown to be preferentially sequestered in subcellular compartments and cell types where they do the least damage to photosynthesis. Hyperaccumulators generally seem to sequester them in the vacuole [86,234,291]. While work with chemically fixed samples suggested a precipitation in the vacuole [291], later work with frozen-hydrated samples or micropipette extraction revealed that the heavy metals are in soluble form and the precipitations were preparation artifacts [86,234]. Most hyperaccumulator plants have been found to accumulate the metals in the epidermis [86,234,291]. A functional differentiation inside the epidermis leading to accumulation mainly in large, non-photosynthetic storage cells and exclusion from the photosynthetic guard cells was found and proposed to further reduce toxicity [86,234–237]. The mechanisms by which this epidermal sequestration is achieved are not known. They likely involve an increased expression of metal transporters in the cytoplasmic and vacuolar membranes of the epidermis cells, but were recently found not to involve an increased transport into the mesophyll cells that are between the veins and the epidermis [292]. The transport of accumulated metals into the vacuole may be related to compartmentation, rather than normal ion pumping, as well. Neumann and De Figueiredo [290] found (in *T. caerulescens*) Zn in vesicles that generally (also in non-accumulator plants) transport silicates into the vacuole. This is an interesting discovery, and should be studied further. The strong sequestration of metals into the vacuoles makes hyperaccumulators inefficient in using these metals as micronutrients; they need much higher concentrations of these metals for normal growth than all other plants [86,234,273] (Fig. 3).

Three noteworthy exceptions from the described pattern of metal compartmentation have been found. In the Cd/Zn hyperaccumulator *A. halleri*, most of the metal is found in the mesophyll vacuoles; probably, this contributes to the lower Cd tolerance of this species compared to *T. caerulescens* [188]. The only highly Cd and Zn accumulating epidermis cells in *A. halleri* are the trichomes of the leaf surface [188,293]. Old leaves of the Ni hyperaccumulator *Berkheya coddii* accumulate most of the metal as Ni silicates in the cell walls [294], and in the Ni-hyperaccumulating tree *Sebertia acuminata*, the latex has by far the highest Ni concentration [252].

Several studies have shown higher expression levels of various metal transporters in both roots and shoots of hyperaccumulator plants than in related non-accumulators. Pence et al. [101] found that the elevated levels of root and shoot Zn uptake in *T. caerulescens* correlated with a much higher expression of ZNT1, a Zn transporter from the ZIP family (Section 2.4) with high similarity to AtZIP4. Very recent studies using microarrays and *A. halleri* revealed a high expression of transporters similar to AtZIP6 in shoots [102] and AtZIP9 in roots [223]. *Thlaspi caerulescens* was found to have elevated expression of ZNT1, the closely related ZNT2, and ZTP1 [105]. The latter is a member of the CDF family, a homolog to

Figure 3 Heavy metal deficiency in heavy metal hyperaccumulator plants. Grown on concentrations of the hyperaccumulated metal that are sufficient or even toxic to normal non-accumulator plants, hyperaccumulators suffer from deficiency of this metal. (A: Top part) Cadmium deficiency in the Cd/Zn hyperaccumulator *Thlaspi caerulescens*, Prayon ecotype. Left pot: four plants grown with 10 μM Zn^{2+} and 10 μM Cd^{2+} in the hydroponic nutrient solution. Right pot: four plants grown with 10 μM Zn^{2+} but no Cd^{2+} in the nutrient solution. The cadmium deficiency manifested itself in an increased vulnerability of the plants to the attack of herbivores. In the example shown here, the plants grown without Cd suffered from damage (particularly visible on the leaf edges) caused by *Thrips palmae*, while the plants grown with Cd were not attacked by these insects and therefore reached a higher biomass than those without Cd. (B: Lower part) Nickel deficiency in the nickel hyperaccumulator *Alyssum lesbiacum*. Front row: plants grown on normal soil. Back row: plants grown on soil with addition of 500 ppm $NiSO_4$. Note that the plants grown on normal soil are much smaller than the high-Ni plants in the back, and even display damaged chlorotic leaves, showing that in this case the heavy metal deficiency did not only result in increased pathogen sensitivity, but also symptoms of a nickel-deficient metabolism as they have been reported for other species under Ni deficiency.

the *A. thaliana* gene ZAT [104] and the *T. goesingense* gene MTP1 [89]; its over-expression leads to elevated accumulation [104] and its homolog was found to be highly expressed also in *A. halleri* [102]. The latter study has additionally shown an elevated expression of a homolog to the CPx-type heavy metal ATPase AtHMA3. Pence et al. [101] presented evidence indicating that the Zn-dependent regulation of ZNT1 expression is altered in *T. caerulescens* as compared with non-accumulator plants. That is, plants needed to be grown on much higher Zn levels to down-regulate ZNT1 expression in *T. caerulescens* as compared with the non-accumulator *T. arvense*. However, it remains unknown in which specific tissues and cell types these transporters are expressed and whether their expression is directly responsible for the metal hyperaccumulation phenotype, or if this elevated expression is only a result of Zn deficiency in the tissue, result-ing from the highly efficient metal sequestration into vacuoles by vesicles or only one/few transporters.

When metal concentrations in hyperaccumulators reach toxic levels, several additional resistance mechanisms may become activated. In the meso-phyll of Cd-stressed Cd-hyperaccumulating *A. halleri* [188] and Ni-stressed Ni-hyperaccumulating *Alyssum* and *Thlaspi* species [86], Küpper et al. found an increase of the Mg content and interpreted it as a defense against the replace-ment of Mg^{2+} in chlorophyll by heavy metals [136,147,151]. Similarly, Zn uptake was found to be enhanced in hyperaccumulator plants stressed by excess Cd, which could be a defense against substitution of Cd in the catalytic centers of enzymes by Cd [189].

The sequestration-based metal resistance strategy seems to leave hyper-accumulators vulnerable to toxicity of those metals which the highly expressed transporters (usually specific for one or a few metals) cannot detoxify. Most likely for this reason, the Cd/Zn hyperaccumulators *Arabidopsis halleri* and *Thlaspi caerulescens* are rather sensitive to Cu toxicity [295,296].

4.3. Use of Hyperaccumulator Plants for Phytomining and Phytoremediation

Many hyperaccumulators have a good potential to be used for phytoremediation, i.e., to extract and remove heavy metals from anthropogenically contaminated soils, which was first proposed by Chaney [297]. Some of them even allow for commercially profitable phytomining, i.e., the extraction of metals from naturally heavy metal-rich soils (that are not directly usable as metal ores) with subsequent burning of the plants, the ash of which can be used as a metal ore (first proposed by Baker and Brooks [298]). These applications of metal phytoextraction have been a subject to extensive research as reviewed in Refs. [18,299–308]. In the following, some reasonably well studied cases are presented in more detail.

For the metalloid arsenic, the fast growing, high-biomass, As-hyperaccumulating fern *Pteris vittata* is a very promising candidate for phytoremediating

As-contaminated areas [258]; the same applies to some related *Pteris* species [309,310]. In aqueous environments, the submerged plant *Ceratophyllum demersum* may accumulate enough arsenic to have a potential for phytoremediation [311,312]; the same may apply to *Amaranthus blitoides* [313]. Both species were found to accumulate relatively large amounts of heavy metals in contaminated natural wetland habitats, but have not yet been tested under controlled conditions.

For Cd, the Cd/Zn hyperaccumulator *T. caerulescens* seems to be the best known candidate for phytoremediation. Although it has a rather small biomass of 2–5 tons/ha [304,314], the extreme bioaccumulation coefficient of its southern French ecotypes [30,315] yields Cd extraction rates high enough for cleaning up Cd-contaminated soils within a few years as tested in the field by Robinson et al. [314] and Hammer and Keller [316]. A major problem, however, may be the high Cu sensitivity of *T. caerulescens*; concentrations that may easily occur in multi-contaminated soils were found to strongly inhibit growth of this plant [296].

Nickel was the first metal for which the economic feasibility of phytomining was shown, and some Ni hyperaccumulators hyperaccumulate the even more valuable cobalt as well [255,317]. Nicks and Chambers [318] yielded a crop of Ni of equal value compared to an average crop of wheat by planting *Streptanthus polygaloides* on a metal-rich soil in California (USA). They furthermore showed that by burning these plants it is possible to yield, with low input of energy, a bio-ore (the plant ash) containing ~15% Ni. *Berkheya coddii* has been known as a high-biomass Ni hyperaccumulator since the work of Anderson et al. [319]. Robinson et al. [253,255] carried out comprehensive studies of metal uptake, and showed that fertilization with sulfur and nitrogen greatly increased Ni and Co hyperaccumulation. Thus, their work has demonstrated that this species is a very promising candidate for both phytoremediation and phytomining. This has been confirmed by field trials in a recent study, which demonstrated that this species easily yields 110 kg Ni/ha and year [320], and even 170 kg should be possible [321]. Similarly, *Alyssum bertolonii* has already been shown to produce high enough Ni yields per hectare for phytomining [254,317,321], which has already been put into commercial operation [304] despite earlier doubts about the suitability of this species for phytomining expressed by Brooks et al. [320]. Several other large plants hyperaccumulate Ni, but their potential for phytoremediation and phytomining has not yet been tested. An extreme case seems to be the widespread tropical genus *Phyllanthus* (Euphorbiaceae). The Cuban species *P. x pallidus* was found to accumulate up to 6% Ni in its dry mass [322], and many others contained between 3% and 4% Ni; only the Australian *Phyllanthus* species are non-hyperaccumulating [323]. *Phyllanthus* species are medium-sized (1–5 m high) shrubs; despite their size and extreme Ni contents (in particular in the Cuban species), no study has been published so far investigating the potential of *Phyllanthus* species for phytoremediation and phytomining [323].

A very unusual phytoremediation strategy has been proposed for mercury [306]. Rugh et al. [324,325] engineered plants that reduce Hg^{2+} to Hg^0, which is then volatilized. This method does not seem very reasonable since it does not actually remove the toxic metal from the environment, but just distributes it over a large area making later cleanup impossible; it is comparable to the industrial strategy in the 1960s–1980s of building tall chimneys to reduce local contamination with various toxic substances.

For Zn, the Chinese plant *Sedum alfredii* may be the most promising candidate for phytoremediation and possibly even commercial phytomining because of its correlation of high Zn accumulation with relatively high biomass [326,327]. In contrast, the model Zn hyperaccumulator plant *Thlaspi caerulescens* [256,257] has a rather low biomass and at high soil Zn concentrations also a low bioaccumulation coefficient [314,315], so that its use in Zn phytoremediation is generally limited to moderate levels of contamination. Indeed, while field trials on moderately contaminated soil by Baker et al. [301] were successful, those on more heavily Zn-contaminated soil failed [316]. In any case, *T. caerulescens* is definitely unsuitable for commercially phytomining Zn. As was the case for Cd, Zn phytoremediation by *T. caerulescens* is further limited by its Cu sensitivity (see above [296]).

In addition to true hyperaccumulator plants, various other plants have been proposed for use in phytoremediation. One idea is to use high-biomass plants for absorbing the metals; it is argued that the much higher biomass will yield higher metal extraction per area of land compared to hyperaccumulators, despite the much lower metal content of non-accumulator plants [305,306,328]. The same has been proposed for various fast-growing macrophytic water plants such as water hyacinths, *Eichhornia crassipes* [329]. Those who argue for such an approach, however, mostly ignore that such a strategy would dilute the extracted metal in a much larger amount of toxic biomass compared to hyperaccumulator plants; this biomass would be too toxic for use as compost and would not contain enough metal to make a recycling of the phytoextracted metal feasible [307,329]. In addition, the bioaccumulation factor of metals in such plants is usually so low (below 1) that hundreds of crops would be required for phytoremediation of even a moderately contaminated soil [301,304,307]. Those who argue for this approach because of the low biomass of many (not all, see above!) hyperaccumulators should also keep in mind the following facts:

(a)   The biomass yield of non-accumulator plants on contaminated soils is reduced by phytotoxicity of the contaminating metal [307,330].

(b)   The biomass of hyperaccumulators can be rather easily improved by selecting suitable ecotypes and individuals within the natural population [331,332], breeding [333], and fertilization (2–3 times increase) [303,320,331,332,334].

(c)   The metal accumulation of hyperaccumulators can further be optimized by selection. Many recent studies pointed out the more than

20-fold variation of bioaccumulation factors for the same metal between ecotypes/populations [30,315,335–340]. Furthermore, the accumulation efficiency is not directly correlated to the metal content of the habitat [337], and strong variation of metal bioaccumulation factors exists even within one population [339]. Finally, accumulation is higher on the average moist agricultural land compared to their dry natural habitats [341], and fertilization increases it further [332].

In summary, presently it is not the phytoremediation by hyperaccumulators that is a "hype" [342], but the use of non-accumulating plants for this task. The only way that non-hyperaccumulating plant species may become a better alternative would be creating (by genetic engineering or traditional breeding) metal-accumulating cultivars.

One major obstacle to phytoremediation is the often low bioavailability of metals in the soil [295,304,306,307,343]. Addition of chelating agents to the soil may increase metal availability to the hyperaccumulators [306], but this method is not recommendable because it causes strong leaking of heavy metals from the contaminated soil into the ground water [344]. A safer way to overcome this problem may be the use of soil microorganisms, some of which have been shown to enhance the Zn uptake by *Thlaspi caerulescens* [345], but results with other organisms were contradictory [343].

A classic example of phytoremediation are algae, which can be used, either alive or as dead biomass, for absorbing heavy metals from contaminated water [116]. Since this passive type of "phytoremediation" is, in contrast to true hyperaccumulation, not metal-specific (cell walls bind a very broad range of metals), it will again yield, in most cases, a toxic biomass that cannot be used but has to be dumped on toxic waste sites, just like conventional, artificial ion exchange resins.

5. CONCLUSIONS

This chapter summarizes recent important developments in the field of heavy metals and their interaction with higher plants, algae, and cyanobacteria. Three major topics have been stressed: (i) the role of heavy metals as micronutrients, (ii) the inhibition of plant metabolism by heavy metals, and (iii) the use of heavy metals in chemical defense against herbivores and pathogens. As expected, these three divergent roles require a complex network of mechanisms to regulate metal uptake as well as metal trafficking in the organisms. The latter includes transport of heavy metals to their targets, their sequestration into compartments used for storage and/or for detoxification, and the development of various strategies for detoxification inside those compartments.

Extensive research has been applied in recent years to unravel the different mechanisms within the network, and to get a deeper insight into the different functions of heavy metals at their target locations. This overview can only cover the most important developments of the past decade. Despite the huge number of impressive results of a rapidly growing field, many questions still remain to be answered. It also seems that some research areas advance much faster than others. For example, while many genes involved in heavy metal metabolism have been identified in recent years, very few of the proteins encoded by these genes have been actually isolated and biochemically and structurally characterized. Thus, in many cases, the physiological roles of these proteins and their mechanism of action have not been identified.

Furthermore, particularly in the field of heavy metal toxicity and hyper-accumulation, only a few of the mechanisms of inhibition and detoxification have been investigated under physiologically relevant conditions. Note that completely different mechanisms may be important under low vs. high concentrations of the heavy metal under investigation.

In conclusion, research on heavy metals and their impact on the environment has become an important and rapidly growing area of research. To understand in depth the many structural and functional aspects of heavy metals and their interplay with the biosphere tremendous joint efforts of biologists, biochemists, and biophysicists will be required in the future.

ACKNOWLEDGMENTS

The authors would like to thank Laura Ort Seib for critical reading of the manuscript, and gratefully acknowledge a fellowship of the Alexander von Humboldt Foundation to H. Küpper and grants of the Deutsche Forschungsgemeinschaft to P. M. H. Kroneck.

ABBREVIATIONS

ATFP	*Arabidobsis thaliana* farnesylated protein
CAX	cation exchanger
CCH	copper chaperone
CDF	cation diffusion facilitator
Chl	chlorophyll
COX	cytochrome oxidase
CPx ATPases	ATPases that have a common conserved intramembranous cysteine–proline–cysteine or cysteine–proline–histidine ("CPx") motif. They usually transport heavy metals and are a separate evolutionary branch of ion-transporting P-type ATPases
ETR	ethylene receptor
F_0	basic fluorescence yield of a dark adapted sample, fluorescence in non-actinic measuring light
F_0'	basic fluorescence yield of a light adapted sample, i.e., fluorescence in non-actinic light after actinic irradiance

F_m	maximum fluorescence yield of a dark adapted sample
F'_m	maximum fluorescence yield of a sample during the exposure to actinic light, and reduced by non-photochemical quenching
F'_t	fluorescence under actinic irradiance immediately before the measurement of F'_m
F_v	variable fluorescence; $F_v = F_m - F_0$
hms-Chl	heavy metal-substituted chlorophyll
IRT	iron regulated transporter
IUB	International Union of Biology
LHC	light harvesting complex
MT	metallothionein
Nramp	natural resistance associated macrophage protein
PC	phytochelatin
PS	photosystem
RC	reaction center
SOD	superoxide dismutase
ZAT	Zn transporter of *Arabidopsis thaliana* (a member of the CDF family)
ZIP	ZRT, IRT-like protein, sometimes also referred to as "zinc inducible protein"
ZNT	zinc transporter of *Thlaspi* (while the mammalian ZnTs belong to the CDF family, the *Thlaspi* ZNTs belong to the ZIP family of micronutrient transporters)
ZRT	zinc-regulated transporter (also these transporters belong to the ZIP family of micronutrient transporters)
ZTP	zinc transport protein (the *Thlaspi* ZTPs are highly similar to the *Arabidopsis* ZAT and belong to the CDF family)

REFERENCES

1. Lane TW, Morel FMM. Proc Natl Acad Sci USA 2000; 97:4627–4631.
2. Hacisalihoglu G, Kochian LV. New Phytol 2003; 159:341–350.
3. Morel FMM, Price, NM. Science 2003; 300:944–947.
4. Bertini I, Sigel A, Sigel H, eds. Handbook on Metalloproteins. New York, Basel: Marcel Dekker Inc., 2001.
5. Messerschmidt A, Huber R, Poulos T, Wieghardt K, eds. Handbook of Metalloproteins. Weinheim: Wiley-VCH Verlag GmbH, 2001.
6. Holm RH, Solomon EI, eds. Chem Rev 2004; 104:347–1200.
7. Cavet JS, Borrelly PM, Robinson NJ. FEMS Microbiol Rev 2003; 7:165–181.
8. Hall JL, Williams LE. J Exp Bot 2003; 54:2601–2613.
9. Cobbett CS, Hussain D, Haydon MJ. New Phytol 2003; 159:315–321.
10. Mann EL, Ahlgren N, Moffet JW, Chisholm, SW. Limnol Oceanogr 2002; 47:976–988.
11. Prasad MNV, Hagemeyer J, eds. Heavy Metal Stress in Plants: From Molecules to Ecosystems. Berlin, Heidelberg: Springer, 1999.

12. van Hoof NALM, Koevoets PLM, Hakvoort HWJ, Ten Bookum WM, Schat H, Verkleij JAC, Ernst WHO. Physiol Plant 2001; 113:225–232.
13. Cobbett CS, Goldsbrough P. Ann Rev Plant Biol 2002; 53:159–182.
14. Krämer U, Cotterhowells JD, Charnock JM, Baker AJM, Smith JAC. Nature 1996; 379:635–638.
15. Maeda S, Sakaguchi T. In: Akatsuka I, ed. Introduction to Applied Phycology. The Hague: SPB Academic Publishing, 1990:109–136.
16. Baker AJM. J Plant Nutr 1981; 3:643–654.
17. Brooks RR. Plant Soil 1977; 48:541–544.
18. McGrath SP, Sidoli CMD, Baker AJM, Reeves RD. In: Eijsackers HJP, Hamers T, eds. Integrated Soil and Sediment Research: A Basis for Proper Protection. Dordrecht: Kluwer Academic Publishers, 1993:673–677.
19. McGrath SP, Zhao FJ, Lombi E. Adv Agron 2002; 75:1–56.
20. Bruland KW. Earth Planet Sci Lett 1980; 47:176–198.
21. Boyle EA, Sclater F, Edmond JM. Nature 1976; 263:42–44.
22. Lee JG, Roberts SB, Morel FMM. Limnol Oceanogr 1995; 40:1056–1063.
23. Morel FMM, Reinfelder JR, Roberts SB, Chamberlain CP, Lee JG, Yee D. Nature 1994; 369:740–742.
24. Cullen JT, Lane TW, Morel FMM, Sherrell RM. Nature 1999; 402:165–167.
25. Meisch HU, Schmitt JA, Scholl AR. Naturwiss 1979; 66: 209–209.
26. Meisch HU, Scholl AR, Schmitt JA. Z Naturforsch C-A J Biosci 1981; 36:765–771.
27. Tumova E, Sofrova D. Photosynthetica 2002; 40:103–108.
28. Lee JG, Morel FMM. Mar Ecol Progr Ser 1995; 127:305–309.
29. Karavaev VA, Baulin AM, Gordienko TV, Dovyd'kov SA, Tikhonov AN. Russ J Plant Physiol 2001; 48:38–44.
30. Lombi E, Zhao FJ, Dunham SJ, McGrath SP. New Phytol 2000; 145:11–20.
31. Lombi E, Zhao FJ, McGrath SP, Young SD, Sacchi GA. New Phytol 2001; 149:43–60.
32. Lombi E, Tearall KL, Howarth JR, Zhao FJ, Hawkesford MJ, McGrath SP. Plant Physiol 2002; 128:1359–1367.
33. Korshunova YO, Eide D, Clark WG, Guerinot ML, Pakrasi HB. Plant Mol Biol 1999; 40:37–44.
34. Kowalewska G, Lotocka M, Latala A. Polsk Arch Hydrobiol 1992; 39:41–49.
35. Raven JA, Evans MCW, Korb RE. Photosyn Res 1999; 60:111–149.
36. Colman PM, Freeman HC, Guss JM, Murata M, Norris VA, Ramshaw JAM, Ventakappa MP. Nature 1978; 272:319–324.
37. Guss JM, Freeman HC. J Mol Biol 1983; 169:521–563.
38. Navarro JA, Hervás M, De la Rosa MA. J Biol Inorg Chem 1997; 2:11–22.
39. Freeman HC, Guss JM. In: Messerschmidt A, Huber R, Poulos T, Wieghardt K, eds. Handbook of Metalloproteins. Weinheim: Wiley-VCH Verlag GmbH, 2001:1153–1169.
40. McCord JM, Fridovich I. J Biol Chem 1969; 244:6049–6055.
41. Bordo D, Pesce A, Bolognesi M, Stroppolo ME, Falconi M, Desideri A. In: Messerschmidt A, Huber R, Poulos T, Wieghardt K, eds. Handbook of Metalloproteins. Weinheim: Wiley-VCH Verlag GmbH, 2001:1284–1300.
42. Getzoff E, Tainer JA, Weiner PK, Kollman PA, Richardson JS, Richardson DC. Nature 1983; 306:287–290.
43. Keilin D, Mann T. Nature 1939; 143:23–24.
44. Malmström BG, Mosbach R, Vänngard T. Nature 1959; 183:321–322.

45. (a) James WO. Proc Roy Soc Lond B 1953; 141:289–299; (b) Alge D, Wastyn M, Mayer C, Jungwirth C, Zimmermann U, Zoder R, Fromwald S, Peschek GA. IUBMB-Life 1999; 48:187–197; (c) Pescheck GA, Wastyn M, Trnka M, Molitor V, Fry IV, Packer L. Biochemistry 1989; 28:3057–3063; (d) Peschek GA. Biochim Biophys Acta 1996; 1275:27–32.

46. Rodriguez FI, Esch JJ, Hall AE, Binder BM, Schaller GE, Bleecker AB. Science 1999; 283:996–998.

47. Herbik A, Bölling C, Buckhout TJ. Plant Physiol 2002; 130:2039–2048.

48. Lanaras T, Moustakas M, Symeonidis L, Diamantoglou S, Karataglis S. Physiol Plant 1993; 88:307–314.

49. Williams LE, Pittman JK, Hall JL. Biochim Biophys Acta 2000; 1465:104–126.

50. Hill KL, Hassett R, Kosman D, Merchant S. Plant Physiol 1996; 112:697–704.

51. Liao MT, Hedley MJ, Woolley DJ, Brooks RR, Nichols MA. Plant Soil 2000; 223:243–252.

52. Kampfenkel K, Kushnir S, Babiychuk E, Inze D, Van Montagu M. J Biol Chem 1995; 270:28479–28486.

53. Sancenón V, Puig S, Mira H, Thiele DJ, Peñarrubia L. Plant Mol Biol 2003; 51:577–587.

54. Hirayama T, Kieber JJ, Hirayama N, Kogan M, Guzman P, Nourizadeh S, Alonso JM, Dailey WP, Dancis A, Ecker JR. Cell 1999; 97:383–393.

55. Woeste KE, Kieber JJ. Plant Cell 2000; 12:443–455.

56. Shikanai T, Müller-Moulé P, Munekage Y, Niyogi KK, Pilon M. Plant Cell 2003; 15:1333–1346.

57. Axelsen KB, Palmgren MG. Plant Physiol 2001; 126:696–706.

58. Tottey S, Rich PR, Rondet SAM, Robinson NJ. J Biol Chem 2001; 276: 19999–20004.

59. Bughio N, Yamaguchi N, Nishizawa NK, Nakanishi H, Mori S. J Exp Bot 2002; 53:1677–1682.

60. Harrison MD, Jones CE, Dameron CT. J Biol Inorg Chem 1999; 4:145–153.

61. Wintz H, Vulpe C. Biochem Soc Trans 2002; 30:732–735.

62. Arnesano F, Banci L, Bertini I, Ciofi-Baffoni S, Molteni E, Huffman DL, O'Halloran TV. Genome Res 2002; 12:255–271.

63. Pufahl R, Singer CP, Peariso KL, Lin SJ, Schmidt PJ, Fahrni CJ, Cizewski Culotta V, Penner-Hahn JE, O'Halloran TV. Science 1997; 278:853–856.

64. Bertini I, Rosato A. Proc Natl Acad Sci USA 2003; 100:3601–3604.

65. Himelblau E, Mira H, Lin SJ, Culotta VC, Penarrubia L, Amasino RM. Plant Physiol 1998; 117:1227–1234.

66. Himelblau E, Amasino RM. J Plant Physiol 2001; 158:1317–1323.

67. Li L, Tutone AF, Drummond RSM, Gardner RC, Luan S. Plant Cell 2001; 13: 2761–2775.

68. Nautiyal N, Chatterjee C, Sharma CP. Comm Soil Sci Plant Anal 1999; 30: 1625–1632.

69. Romera FJ, Frejo VM, Alcántara E. Plant Physiol Biochem 2003; 41:821–827.

70. Buchanan-Wollaston V. Plant Physiol 1994; 105:839–846.

71. Guo WJ, Bunditha W, Goldsbrough PB. New Phytol 2003; 159:369–381.

72. Dixon NE, Gazzola C, Blakeley RL, Zerner B. J Am Chem Soc 1975; 97: 4131–4133.

73. Fishbein WN, Smith MJ, Nagarajan K, Sarz W. Fed Proc 1976; 35:1680.

74. Sumner JB. J Biol Chem 1926; 69:435–441.
75. Eskew DL, Welch RM, Carey EE. Science 1983; 222:621–623.
76. Eskew DL, Welch RW, Norvall WA. Plant Physiol 1984; 76:691–693.
77. Brown PH, Welch RM, Cary EE. Plant Physiol 1987; 85:801–803.
78. Jabri E, Carr MB, Hausinger RP, Karplus PA. Science 1995; 268:998–1004.
79. Dixon NE, Riddles PW, Gazzola C, Blakeley RL, Zerner B. Can J Biochem 1980; 58:1335–1344.
80. Hausinger RP, Karplus PA. In: Messerschmidt A, Huber R, Poulos T, Wieghardt K, eds. Handbook of Metalloproteins. Weinheim: Wiley-VCH Verlag GmbH, 2001: 867–879.
81. Gerendás J, Polacco JC, Freyermuth SK, Sattelmacher B. J Plant Nutr Soil Sci 1999; 162:241–256.
82. Asher CJ. In: Micronutrients in Agriculture. 2d ed. SSSA Book Series No 4. Madison, USA: Soil Science Society of America, 1991:703–723.
83. Brown PH, Welch RM, Cary EE, Checkai RT. J Plant Nutr 1987; 10:2125–2135.
84. Brown PH, Welch RM, Madison JT. Plant Soil 1990; 125:19–27.
85. Benchemski-Bekkari N, Pizelle G. Plant Physiol Biochem 1992; 30:187–192.
86. Küpper H, Lombi E, Zhao FJ, Wieshammer G, McGrath SP. J Exp Bot 2001; 52:2991–2300.
87. Cammak R. Nature 1995; 373:556–557.
88. Papen H, Kentemich T, Schmulling T, Bothe T. Biochimie 1986; 68:121–132.
89. Persans MW, Nieman K, Salt DE. Proc Natl Acad Sci USA 2001; 98:9995–10000.
90. Price NM, Morel FMM. Limnol Oceanogr 1991; 36:1071–1077.
91. Bekheet IA, Syrett PJ. Br Phycol J 1977; 12:137–143.
92. Raulin J. Etudes cliniques sur la vegétation. Ann Sci Nat IX 1869; 93.
93. Auld DS. In: Bertini I, Sigel A, Sigel H, eds. Handbook on Metalloproteins. New York, Basel: Marcel Dekker Inc., 2001:881–959.
94. (a) Keilin D, Mann T. Biochem J 1940; 34:1163–1176; (b) Vallee BL, Auld DS. Proc Natl Acad Sci USA 1990; 87:220–224.
95. Folkers GE, Hanzawa H, Boelens R. In: Bertini I, Sigel A, Sigel H, eds. Handbook on Metalloproteins. New York, Basel: Marcel Dekker Inc., 2001:961–1000.
96. Sträter N, Klabunde T, Tucker P, Witzel H, Krebs B. Science 1995; 268:189–1492.
97. Grotz N, Fox T, Conolly E, Park W, Guerinot ML, Eide D. Proc Natl Acad Sci USA 1998; 95:7220–7224.
98. (a) Guerinot ML, Eide D. Curr Opin Plant Biol 1999; 2:244–249; (b) Guerinot ML. Biochim Biophys Acta 2000; 1465:190–198.
99. Connolly EL, Fett JP, Guerinot ML. Plant Cell 2002; 14:1347–1357.
100. Ramesh SA, Shin R, Schachtman DP. Plant Physiol 2003; 133:126–134.
101. Pence NS, Larsen PB, Ebbs SD, Lasat MM, Letham DLD, Garvin DF, Eide D, Kochian LV. Proc Natl Acad Sci USA 2000; 97:4956–4960.
102. Becher M, Talke IN, Krall L, Krämer U, Plant J 2004; 37:251–268.
103. Gaither LA, Eide DJ. Biometals 2001; 14:251–270.
104. Van der Zaal BJ, Neuteboom LW, Pinas JE, Chardonnens AN, Schat H, Verkleij JAC, Hooykaas PJJ. Plant Physiol 1999; 119:1047–1055.
105. Assunção AGL, Costa Martins PDA, De Folter S, Vooijs R, Schat H, Aarts MGM. Plant Cell Environ 2001; 24:217–226.
106. Graham RD, Asher JS, Hynes SC. Plant Soil 1992; 38:77–80.
107. Kim T, Wetzstein HY. J Am Soc Horticult Sci 2003; 128:171–175.

108. Graham RD, Rengel Z. In: Robson D, ed. Zinc in Soils and Plants. Dordrecht, The Netherlands: Kluwer Academic Publishers, 1993:107–114.
109. Rengel Z. Ann Bot 2000; 86:1119–1126.
110. Hacisalihoglu G, Hart JJ, Kochian LV. Plant Physiol 2003; 131:595–602.
111. Hajiboland R, Yang XE, Römheld V. Plant Soil 2003; 250:349–357.
112. Singh SP, Westermann DT. Crop Sci 2002; 42:1071–1074.
113. Genc Y, Shepherd KW, McDonald GK, Graham RD. Plant Breed 2003; 122: 283–284.
114. Kawashima I, Kennedy TD, Chino M, Lane BG. Eur J Biochem 1992; 209:971–976.
115. Wood JM, Wang H. Environ Sci Technol 1983; 17:582A–590A.
116. Lytle JS, Lytle TF. Environ Toxicol Chem 2001; 20:68–83.
117. Gallagher DL, Johnston KM, Dietrich AM. Water Res 2001; 35:2984–2994.
118. Patra M, Sharma A. Bot Rev 2000; 66:379–422.
119. Morel FMM. Principles in Aquatic Chemistry. New York, USA: John Wiley and Sons, 1983.
120. Morel FMM, Hering JG. Principles and Applications of Aquatic Chemistry. New York: John Wiley and Sons, 1993.
121. Hudson RJM. Crit Rev Anal Chem 1998; 28:19–26.
122. Tubbing DM, Admiraal W, Cleven RFMJ, Iqbal M, van de Meent D, Verweij W. Water Res 1994; 28:37–44.
123. Errecalde O, Campbell PGC. J Phycol 2000; 36:473–483.
124. Stobart AK, Griffiths WT, Ameen-Bukhari I, and Sherwood RP. Physiol Plant 1985; 63:293–298.
125. Barón, M. Arellano JB, Gorge JL. Physiol Plant 1995; 94:174–180.
126. Pätsikkä E, Aro EM, Tyystjärvi E. Plant Physiol 1998; 117:619–627.
127. Vavilin DV, Polynov VA, Matorin DN, Venediktov PS. J Plant Physiol 1995; 146:609–614.
128. Luna CM, González CA, Trippi VS. Plant Cell Physiol 1994; 35:11–15.
129. Okamoto OK, Asano CS, Aidar E, Colepicolo P. J Phycol 1996; 32:74–79.
130. Weckx JEJ, Clijsters HMM. Physiol Plant 1996; 96:506–512.
131. Clijsters H, Cuypers A, Vangronsveld J. Z Naturforsch C 1999; 54:730–734.
132. Kimimura M, Katoh S. Biochim Biophys Acta 1972; 283:279–292.
133. Sandmann G, Böger P. Plant Physiol 1980; 66:797–800.
134. Siedlecka A, Krupa Z. Photosynthetica 1999; 36:321–331.
135. Pätsikkä E, Kairavuo M, Sersen F, Aro EM, Tyystjärvi E. Plant Physiol 2002; 129:1359–1367.
136. Küpper H, Setlík I, Spiller M, Küpper FC, Prásil O. J Phycol 2002; 38:429–441.
137. Sheoran IS, Singal HR, Singh R. Photosyn Res 1990; 23:345–351.
138. Yruela I, Montoya G, Alonso PJ, Picorel R. J Biol Chem 1991; 266:22847–22850.
139. Stokes PM, Dreier SI. Can J Bot 1981; 59:1817–1823.
140. Punshon T, Dickinson N. New Phytol 1997; 137:303–314.
141. Carrier P, Baryla A, Havaux M. Planta 2003; 216:939–950.
142. Küpper H, Setlík I, Setliková E, Ferimazova N, Spiller M, Küpper FC. Funct Plant Biol 2003; 30:1187–1196.
143. Prasad MNV, Strzalka K. In: Prasad MNV, Hagemeyer J, eds. Heavy Metal Stress in Plants: From Molecules to Ecosystems. Berlin, Heidelberg: Springer, 1999: 117–128.
144. Gross RE, Pugno P, Dugger WM. Plant Physiol 1970; 46:183–185.

145. Clijsters H, Van Assche F. Photosyn Res 1985; 7:31–40.
146. Atal N, Saradhi PP, Mohanty P. Plant Cell Physiol 1991; 32:943–951.
147. Küpper H, Küpper F, Spiller M. J Exp Bot 1996; 47:259–266.
148. Kowalewska G, Falkowski L, Hoffmann SK, Szczepaniak LS. Acta Physiol Plant 1987; 9:43–52.
149. Kowalewska G, Hoffmann SK. Acta Physiol Plant 1989; 11:39–50.
150. Kowalewska G. Polsk Arch Hydrobio 1990; 37:327–339.
151. Küpper H, Küpper F, Spiller M. Photosyn Res 1998; 58:123–133.
152. Küpper H, Dedic R, Svoboda A, Hála J, Kroneck PMH. Biochim Biophys Act 2002; 1572:107–113.
153. Küpper H, Küpper FC, Spiller M. In: Grimm B, Porra R, Rüdiger W, Scheer H, eds. Chlorophylls and Bacteriochlorophylls: Biochemistry, Biophysics and Biological Functions, Advances in Photosynthesis (Series editor: Govindjee), Dordrecht: Kluwer Academic Publishers, 2004, in press.
154. Prasad MNV, Malec P, Waloszek A, Bojko M, Strzalka K. Plant Sci 2001; 161: 881–889.
155. Cedeno-Maldonado A, Swader JA, Heath RL. Plant Physiol 1972; 50:698–701.
156. Boucher LJ, Katz JJ. J Am Chem Soc 1967; 89:4703–4308.
157. Rebeiz CA, Belanger FC. Spectrochim Acta 1984; 40A:793–806.
158. Paulsen H, Finkenzeller B, Kühlein N. Eur J Biochem 1993; 215:809–816.
159. Watanabe T, Kobayashi M. Special Articles on Coordination Chemistry of Biologically Important Substances 1988; 4:383–395.
160. Fiedor L, Leupold D, Teuchner K, Voigt B, Hunter CN, Scherz A, Scheer H. Biochemistry 2001; 40:3737–3747.
161. Wu JT, Lorenzen H. Bot Bull Acad Sin 1984; 25:125–132.
162. La Fontaine S, Quinn JM, Nakamoto SS, Page MD, Göhre V, Moseley JL, Kropat J, Merchant S. Eukar Cell 2002; 1:736–757.
163. Lidon FC, Ramalho JC, Henriques FS. J Plant Physiol 1993; 142:12–17.
164. Sgardelis S, Cook CM, Pantis JD, Lanaras T. Sci Total Environ 1994; 158:157–164.
165. Nielsen HD, Brownlee C, Coelho SM, Brown MT. New Phytol 2003; 160:157–165.
166. Pietrini F, Iannelli MA, Pasqualini S, Massacci A. Plant Physiol 2003; 133:829–837.
167. Yruela I, Alfonso M, Barón M, Picorel R. Physiol Plant 2000; 110:551–557.
168. Horváth G, Droppa M, Oravecz A, Raskin VI, Marder JB. Planta 1996; 199: 238–243.
169. Somashekaraiah BV, Padmaja K, Prasad ARK. Physiol Plant 1992; 85:85–89.
170. Jegerschöld C, Arellano JB, Schröder WP, van Kan PJM, Barón M, Styring S. Biochemistry 1995; 34:12747–12754.
171. Yruela I, Gatzen G, Picorel R, Holzwarth AR. Biochemistry 1996; 35:9469–9474.
172. Samson G, Morissette JC, Popovic R. Photochem Photobiol 1988; 48:329–332.
173. Yruela I, Alfonso M, de Zarate, IO, Montoya G, Picorel R. J Biol Chem 1993; 268:1684–1689.
174. Schröder WP, Arellano JB, Bittner T, Baron M, Eckert HJ, Renger G. J Biol Chem 1994; 269:32865–32870.
175. Boucher N, Carpentier R. Photosyn Res 1999; 59:167–174.
176. Burda K, Kruk J, Schmid GH, Strzalka K. Biochem J 2003; 371:597–601.
177. Li EH, Miles CD. Plant Sci Lett 1975; 5:33–40.
178. Hsu BD, Lee JY. Plant Physiol 1988; 87:116–119.
179. Yruela I, Montoya G, Picorel R. Photosyn Res 1992; 33:227–233.

180. Brown NT, Newman JE. Aquat Toxicol 2003; 64:201–213.
181. Pätsikkä E, Aro EM, Tyystjärvi E. Physiol Plant 2001; 113:142–150.
182. Yruela I, Pueyo JJ, Alonso PJ, Picorel R. J Biol Chem 1996; 271:27408–27415.
183. Ouzounidou G, Moustakas M, Strasser RJ. Aust J Plant Physiol 1997; 24:81–90.
184. Krupa Z, Öquist G, Huner NPA. Physiol Plant 1993; 88:626–630.
185. Krupa Z, Siedlecka A, Maksymiec W, Baszynski T. J Plant Physiol 1993; 142: 664–668.
186. Xyländer M, Hagen C, Braune W. Bot Acta 1996; 109:222–228.
187. Murthy SDS, Mohanty P. Plant Cell Physiol 1991; 32:231–237.
188. Küpper H, Lombi E, Zhao FJ, McGrath SP. Planta 2000; 212:75–84.
189. Küpper H, Mijovilovich A, Meyer-Klaucke W, Kroneck PMH. Plant Physiol 2004; 134:748–757.
190. Liao MT, Hedley MJ, Woolley DJ, Brooks RR, Nichols MA. Plant Soil 2000; 221:135–142.
191. DeVos CHR, Schat H, DeWaal MAM, Vooijs R, Ernst WHO. Physiol Plant 1991; 82:523–528.
192. Ouzounidou G, Lannoye R, Karataglis S. Plant Sci 1993; 89:221–226.
193. Cuypers A, Vangronsveld J, Clijsters H. J Plant Physiol 2002; 159:869–876.
194. Pinto E, Sigaud-Kutner TCS, Leitao MAS, Okamoto OK, Morse D, Colepicolo P. J Phycol 2003; 39:1008–1018.
195. Sandalio LM, Dalurzo HC, Gómez M, Romero-Puertas MC, del Río LA, J Exp Bot 2001; 52:2115–2126.
196. De Vos CHR, Schat H. In: Rozema J, Verkleij JAC, eds. Ecological Responses to Environmental Stresses. Dordrecht, The Netherlands: Kluwer Academic Publishers, 1991:22–30.
197. Clarkson TW. Annu Rev Pharmacol Toxicol 1993; 33:545–571.
198. Demidchik V, Sokolik A, Yurin V. Plant Physiol 1997; 114:1313–1325.
199. Demidchik V, Sokolik A, Yurin V. Planta 2001; 212:583–590.
200. Stohs SJ, Bagchi D. Free Radic Biol Med 1995; 18:321–336.
201. Krupa Z, Baszynski T. Acta Physiol Plant 1989; 11:111–116.
202. Maksymiec W, Russa R, Urbaniksypniewska T, Baszynski T. J Plant Physiol 1992; 140:52–55.
203. Berglund AH, Quartacci MF, Liljenberg C. Biochem Soc Trans 2000; 28:905–907.
204. Quartacci MF, Pinzino C, Sgherri CLM, F. Dalla Vecchia, Navari-Izzo F. Physiol Plant 2000; 108:87–93.
205. Jemal F, Zarrouk M, Ghorbal MH. Biochem Soc Trans 2000; 28:907–910.
206. Gushina IA, Harwood JL. Biochem Soc Trans 2000; 28:910–912.
207. Cuypers A, Vangronsveld J, Clijsters H. Plant Physiol Biochem 2001; 39:657–664.
208. Grill E, Winnacker EU, Zenk MH. Science 1985; 230:674–676.
209. Grill E, Löffler S, Winnacker EL, Zenk MH. Proc Natl Acad Sci USA 1989; 86:6838–6842.
210. Vatamaniuk OK, Mari S, Lu YP, Rea PA. J Biol Chem 2000; 275:31451–31459.
211. Salt DE, Rauser WE. Plant Physiol 1995; 107:1293–1301.
212. Howden R, Goldsbrough PB, Andersen CR, Cobbett CS. Plant Physiol 1995; 107:1059–1066.
213. Clemens S, Simm C. New Phytol 2003; 159:323–330.
214. Vatamaniuk OK, Bucher EA, Ward JT, and Rea PA. Trends Biotechnol 2002; 20:61–64.

215. Hartley-Whitaker J, Ainsworth G, Vooijs R, Ten Bookum W, Schat H, Meharg AA. Plant Physiol 2001; 126:299–306.
216. Maitani T, Kubota H, Sato K, Yamada T. Plant Physiol 1996; 110:1145–1150.
217. Satofuka H, Fukui T, Takagi M, Atomi H, Imanaka T. J Inorg Biochem 2001; 86:595–602.
218. Schat H, Llugany M, Vooijs R, Hartley-Whitaker J, Bleeker PM. J Exp Bot 2002; 53:2381–2392.
219. García-Hernández M, Murphy A, Taiz L. Plant Physiol 1998; 118:387–397.
220. Ahner BA, Wei L, Oleson JR, Ogura N. Mar Ecol Progr Ser 2002; 232:93–103.
221. Salt DE, Prince RC, Baker AJM, Raskin I, Pickering IJ. Environ Sci Technol 1999; 33, 712–717.
222. Scholz G, Becker R, Pich A, Stephan UW. J Plant Nutr 1992; 15:1647–1665.
223. Weber M, Harada E, Vess C, von Roepenack-Lahaye E, Clemens S, Plant J 2004; 37:269–281.
224. Bassi R, Sharma SS. Phytochemistry 1993; 33:1339–1342.
225. Wu JT, Hsieh MT, Kow LC. J Phycol 1999; 34:113–117.
226. Mehta SK, Gaur JP. New Phytol 1999; 143:253–259.
227. Schat H, Sharma SS, Vooijs R. Physiol Plant 1997; 101:477–482.
228. Dykema PE, Sipes PR, Marie A, Biermann BJ, Crowell DN, Randall SK. Plant Mol Biol 1999; 41:139–150.
229. Hale KL, McGrath SP, Lombi E, Stack SM, Terry N, Pickering IJ, George GN, Pilon-Smits AH, Plant Physiol 2001; 126:1391–1402.
230. Leal MFC, Vasconcelos MTSD, van den Berg CMG. Limnol Oceanogr 1999; 44:1750–1762.
231. Ernst WHO, Verkleij JAC, Schat H. Acta Bot Neerland 1992; 41:229–248.
232. De DN. Plant cell vacuoles. Collingwood, Australia: CSIRO Publishing, 2000.
233. Chardonnens AN, ten Bookum WM, Kuijper LDJ, Verkleij JAC, Ernst WHO. Physiol Plant 1998; 104:75–80.
234. Küpper H, Zhao F, McGrath SP. Plant Physiol 1999; 119:305–311.
235. Frey B, Keller C, Zierold K, Schulin R. Plant Cell Environ 2000; 23:675–687.
236. Psaras GK, Constantinidis TH, Cotsopoulos B, Manetas Y. Ann Bot 2000; 86: 73–78.
237. Psaras GK, Manetas Y. Ann Bot 2001; 88:513–516.
238. Karley AJ, Leigh RA, Sanders D. Trends Plant Sci 2000; 5:465–470.
239. Salt DE, Wagner GJ. J Biol Chem 1993; 268:12297–12302.
240. Li L, Zengyong H, Pandey GK, Tsuchiya T, Luan S. J Biol Chem 2002; 277: 5360–5368.
241. Dudev T, Lim C. Chem Rev 2003; 103:773–787.
242. Sunda WG, Huntsman SA. Limnol Oceanogr 1998; 43:1055–1064.
243. Siedlecka A, Baszynski T. Physiol Plant 1993; 87:199–202.
244. Morel NML, Rueter JG, Morel FMM. J Phycol 1978; 14:43–48.
245. Rueter JG Jr, Chisholm SW, Morel FMM. J Phycol 1981; 17:270–278.
246. Vassilev A, Yordanov I, Tsonev T. Photosynthetica 1997; 34:293–302.
247. Baumann A. Landwirtsch Verss 1885; 31:1–53.
248. Ingrouille MJ, Smirnoff N. New Phytol 1986; 102:219–33.
249. Brooks RR, Lee J, Reeves RD, Jaffre T. J Geochem Explor 1977; 7:49–57.
250. Brooks RR. In: Brooks RR, ed. Plants that Hyperaccumulate Heavy Metals. Wallingford, UK: CAB International, 1998:55–94.

251. Baker AJM, McGrath SP, Reeves RD, Smith JAC. In: Terry N, Bañuelos G, Vangronsveld J, eds. Phytoremediation of Contaminated Soil and Water. Boca Raton, Florida: Lewis Publishers. 2000:85–107.

252. Sagner S, Kneer R, Wanner G, Cosson JP, Deus-Neumann, B, Zenk MH. Phytochemistry 1998; 47:339–347.

253. Robinson BH, Brooks RR, Howes AW, Kirkman JH, Gregg PEH. J Geochem Explor 1997; 60:115–126.

254. Robinson BH, Chiarucci A, Brooks RR, Petit D, Kirkman JH, Gregg PEH, De Dominicis V. J Geochem Explor 1997; 59:75–86.

255. Robinson BH, Brooks RR, Clothier BE. Ann Bot 1999; 84:689–694.

256. Assunção AGL, Schat H, Aarts MGM. New Phytol 2003; 159:351–360.

257. Peer WA, Mamoudian M, Lahner B, Reeves RD, Murphy AS, Salt DE. New Phytol 2003; 159:421–430.

258. Ma LQ, Komar KM, Tu C, Zhang WH, Cai Y, Kennelley ED. Nature 2001; 409:579.

259. Paton A, Brooks RR. J Geochem Explor 1996; 56:37–45.

260. Boyd RS, Martens SN. Oikos 1994; 70:21–25.

261. Martens SN, Boyd RS. Oecologia 1994; 98:379–384.

262. Boyd RS, Davis MA, Wall MA, Balkwill K. Chemoecology 2002; 12:91–97.

263. Hanson B, Garifullina GF, Lindblom SD, Wangeline A, Ackley A, Kramer K, Norton AP, Lawrence CB, Pilon-Smits EAH. New Phytol 2003; 159:461–469.

264. Pollard JA, Baker AJM. New Phytol 1997; 135:655–658.

265. Jhee EM, Dandridge KL, Christy AM, Pollard AJ. Chemoecology 1999; 9:93–95.

266. Davis MA, Boyd RS. New Phytol 2000; 146:211–217.

267. Ghaderian SM, Lyon AJE, Baker AJM. New Phytol 2000; 146:219–224.

268. Huitson SB, Macnair MR. New Phytol 2003 159:453–459.

269. Davis MA, Murphy JF, Boyd RS. J Environ Qual 2001; 30:85–90.

270. Peterson LR, Trivett V, Baker AJM, Aguiar C, Pollard AJ. Chemoecology 2003; 13:103–108.

271. Pollard AJ. New Phytol 2000; 146:179–181.

272. Boyd RS, Jaffre T. South Africa J Sci 2001; 97:535–538.

273. Shen ZG, Zhao FJ, McGrath SP. Plant Cell Environ 1997; 20:898–906.

274. Perronnet K, Schwartz C, Gérard E, Morel L. Plant Soil 2000; 227:257–263.

275. Baker AJM, Proctor J, van Balgooy, MMJ, Reeves RD. In: Baker AJM, Proctor J, Reeves RD, eds. The Vegetation of Ultramafic (Serpentine) Soils. Dordrecht: Kluwer Academic Publishers, 1992:Chap. 22, 291–304.

276. Krämer U, Smith RD, Wenzel WW, Raskin I, Salt DE. Plant Physiol 1997; 115:1641–1650.

277. Whiting SN, Neumann PM, Baker AJM. Plant Cell Environ 2003; 26:351–360.

278. Perronnet K, Schwartz C, Morel JL. Plant Soil 2003; 249:19–25.

279. Whiting SN, Leake JR, McGrath SP, Baker AJM. New Phytol 2000; 145:199–210.

280. Pollard AJ, Powell KD, Harper FA, Smith JAC. Crit Rev Plant Sci 2002; 21: 539–566.

281. Macnair MR, Bert V, Huitson SB, SaumitouLaprade P, Petit D. Proc Roy Soc Lon B 1999; 266:2175–2179.

282. Bert V, Meerts P, Saumitou-Laprade P, Salis P, Gruber W, Verbruggen N. Plant Soil 2003; 249:9–18.

283. Whiting SN, De Souza MP, Terry N. Environ Sci Technol 2001; 35:3144–3150.

284. Lasat MM, Baker AJM, Kochian LV. Plant Physiol 1996; 112:1715–1722.

285. Lasat MM, Baker AJM, Kochian LV. Plant Physiol 1998; 118:875–883.
286. Kerkeb L, Krämer, U. Plant Physiol 2003; 131:716–724.
287. Wang J, Zhao FJ, Meharg AA, Raab A, Feldmann J, McGrath SP, Plant Physiol 2002; 130:1552–1561.
288. Webb SM, Gaillard JF, Ma LQ, Tu C. Environ Sci Technol 2003; 37:754–760.
289. Ebbs S, Lau I, Ahner B, Kochian LV, Planta 2002; 214:635–640.
290. Neumann D, De Figueiredo C. Protoplasma 2002; 220:59–67.
291. Vázquez MD, Poschenrieder Ch, Barceló J, Baker AJM, Hatton P, Cope GH. Bot Acta 1994; 107:243–250.
292. Cosio C, Martinoia E, Keller C. Plant Physiol 2004; 134:716–725.
293. Zhao FJ, Lombi E, Breedon T, McGrath SP. Plant Cell Environ 2000; 23:507–514.
294. Küpper H. Physiology of Hyperaccumulator- and Non-accumulator Plants: Heavy Metal Uptake, Transport, Compartmentation, Stress and Resistance, Allensbach, Germany: UFO-Verlag, 2001.
295. Dahmani-Muller H, van Oort F, Balabane M. Environ Pollut 2001; 114:77–84.
296. Walker DJ, Bernal MP. Water Air Soil Pollut 2004; 151:361–372.
297. Chaney RL. In: Parr JE, Marsh PB, Kla JM, eds. Land Treatment of Hazardous Wastes. Park Ridge, IL USA: Noyes Data Corp., 1983:50–76.
298. Baker AJM, Brooks RR. Biorecovery 1989; 1:81–126.
299. Baker AJM, Brooks RR, Reeves R. New Sci 1988; 117:44–48.
300. Baker AJM, Reeves RD, McGrath SP. In: Hinchee RE, Olfenbuttel RF, eds. In Situ Bioreclamation. Stoneham, MA, USA: Butterworth-Heinemann, 1991:539–44.
301. Baker AJM, McGrath SP, Sidoli CMD, Reeves RD. Resourc Conserv Recycl 1994; 11:41–49.
302. McGrath SP. In: Brooks RR, ed. Plants that Hyperaccumulate Heavy Metals. Wallingford, UK: CAB International, 1998:261–287.
303. McGrath SP, Dunham SJ, Correll RL. In: Terry N, Bañuelos G, eds., Phytoremediation of Contaminated Soil and Water. Boca Raton, Florida, USA: Lewis Publishers, 2000:109–128.
304. McGrath SP, Zhao FJ. Curr Opin Biotechnol 2003; 14:277–282.
305. Salt DE, Blaylock M, Nanda Kumar PBA, Dushenkov V, Ensley BD, Chet I, Raskin I. Biotechnology 1995; 13:468–474.
306. Salt DE, Smith RD, Raskin I. Annu Rev Plant Physiol Plant Mol Biol 1998; 49:643–668.
307. Chaney RL, Malik M, Li YM, Brown SL, Brewer EP, Angle JS, Baker AJM. Curr Opin Biotechnol 1997; 8:279–284.
308. Chaney RL, Li YM, Brown SL, Homer FA, Malik M, Angle JS, Baker AJM, Reeves RD, Chin M. In: Terry N, Banuelos G, eds. Phytoremediation of Contaminated Soil and Water. Boca Raton, FL: CRC Press, 2000:129–158.
309. Zhao FJ, Dunham SJ, McGrath SP. New Phytol 2002; 156:27–31.
310. Meharg AA. New Phytol 2003; 157:25–31.
311. Reay PF. J Appl Ecol 1972; 9:557–565.
312. Robinson B, Outred H, Brooks R, Kirkman J. Chem Speciat Bioavail 1995; 7:89–96.
313. Del Ríoa M, Fonta R, Almelab C, Vélezb, D, Montorob R, De Haro Bailón A. J Biotechnol 2002; 98:125–137.
314. Robinson BH, Leblanc M, Petit D, Brooks RR, Kirkman JH, Gregg PEH. Plant Soil 1998; 203:47–56.
315. Zhao FJ, Lombi E, McGrath SP. Plant Soil 2003; 249:37–43.

316. Hammer D, Keller C. Soil Use Manage 2003; 19:144–149.
317. Brooks RR, Robinson BH. In: Brooks RR, ed. Plants that Hyperaccumulate Heavy Metals. Wallingford, UK: CAB International, 1998:203–226.
318. Nicks L, Chambers MF. Mining Environ Manage September, 1995; 15–18.
319. Anderson TR, Howes AW, Slater K, Dutton MF, In: Jaffre T, Reeves RD, Becquer T, eds. The Ecology of Ultramafic and Metalliferous Areas, Proc. Second Intern. Conf. on Serpentine Ecology, Noumea, New Caldonia, July 31–Aug. 5, 1995. New Caledonia: OSTROM, 1996:261–266.
320. Brooks RR, Robinson BH, Howes AW, Chiarucci A. South Africa J Sci 2001; 97:558–560.
321. Brooks RR, Chambers MF, Nicks LJ, Robinson BH. Trends Plant Sci 1998; 3: 359–362.
322. Reeves RD, Baker AJM, Borhidi A, Berazaín R. New Phytol 1996; 133:217–224.
323. Reeves RD. Plant Soil 2003; 249:57–65.
324. Rugh CL, Wilde HD, Stack NM, Thompson DM, Summers AO, Meagher RB. Proc Natl Acad Sci USA 1996; 93:3182–3187.
325. Rugh CL, Senecoft JF, Meagher RB, Merkle SA. Nat Biotechnol 1998; 33: 616–621.
326. Long XX, Yang XE, Ye ZQ, Ni QZ, Shi WY. Acta Bot Sinica 2002; 44: 152–157.
327. Ye HB, Yang XE, He B, Long XX, Shi WY. Acta Bot Sinica 2003; 45: 1030–1036.
328. Pulford ID, Watson C. Environ Int 2003; 29:529–540.
329. Williams JB. Crit Rev Plant Sci 2002; 21:607–635.
330. Ebbs SD, Kochian LV. J Environ Qual 1997; 26:776–781.
331. Li YM, Chaney R, Brewer E, Roseberg R, Angle JS, Baker A, Reeves R, Nelkin J. Plant Soil 2003; 249:107–115.
332. Schwartz C, Echevarria G, Morel JL. Plant Soil 2003; 249:27–35.
333. Brewer EP, Saunders JA, Angle JS, Chaney RL, McIntosh MS. Theor Appl Genet 1999; 99:761–771.
334. Bennett FA, Tyler EK, Brooks RR, Gregg PEH, Stewart RB. In: Brooks RR, ed. Plants that Hyperaccumulate Heavy Metals. Wallingford, UK: CAB International, 1998:249–259.
335. Meerts P, Van Isacker N. Plant Ecol 1997; 133:221–231.
336. Bert V, Macnair MR, de Lagúrie P, Saumitou-Laprade P, Petit D. New Phytol 2000; 146:225–233.
337. Bert V, Bonnin I, Saumitou-Laprade P, de Laguérie P, Petit D, New Phytol 2002; 155:47–57.
338. Escarré J, Lefèbvre C, Gruber W, Leblanc M, Lepart J, Rivière Y, Delay B, New Phytol 2000; 145:429–437.
339. Macnair MR, New Phytol 2002; 155:59–66.
340. Roosens N, Verbruggen N, Meerts P, Ximénez-Embun P, Smith JAC. Plant Cell Environ 2003; 26:1657–1672.
341. Angle JS, Baker AJM, Whiting SN, Chaney RL. Plant Soil 2003; 256:325–332.
342. Ernst WHO. New Phytol 2000; 146:357–358.
343. Lasat MM. J Environ Qual 2002; 31:109–120.
344. Lombi E, Zhao FJ, Dunham SJ, McGrath SP. J Environ Qual 2003; 30: 1919–1926.

345. Whiting SN, Leake JR, McGrath SP, Baker AJM. Plant Soil 2001; 237: 147–156.
346. Piontek K, Antorini M, Choinowski T. J Biol Chem 2002; 277:37663–37669.
347. Richter OMH, Ludwig B. Rev Physiol Biochem Pharmacol 2003; 147:47–74.
348. Askwith C, Eide D, Van Ho A, Bernard PS, Li L, Davis-Kaplan S, Sipe DM, Kaplan J. Cell 1994; 76:403–410.
349. Lindley PF, Card G, Zaitseva I, Zaitsev V, Reinhammar B, Selin-Lindgren E, Yoshida K. J Biol Inorg Chem 1997; 2:454–63.

6

Arsenic: Its Biogeochemistry and Transport in Groundwater

Charles F. Harvey[1] and Roger D. Beckie[2]

[1]*Parsons Laboratory, Department of Civil and Environmental Engineering, Massachusetts Institute of Technology, 48-321, Cambridge, MA 02139, USA*
[2]*Department of Earth and Ocean Sciences, University of British Columbia, 6339 Stores Road, Vancouver, British Columbia, V6T 1ZA Canada*

1. INTRODUCTION

Arsenic is dissolved in groundwater throughout the world at levels believed to be dangerous. Arsenic is clearly poisonous at high concentrations—a lethal dose of sodium arsenite is $\sim 15-30$ mg [1]. However, the cancer risks of low doses (e.g., 50 μg/L, typical in contaminated groundwater) are more difficult to quantify and appropriate drinking water standards continue to be debated, particularly in the USA. Smith et al. [2] studied the epidemiology of arsenic exposures to humans in Taiwan, Japan, and Argentina and estimated a combined cancer mortality risk of 1 in 100 for the current US maximum contaminant level (MCL) of 50 μg/L arsenic in drinking water [2]. This risk is 100 times greater than for any other drinking water contaminant that has a United States Environmental Protection Agency (USEPA) defined MCL [2]. Even though evidence from as early as the 1960s suggested that arsenic was a severe carcinogen, the US will not lower its MCL from 50 μg/L to the World Health Organization's recommendation of 10 μg/L until 2006. The state of California recently set its public health goal for arsenic in drinking water to 0.004 μg/L, much lower than can currently be measured. Smith attributes the delay in reducing standards to uncertainty about the role of other factors such as diet, and threshold levels for toxicity, and many publications (e.g., [3,4]) discuss the uncertainties in current epidemiological models.

In the next section, we review the fundamentals of low-temperature arsenic geochemistry that control arsenic concentrations in groundwater including arsenic abundance, aqueous and gas-phase speciation, arsenic mineralogy, and arsenic sorption behavior. We then discuss two important arsenic problems: arsenic in mining, which probably represents the most significant dispersal mechanism of arsenic into the environment, and arsenic in Bangladesh, the most significant human impact of arsenic in the world. The dispersal of arsenic by mining is anthropogenic, whereas the arsenic in Bangladeshi groundwater is naturally occurring, and the role of human activities on arsenic levels is subtle and indirect, illustrating the complex controls that govern arsenic mobilization and transport in groundwater. A considerable number of reviews, compendiums and special volumes have been published on arsenic, and we refer interested readers to these for more information [5–14].

2. GLOBAL ARSENIC ABUNDANCE AND FLUX ESTIMATES

Several tabulations of arsenic abundance in various earth materials have been collected [8,9,12,15,16]. As Cullen and Reimer note in their comprehensive 1989 review [5], it is difficult to estimate "typical" or representative numbers for arsenic abundance in nature because of its great variability. The reservoir of As in the Earth's crust is vast, on the order of 10^{13} ton, compared to the hydrosphere (oceans, rivers, groundwaters, and icecaps) on the order of 10^9 ton and the atmosphere, on the order of 10^3 ton. A recent study of over

280,000 reconnaissance surface samples (mostly sediments from streams, lakes, and surficial soils) collected in the USA and Canada [16] shows solid-phase arsenic concentrations range from <1 parts per million (ppm) in large regions of the US southeast and continental shield to 50–100 ppm in the Selwyn basin in the Yukon and much of Alaska to 100–200 ppm on Baffin Island in the Canadian arctic. Arsenic concentrations are strongly tied to rock and sediment type, and are particularly high in pyrite-rich shales and sediments. Arsenic concentrations in natural groundwaters are typically below 10 μg/L, but naturally reach as high as 1000 μg/L in areas such as Bangladesh and may rise to several hundred milligrams per liter in tailings from mine wastes.

Several authors have attempted to quantify the natural and anthropological fluxes of arsenic through the environment [8,12,14,17]. While the estimates are uncertain, it is clear that human activities have had a very large impact on the global arsenic cycle. The following discussion, and Fig. 1, is based upon estimates from [8,12].

Arsenic is a volatile supplied to the atmosphere naturally by volcanoes, which release ∼17,000 ton/year, and anthropologically at a rate of ∼25,000 ton/year by smelting of non-ferrous ores, principally copper, zinc and gold, and burning of coal. Matschullat [8] estimates 125,000 ton/year discharged to the environment world-wide as dissolved arsenic from human sources, but does not report arsenic disposed on land. Data compiled on the USEPA toxic release inventory (TRI) shows that in 2001, 174,000 ton of As was released on land in the USA alone (down from 267,000 ton in 1999). Almost all of this arsenic was in mining-derived waste rock and tailings, of which 89% was from gold ores, 4% from copper ores and 3% from silver ores. The State of Nevada, where there are a substantial number of large gold mines and extensive shale deposits, accounted for 87% of all US arsenic waste. Arsenic production associated with gold mining in the rest of the world is difficult to estimate, but is probably of at least the same order of magnitude as that reported in the USA, and likely the most important mechanism by which arsenic is dispersed in the environment.

3. ARSENIC MINERALS AND THE GEOCHEMICAL ARSENIC CYCLE

We next present an overview of the forms and cycling of arsenic at the Earth's surface. Arsenic is commonly found in four oxidation states: As(−III) as in AsH_3, arsine gas, As(0), as in native or metallic arsenic, As(III), as in As_2O_3, arsenic trioxide, also known as either claudetite and arsenolite (two distinct polymorphs), and As(V), as in $FeAsO_4 \cdot 2H_2O$, scorodite. Arsenic is rarely a major species in solution, and therefore arsenic geochemistry is largely dictated by the prevailing pH and redox conditions.

In the aqueous phase, arsenic is almost always in the As(III) and As(V) oxidation states, dominantly as inorganic species. The As(III) and As(V) ions have a high ionic potential (IP), defined as the valance of the ion divided by

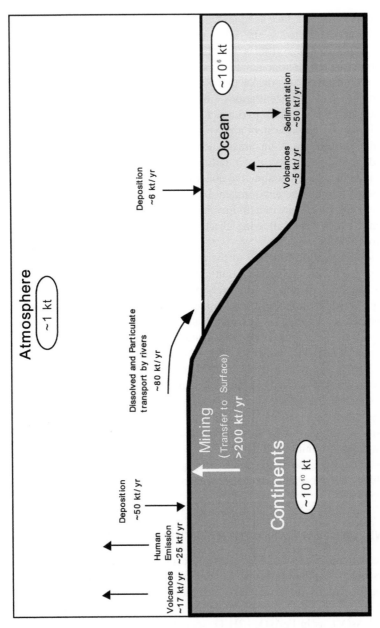

Figure 1 Simplified description of rough estimates of global arsenic fluxes (in kilotons per year, kton/year) based largely on the work of Ref. [8] that provides a detailed overview of the substantial uncertainty in these values. In the USA ~200 kton/year of arsenic waste are produced by mining each year [23], so the global value is substantially larger. The mismatch between input and output fluxes represents uncertainty in the flux values, each of which may be an order of magnitude different than shown above.

the crystallographic radius [18] and therefore exist in solution as the oxyanions arsenite As(III), AsO_3^{2-}, and arsenate As(V), AsO_4^{3-}. Other ions with similarly high IP also exist as oxyanions, for example, phosphate P(V), PO_4^{3-}, chromate Cr(VI), CrO_4^{4-}, selenate Se(VI), SeO_4^{2-}, and carbonate C(IV), CO_3^{2-}. Most trace metals, in contrast, have lower IPs and therefore exist in solution as hydrated cations (e.g., Ni^{2+}, Cu^{2+}, Zn^{2+}). Oxyanions, such as arsenite and arsenate, tend to desorb from mineral surfaces as the pH rises above 6, while cations such as nickel, copper, and zinc tend to sorb more strongly as pH increases. Arsenite exists as a neutral complex H_2AsO_3 over a wide pH range; only at pH 9.17 does the complex $HAsO_3^-$ become dominant. Arsenate is principally a negatively charged anion at pH above 2.3 [19].

Methylated arsenic species, most commonly monomethyl and dimethyl As(V) species, can be formed inorganically or at pH > 5 through microbial action [5,11,20]. Under stronger reducing conditions, volatile methyl-arsines [As(III)] can form, the most famous of which is Gosio gas, trimethylarsine, which was discovered at the turn of the last century to be produced in molds growing in wall paper pigments made from arsenic compounds [5]. Arsine gas, AsH_3 [As(−III)] is formed under the strongest reducing conditions.

Over 99% of arsenic is estimated to be in rocks [21], and arsenic is found at some level in most rock types, but is enriched in minerals derived from hydrothermal solutions of magmatic origin. As magma cools and crystallizes, arsenic and metals such as gold, silver, copper, and uranium tend to partition into the solution phase because their size or charge does not allow them to be easily incorporated into igneous minerals [22]. When these fluids rise to the near surface environment and cool to between 50°C and 300°C, a suite of sulfide minerals is precipitated, often forming hydrothermal ore deposits. Arsenic is commonly found in association with these ore deposits, and particularly with those of gold.

The most important arsenic minerals found in nature fall into four groups shown in Table 1 [23]. Arsenian pyrite, $Fe(As,S)_2$, is probably the most significant reservoir of mineral arsenic in nature [24]. Pyrite can incorporate substantial amounts of trace metal impurities into its structure and a maximum of 8 wt.% arsenic, although it is now speculated that pyrite forms a solid solution with arsenopyrite, FeAsS [25–27]. Pyrite formation is favorable at temperatures <300°C, and is observed in modern wetland, lake, and ocean sediments [28,29]. So-called authigenic pyrite is typically formed when organic matter is introduced into an iron and sulfur rich system, which causes redox potentials to drop. Highly insoluble ferric iron (iron oxide), is first reduced by the organic matter to soluble ferrous iron, and then sulfate is reduced to sulfide. The dissolved arsenic concentrations then decline as pyrite forms through a complex process [29], co-precipitating arsenic. This mechanism explains the pyrite and therefore high arsenic found in organic-rich shales, mudstones, and coal deposits. However, the introduction of organic matter and the onset of reducing conditions do not unambiguously lead to the sequestration of arsenic in sulfide minerals. As we describe later, a decrease in redox potentials may be

Table 1 Arsenic Mineral Families

Group name	General formula	M	X
Sulfides	MAsS	Fe, Cu, Ni, Sb, Pb, Ag, Hg, Co	
Arsenides	MAs	Fe, Cu, Pb, Ni, Sb, Pd, Co	
Arsenites	$MAsO_3X$	Fe, Ca, Mn, Pb, Mg, Zn, Si, Cu	OH
Arsenates	$MAsO_4X$	Fe, Ca, Zn, Cu, Mn, Pb, Mg	OH, H_2O

associated with an increase in dissolved arsenic concentrations as arsenic is liberated from dissolving iron oxide minerals.

The thermodynamics of arsenopyrite, a much rarer mineral, has recently been re-evaluated [30]. It had previously been considered a high-temperature mineral 250–500°C [31], but appears to be stable at much lower near-surface temperatures. There is field evidence to suggest that it can persist on geologic time scales in anoxic near-surface environments [32] and Rittle et al. [33] report an amorphous authigenic iron-arsenic-sulfide phase formed in their microcosms derived from mining-impacted sediments.

Orpiment, As_2S_3, is found in hydrothermal ore deposits, in hot-spring precipitates, and as a secondary mineral (weathering product) derived from arsenopyrite and realgar AsS. The temperature range under which realgar is stable is narrower than that of orpiment and although amorphous orpiment has been documented using X-ray diffraction methods, amorphous realgar has never been reported [19]. Both orpiment and realgar can be precipitated by bacteria [34,35].

Under oxidizing conditions, arsenic concentrations in water are most often controlled by adsorption rather than by mineral solubility constraints. Indeed, with the exception of environments rich in arsenic such as mine wastes, in the presence of oxygen most arsenic mineral phases are too soluble to control arsenic concentrations in water. Oxides of iron, aluminum, and manganese are the most important sorbents in the environment because of their ubiquity, tendency to appear as coatings on mineral surfaces, and corresponding high surface area [36]. Synchrotron studies demonstrate that arsenite and arsenate ions sorb to iron oxides as both bidendate and to a lesser degree monodendate inner-sphere complexes [37]. Generally, the adsorption capacity of oxide minerals decreases as the degree of mineral crystallinity increases [38,39], and appears to be tied to the decrease in surface site density and not to a change in site binding strength [40]. While arsenic adsorption is strongest on oxide mineral surfaces, arsenic also adsorbs onto clay minerals, silicates, and magnetite [40,41]. Green rusts (amorphous ferric–ferrous oxides), are an important corrosion product of native (zero valent) iron and are reported to control dissolved arsenic in iron-based permeable reactive barriers [42].

The degree of arsenic sorption is strongly controlled by solution pH and composition. Arsenate and to a lesser extent arsenite tend to desorb as pH increases. This is due to both an increase in negative surface charge with increasing pH as well as deprotonation of the arsenate ion at pH 7 [43]. Phosphate strongly competes with arsenic and sorbs by the same bidendate mechanism [37]. Natural dissolved organic matter tends to form soluble complexes with arsenic, particularly with As(III), which dramatically decreases the amount of arsenic sorption on hematite [44]. Silicate [45] and sulfate [46] are also known to compete with arsenic for surface sites but are weaker than phosphate. Recently, the effect of carbonate ions on arsenic adsorption has received attention, and has been proposed as an important control on arsenic mobility in sorption-dominated systems [47,48]. The effects of carbonate are however not clear at this point, as some experiments show only a very small effect of carbonate on As(III) and As(V) sorption onto ferrihydrite [49,50].

The arsenic sorbed onto iron oxides is vulnerable to release by changing geochemical conditions, in particular a decrease in redox potential. When redox potentials drop, ferric iron is reduced to the highly soluble ferrous iron. In the process, the iron oxide solid phase dissolves and releases sorbed ions such as arsenic into solution. We believe that the reductive dissolution of iron oxides is the dominant process by which arsenic has been mobilized into Bangladeshi groundwater, although the timing and location of this process remains uncertain. The reduction is driven by the introduction of organic matter into the aquifer. The extent of iron reduction and corresponding oxide dissolution depends upon local redox buffering, the flux of organic matter and the stability of iron oxide phases [51].

4. ARSENIC AND MINING

Economic geologists have a long-standing interest in arsenic because gold and arsenic are strongly correlated in many deposits [15,52]. For this reason, and the relatively low concentration of arsenic in most crustal rocks, prospectors use arsenic as a pathfinder element when exploring for gold. This strong association also explains why mining industries are a major source of arsenic waste, and mine sites are major sources of arsenic to the wider environment. The arsenic-bearing minerals in ore deposits are usually stable over geologic time in their undisturbed environment. The disturbance to the natural environment from mining and mineral processing often leaves arsenic in a less stable state, which creates opportunities for off-site transport, usually by water.

Acid rock drainage (ARD) represents a significant source of arsenic to the environment, particularly in North America. ARD occurs where sulfide minerals are exposed to the atmosphere and subsequently oxidize, typically with microbial catalysis, releasing acid, sulfate, and metals. While ARD can occur naturally, it is most common at mine sites, issuing from abandoned mine workings, and from mine wastes such as tailings and waste rock [53]. Arsenic sulfide minerals are believed to be more reactive than the much more abundant pyrite [27,54]. McGuire et al. [54] report that arsenopyrite oxidizes at a rate approximately 10 times faster than pyrite, both abiotically and with microbial acceleration. Arsenic tends to diffuse to the mineral surface as arsenopyrite weathers and is therefore readily available for leaching [55]. ARD can lead to very high and long-duration dissolved arsenic concentrations. Indeed, the highest reported arsenic concentration of 850 mg/L was observed in solutions derived from the oxidation of pyrite in abandoned mine workings at Iron Mountain, USA [56]. Without remedial measures, the Iron Mountain mine is expected to produce acidic drainage for the next 3000 years.

Mineral processing usually produces arsenic wastes that do not occur naturally and are susceptible to dispersal in the environment. For example, highly pure arsenic trioxide is produced during roaster processing of sulfide-rich refractory ores. Roasting, or pyrometallurgy, involves oxidizing the ore at high temperatures, driving off gas-phase sulfur and arsenic. Arsenic trioxide dust is precipitated by cooling the flue gas and is usually collected for disposal using large "bag houses" or electrostatic filters. At the Giant Mine located in Yellowknife, Northwest Territories, Canada, roasting of sulfide gold ores produced 10–15 ton of 70% pure arsenic trioxide dust each day for over 50 years of operations between 1949 and 1999 [57]. The 267,000 ton of arsenic trioxide presently stored in underground chambers at the mine probably represents one of the largest localized sources of highly soluble arsenic in the environment. Groundwater flowing through fracture bedrock threatens to dissolve the arsenic and transport it off site into nearby Great Slave Lake. Current plans are to permanently freeze the material in place at a cost of approximately \$100–\$200 million Canadian dollars (CAD).

Tailings are typically finely ground and contain arsenic in primary gangue minerals such as sulfides as well as altered and secondary minerals produced during processing. Roasting of pyrite tends to produce hematite (α-Fe_2O_3) and maghemite (γ-Fe_2O_3) with a typical spongy-microporous texture. These iron oxides may contain up to 60,000 ppm As and while they are generally stable under oxidizing conditions, they are subject to reductive dissolution, especially in boggy sites or below water bodies with high productivity [58,59]. For example, McCreadie et al. [58], report arsenic concentrations of over 100 mg/L at the base of a tailings deposit that rests upon an organic layer of peat.

Secondary minerals produced in ore processing include a spectrum of ferric–arsenate minerals formed by ferric iron addition and a number of calcium–arsenate minerals that form after lime neutralization of mill raffinates [39,60–63]. Scorodite (Fig. 2) formed from solutions high in both ferric iron and arsenate, has a solubility minimum at pH 2.2, but tends to dissolve incongruently at higher pH to release arsenate and precipitate ferrihydrite. The most efficient removal of arsenic from solutions occurs when arsenic and ferric iron coprecipitate as arsenical ferrihydrite [37,43,64]. Arsenic can be maintained below drinking water standards by this method as long as the Fe/As molar

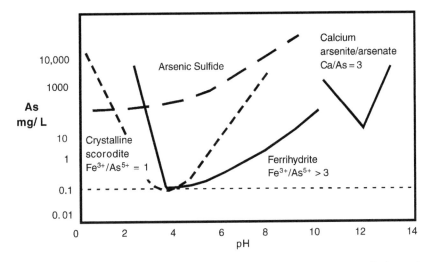

Figure 2 Solubilities of arsenic minerals as a function of pH. At equilibrium, the solubility of the least soluble phase under the prevailing geochemical conditions controls the maximum concentration of arsenic in solution. Adapted from Ref. [74].

ratio is maintained above 4. Arsenic concentrations in equilibrium with ferrihydrite rise steeply as solid-phase Fe/As falls below 4. High As in ferrihydrite inhibits the polymerization of iron oxides, which will mature over time releasing arsenate to solution [37,39].

Lime neutralization which is often used to precipitate heavy metals from solution also produces calcium–arsenate minerals such as arseniosiderite, $Ca_2Fe_3(AsO_4)_3O_2 \cdot 3H_2O$, and yukonite, $Ca_2Fe_3(AsO_4)_4(OH) \cdot 12H_2O$ [63], and arsenate apatite, $Ca_5(AsO_4)_3 \cdot OH$, [65]. The stability of these calcium–arsenate minerals is not well understood, and they may over time convert to calcite in the presence of CO_2 and thus release arsenate [65].

5. ARSENIC IN THE GANGES DELTA

As an example of biogeochemistry of arsenic in groundwater systems, we describe the arsenic contamination problem in the Ganges delta and compare the geochemical characteristics at our field site in central Bangladesh to a national data set. The causes of high arsenic concentrations in the groundwater of this anoxic sedimentary system contrast with the processes described in the previous section that produce high-arsenic groundwater by oxidative dissolution of minerals.

Over the last several decades, much of the population of Bangladesh and West Bengal switched their water supply from surface water to groundwater. As many as 10 million new domestic wells were installed, providing drinking water for over 100 million people. This large-scale transition was motivated by the need for pathogen-free drinking water—diarrheal diseases such as cholera were infecting millions of people—and the transition to well water was readily adopted because of the convenience of water supplies in close proximity to homes and the ease of drilling in the region's high-yielding aquifers. At the same time as the transition of domestic supplies to well water, irrigation wells were also installed across the country. Groundwater pumping for irrigation has greatly increased food production enabling Bangladesh to become self-sufficient in food, even though the population nearly tripled over the last four decades. Irrigation has enabled production of dry-season rice called "Boro" that now provides more yield than the traditional rice grown during the wet season (Fig. 3). Thus, issues of groundwater quality and quantity have become vital for both the supply of drinking water and the production of food in Bangladesh.

Tragically, much of the region's groundwater is dangerously contaminated by arsenic, and consumption of this water has already created severe health effects. About half of the wells in the country have concentrations >10 μg/L, now a common standard. Much of our understanding of the distribution of arsenic across Bangladesh's groundwater comes from the comprehensive work of the British Geological Survey (BGS) [66,67]. Their work shows that high arsenic concentrations may be found throughout the flood plains and delta of the Ganges, Brahmaputra, and Mehgna rivers, but the Delta region of the southern half of the country is the most contaminated. Yu et al. [3] combine the BGS's database with dose–response models to estimate that, if consumption of contaminated water continues, the prevalence of arsenicosis and skin cancer in Bangladesh will be approximately 2,000,000 and 100,000 cases per year, respectively, and the incidence of death from cancer induced by arsenic will be approximately 3000 cases per year. Because detailed health records are not

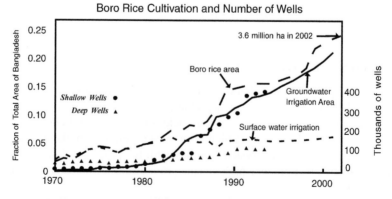

Figure 3 Cultivation of high yielding boro rice has greatly expanded over the last several decades to cover ~25% of Bangladesh, or ~45% of the cultivatable area. Most boro is irrigated by groundwater so extraction has also risen. Adapted from Ref. [74].

kept in Bangladesh, these estimates were made using dose–response curves from the literature. How well these dose–response relationships apply to the broad population of Bangladesh remains an open question and existing epidemiological surveys show a wide spread of arsenicosis prevalence estimates [3].

5.1. Arsenic Mobilization in Ganges-Delta Aquifers

Researchers agree that the high concentrations of arsenic in the groundwater of Bangladesh is not caused by pollution (e.g., not pesticides or industrial pollution) but rather has been mobilized *in situ* from the sediments. However, there is no evidence of widespread levels of solid phase arsenic in the aquifer material that are unusually high—concentrations are typically <10 ppm in sandy sediment and <100 ppm in clays and peats [68]. High concentrations have been reported in the soils of irrigated fields, but these may be the result of arsenic input from groundwater irrigation and sorption to the soils [69,70]. Thus, it appears that high dissolved arsenic concentrations are the result of particular hydrologic and biogeochemical conditions that partition arsenic from the solid to aqueous phase, but have not yet flushed dissolved arsenic from the subsurface.

The reducing conditions of almost all groundwater in Bangladesh (demonstrated by high levels of dissolved ferrous iron and methane, and low measurements of Eh) suggest that most arsenic is liberated by dissolution of iron oxyhydroxides, or perhaps desorption of arsenic after reduction from arsenate to arsenite. The weak but positive correlation of dissolved arsenic to iron and bicarbonate suggest that arsenic may have been liberated concurrently with the reductive dissolution of iron oxyhydroxides by microbial organic carbon oxidation [66,68,71]. The low concentrations of sulfate (and in some areas the negative correlation between arsenic and sulfate) as well as the generally reducing conditions argue against the widespread mobilization of arsenic from sulfide minerals (e.g., [72]). However, it remains a possibility that at the surface, where oxygen is introduced as the water table rises and falls, sulfide minerals could be oxidized and dissolved thereby liberating arsenic. This process would only occur near the surface, and oxide surfaces, also forming under oxic conditions, would likely quickly absorb the mobilized arsenic. The long-term implications of such cyclic processes in a rapidly accreting (~ 1 cm/year) aquifer remain to be studied.

High concentrations of other anions that compete with arsenic for surface sorption sites also contribute to high dissolved arsenic concentrations. Silicate and phosphate are prevalent in groundwater throughout most of the arsenic affected areas, but there is no convincing correlation between these anions and arsenic to suggest that they explain the spatial pattern of dissolved arsenic. Appelo et al. [48] have argued that competition by bicarbonate, which correlates better with arsenic over the country, might explain the distribution of dissolved arsenic. By this scenario oxidation of organic carbon liberates arsenic indirectly through desorption by its byproduct, bicarbonate, rather than directly by reduction of iron oxides or arsenate. However, equilibrium chemical modeling using the parameters measured at our site indicates that the effect of bicarbonate on arsenic sorption is less than that of silicate and no more than phosphate [73]. These conceptual geochemical models are further complicated by the fact that arsenic likely adsorbs to surfaces of many solid phases other than oxyhydroxides, such as magnetite, green rust, and potentially siderite and apatite. Arsenic is known to sorb to magnetite [40] and the results of our density and magnetic separations show that the magnetite fraction of our sediment has the highest arsenic concentration [73].

Figure 4 compares depth profiles of geochemical characteristics measured at our field site in Munshiganj [71,73] with averaged values from the BGS data set within depth intervals chosen by the BGS [67]. Our study site in the Munshiganj district is located 30 km south of Dhaka and 7 km north of the Ganges. It contains a small intensive-study area where 25 sampling wells, located within a 15 m^2 area, extract water from depths ranging between 5 and 165 m of the land surface [intensive-study site in Fig. 6(A) in Section 5.3]. We also monitor water levels at 87 other locations in the surrounding 16 km^2 region. We find surprising agreements between results at our single site and the averaged national data set that suggest the existence of general characteristics of geochemical evolution and transport across the region.

5.1.1. Arsenic as a Function of Depth

Both data sets show peaks of arsenic in the range of 30 m depth. At our site, we find no chemical characteristic of the sediment to explain this pattern [71,73]. However, several types of data, when taken together, suggest a relation between the arsenic peak and groundwater flow patterns. Figure 5(left) shows estimated hydraulic conductivities from pumping tests in each of the wells; Fig. 5(middle) shows relative head differences between the wells and; Fig. 5(right) shows oxygen-18 isotope ratios. The hydraulic conductivity data indicate only mild variations, but suggests that a lower conductivity layer at 22 m may work to separate horizontal flow paths, and the head data (middle part of Fig. 5) show that, at least at some times of the year, there is convergent vertical flow that would mix water from above and below 30 m, thus accelerating horizontal flow. Such mixing is supported by the O-18 data that are consistent with mixing of lighter water from above and heavier water from below at 30 m. The heavier water below 30 m could represent infiltrated pond or river water that has been subject to evaporation. Measurable tritium values are found to 60 m depth [71] indicating the presence of at least a component of water <40 years old throughout this depth interval.

Kinniburg et al. [67] describe the profile of average arsenic concentrations across the country as "bell shaped" (Fig. 4), a pattern that has previously been suggested for both Bangladesh and West Bengal. Although this pattern is evident by averaging over particular depth intervals, and resembles the pattern

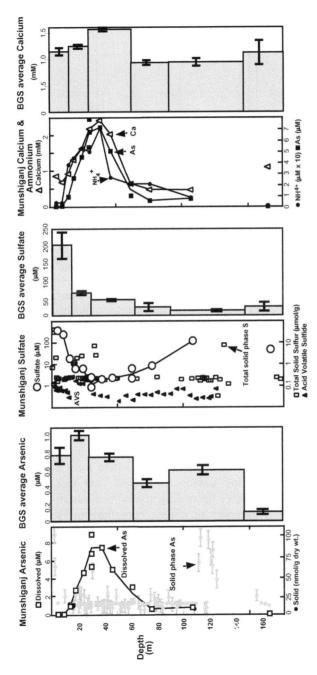

Figure 4 Comparison of vertical profiles of groundwater chemical characteristics at our intensive study site in Munshiganj (yellow triangle in Fig. 6) with average values from the BGS national data set [5,11]. The error bars on the BGS data set represent the sample standard deviation divided by the square root of the sample size.

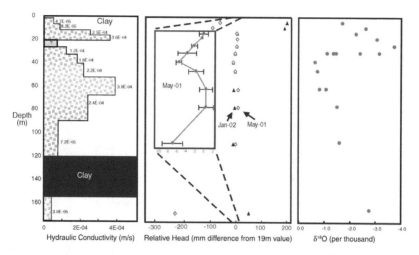

Figure 5 (Left) Horizontal hydraulic conductivity as estimated from pump test at the intensive-study site. The conductivity at 22 m is uncertain because of well silting. (Middle) Hydraulic heads relative to the 20 m piezometer during the dry season, early irrigation (January) and late irrigation (May). The inset represents heads in May measured with a manometer that obtains relative differences to within 1 mm. (Right) Oxygen-18 isotope ratios relative to standard mean ocean water (SMOW).

observed at our field site, there are other locations where the pattern of dissolved arsenic with depth is different [66] and the appearance of bell shape in the smoothed data is not entirely robust with the choice of depth intervals for binning the data. The lower concentration averages near the surface disappear for some other choices of bin intervals, or other moving average algorithms. Depth trends can also be considered within different geologic regions. Yu et al. [3] tabulate the geologic regions of Bangladesh where there is a statistically significant trend of decreasing arsenic with depth, and their geostatistical analysis shows that the trends of decreasing arsenic concentrations with depth explains much of the differences in arsenic concentrations between nearby wells—because the wells withdraw water from different depths the arsenic concentrations differ due to the regional depth trend.

The bell shaped patterns of solutes (arsenic, ammonium, calcium, and dissolved carbon) with depth is typical of vertical profiles of contaminant plumes that originate from surface sources. Such profiles are typically seen on groundwater contamination sites in North America and Europe where release of contaminants at distinct locations contaminates groundwater as laterally migrating plumes. Groundwater flow is primarily horizontal, so plumes quickly move horizontally away from their source, and as recharge enters the aquifer above the contaminated water, these plumes are also pushed deeper into the aquifer. Thus, the vertical component of flow is important for calculating the chemical flux from the surface down into the aquifer, but of course the horizontal component of groundwater flow throughout most of the aquifers is larger, as it must be in any low-topography environment. These basic hydrologic processes create vertical profiles of contamination with this typical bell shape. While it is easy to postulate that the bell-shaped pattern at our site results from a single local source of organic carbon (i.e., a nearby pond, river, or rice field), it remains an open question whether the much less distinct bell-shaped pattern of the national BGS data site results from the same mechanisms. It is possible that such a pattern could also develop from a different sediment geochemistry at the arsenic peak, such as the presence of a peat that could release organic carbon into the groundwater. However, the young radiocarbon ages of dissolved carbon at our site discussed below argue against this hypothesis.

It is intriguing that the bell-shape pattern of arsenic is mirrored by the distribution of well depths (not shown) in the BGS data set. Most wells withdraw water from depths near 25 m. Similarly, at our site in Munshiganj nearly all wells (drinking and irrigation) are completed at the same depth where we find the arsenic peak, 30 m. This correspondence may suggest a hydraulic component to the cause of arsenic mobility with depth. Because installing wells deeper is more expensive, villagers only complete wells to a depth sufficient to provide adequate yield. Thus, a relationship may exist between the depth at which the aquifer is more conductive to groundwater flow, and the depth of maximum arsenic concentration. However, such speculation has not been confirmed by hydrogeologic studies.

5.1.2. Sulfate as a Function of Depth

At our site, the inverse relation of dissolved sulfate with As (Fig. 4), and the presence of acid volatile sulfide (AVS) in the sediments near the dissolved As peak, suggest that arsenic is not currently being liberated from sulfide minerals. Instead, low dissolved sulfur levels appear to limit the precipitation of arsenic sulfides near the arsenic peak.

The BGS data set also shows a distinct pattern of decreasing sulfate with depth. Little is known about the source of this sulfate. However, the distinct pattern of decreasing sulfate with depth from the countrywide data set is consistent with the previously described scenario of sulfide oxidization at the surface, followed by sulfide precipitation at depth. How the genesis of this sulfate pattern may relate to the pattern of arsenic mobilization with depth remains an open question.

5.1.3. Calcium and Ammonium as a Function of Depth

At our site, peaks in ammonium and calcium mirror the sharp peak in arsenic (Fig. 4), and these solutes suggest inflow and oxidation of organic carbon. Ammonium is a product of oxidation of natural organic matter, and calcium may be released from solid carbonate after organic carbon oxidation [73]. The

BGS do not report ammonium for their countrywide data set, but the profile of calcium with depth shows a distinct peak similar to that found at our site.

The correspondence of the depth profiles measured at our site to the average BGS profiles should not be interpreted to mean that the same processes are occurring everywhere. There is great spatial heterogeneity in aqueous chemical characteristics, much of which is likely induced by the complex mosaic of recharge and discharge areas at the surface with variable water chemistry. However, the correspondence does suggest the possibility that some general reactive-transport processes may affect the subsurface biogeochemistry across the country. Below, we discuss possible general characteristics of groundwater flow.

5.2. The Relation of Groundwater Flow and Chemical Transport to Arsenic Concentrations

Relative to the extent of research on biogeochemical processes of arsenic mobilization, little work has been conducted in Bangladesh to understand groundwater flow and how it transports arsenic and the solutes that control arsenic mobility. Much of the existing literature on groundwater flow focuses on isotopic inference, and not on physical understanding of groundwater flow and solute transport. However, several simple lines of reasoning, and some isotope data, indicate that flow and transport play important roles in subsurface arsenic concentrations:

(a) Arsenic concentrations are extremely patchy over small spatial scales. At least in the vertical dimension, high concentrations can be found within tens of meters of low concentrations. This means that, if there is any groundwater flow across these arsenic gradients, arsenic will be transported from areas of high concentrations to areas of low concentration, and low-arsenic groundwater will flow into areas of high concentration. If flow does not cross these arsenic gradients, then this indicates that the spatial pattern of arsenic corresponds to the spatial pattern of groundwater flow paths. In either case, understanding the effects of flow and transport appear important for understanding the behavior of dissolved arsenic.

(b) Measured tritium values indicate that groundwater flow through at least the upper 30 m is often rapid. The largest set of tritium data has been gathered by the International Atomic Energy Agency (IAEA) [76] over the last 30 years, and their final report concludes that groundwater ages in the upper 100 m are generally <100 years. Their recent plot [77] shows a somewhat more complex picture, with tritium values >1 tritium unit (TU) penetrating below 25 m in 1999, but not 1979 prior to greatly increased irrigation pumping. At our site, we also find tritium values above 0.2 to 60 m indicating a component of water <40 years. Clearly some areas of stagnant water exist in the region, either because of low hydraulic conductivity (e.g., clay layers) or because of patterns of local recharge and discharge. However, the common occurrence of tritium in groundwater indicates that groundwater flow is sufficient to rapidly transport arsenic, or solutes that interact with arsenic, through many regions where aquifers are contaminated. The presence of young water also indicates that arsenic is being flushed from the system. Indeed, the combination of these estimated groundwater ages, and the rates of pumping-induced circulation described above, with estimated retardation coefficients for arsenic raise the question of how can such high concentrations of dissolved arsenic remain. Kinniburgh et al. [67] estimate the retardation factor for arsenic to be as low as two. We also find that, where arsenic is high the retardation factors are <10 [71,73]. These values imply a residence time for arsenic of decades to centuries in aquifers that are thousands of years old—so why is the arsenic still there?

(c) Irrigation data (Fig. 3) indicate that, at a national level, pumping alone drives significant groundwater flow. Boro rice requires 1 m or more of irrigation annually [74,75], and Fig. 3 shows that roughly 20% of the country has groundwater irrigated boro. So, assuming a porosity of 20% and 1 m of irrigation, this withdrawal cycles 5 m of groundwater flow annually below rice fields, which averaged over the country gives 1 m of vertical groundwater circulation nationally. Thus, pumping induced groundwater flows reach the depth of the arsenic peak within two or three decades on average. This calculated average travel time is consistent with the tritium data described above which also gives travel times to 30 m on the order of decades, thus providing a possible explanation for why mobile arsenic has not been flushed from the system—the flushing has not been occurring over centuries, but is simply the result of the recent advent of massive irrigation.

Using 1996 irrigation data, Ali [74] estimates that groundwater irrigation applies 1 kg arsenic per hectare of irrigated area each year. Thus, in 2001 (Fig. 3) irrigation pumping would have extracted over a million kilograms of arsenic from the aquifers and moved it into rice fields. (This large flux is still several orders of magnitude less than the flux of arsenic liberated into the environment from acid mine drainage.) Irrigation pumping can then be viewed as analogous to "pump-and-treat" groundwater remediation methods employed in North America and Europe, but without the "treat".

(d) At our site, the radiocarbon data show that detrital organic carbon has not driven recent biogeochemical reactions—the byproducts of microbial activity, both inorganic carbon and methane, have much younger dates than the dissolved organic carbon or the sediment, and the concentration of this inorganic carbon is much larger than that of the older organic carbon. Dissolved carbon with radiocarbon ages much younger than the sediment age [71] must have been transported by flowing groundwater. Tree roots and burrowing animals are unlikely to penetrate below 10 m because the aquifer remains saturated all year. Carbon dating of dissolved inorganic carbon by the BGS shows that it is generally radiocarbon young relative to the sediment age. Since inorganic carbon is the byproduct of organic carbon oxidation, and the sediments are much older than this inorganic carbon, these radiocarbon ages indicate that carbon has been transported into aquifers elsewhere in Bangladesh. McArthur et al. [78] argue that organic carbon has been transported from layers of buried peat, but do not attempt to reconcile the different radiocarbon ages of organic carbon and dissolved inorganic carbon. Although peat deposits certainly exist in many areas of the delta and may have supplied labile dissolved organic carbon to groundwater in the past, the radiocarbon age for the dissolved inorganic carbon, or indeed the organic/inorganic mixture is too young for buried peat at our site. Indeed, at 20 m depth the inorganic carbon has levels of carbon-14 at our site that can only be explained as the results of modern bomb testing.

5.3. The Annual Cycle of Groundwater Flow in Central Bangladesh

Here, we present some hydrologic data that characterizes the annual pattern of groundwater circulation at our field site and indicates several important features of groundwater flow for arsenic mobilization and transport. Figure 6(A) shows a map of a region within our study area including irrigation wells, ponds and a river. The water levels are graphed in Fig. 6(B) and (C) and show that: (1) During monsoon flooding all water levels are nearly the same so groundwater flow essentially ceases. (2) As the flood recedes, but before irrigation begins, the river levels drop much more quickly than groundwater or pond levels. The aquifer is partially confined by the overlying clay, and inflow from the clay appears to maintain the head in the aquifer above the river. (3) When irrigation begins in February, the river stops declining but groundwater decline accelerates so that the piezometric level falls below the river. (Also, the two river channels become disconnected when mud bars become exposed separating the branches.) (4) When monsoon flooding begins in June, all water levels rise rapidly together as both rain and river flooding inundate the land, and the aquifer is recharged by both inflow from the clay above and the river. (5) The ponds, which cover ~15% of the land and are not directly connected to the river, and apparently exchange little with the aquifer as their water levels remain several meters above the water table, and the rate of decline is much slower than the groundwater level.

The drop in water levels in the aquifer after flooding recedes in December can be largely explained by the rate of evapotranspiration [71] which is maintained by irrigation of the rice fields. The groundwater levels recorded every hour at our intensive study area [Fig. 6(B)] show the effects of irrigation, both as a doubling of the rate of drawdown, and as dramatic diurnal oscillations as pumps are turned on during the day, and off at night. The increase in the rate of drawdown occurs simultaneously with the onset of irrigation and is also concurrent with the time that the rivers stop declining (e.g., when the gradient driving flow to the river decreases). Thus, the increased drawdown in January appears to be driven by pumping, not discharge to the river. Indeed, the river changes direction during the second half of December, so that flow is no longer discharging to the Ganges, but rather flows from the Ganges to the field area. Furthermore, the hydraulic gradient between wells perpendicular to the river drives flow away from the river during the irrigation season, even though the water level in the

river is initially below the aquifer heads. The slow decline in pond levels relative to groundwater levels imply that pond-bottom sediments inhibit flow into
the aquifer. However, the head gradients between surface water (rivers and
ponds) and the groundwater must drive some exchange. Finally, the paths by
which recharging water enters the aquifer, as both the rivers rise and heavy
rains begin, is not readily apparent from the water level data in Fig. 6(C),
however it appears that the aquifer is poorly connected with surface water
and that at least a component of the recharge is leakage from the overlying
clay.

Quantifying the source of recharge is a focus of current numerical modeling and is of vital importance because this recharge transports important chemical
loads into the aquifer—recharge introduces modern organic carbon, which may
mobilize arsenic, or oxidants, which may immobilize arsenic, into the aquifer.
Furthermore, the rate of flow along these paths controls how quickly arsenic is
flushed from the aquifer. It is likely that where flow is fast, near the surface or
near irrigation wells, arsenic is being flushed from aquifers, so concentrations
will decrease over time, but deeper in the aquifer arsenic concentrations could
be increasing as organic carbon, or arsenic itself, is transported downward.
Indeed, the "bell-shaped" curve of arsenic with depth may be explained as the
interaction of these two processes. Near the surface arsenic is low because
rapid flow has already begun to flush arsenic from the aquifer, and deeper in
the aquifer arsenic remains high. This explanation for the arsenic profile raises
questions about arsenic concentration changes over the time-scale of aquifer
deposition—is it possible for this "bell shaped" pattern to have existed within
the aquifer over the time-scale of deposition, or does this explanation imply
that the currently observed pattern is transient; the result of the recent advent
of massive irrigation pumping? A simple analysis supports the transient hypothesis. If the near-surface sediment had always been flushed of dissolved arsenic by
groundwater flow, then as the near surface sediment is buried by deposition of
new sediment it would be "cleaned" of both dissolved arsenic and arsenic
that is readily mobilized, so there would not be higher arsenic concentrations
at depth. However, if the rapid flushing of the surface sediment is a new phenom-
ena, then the higher arsenic at depth could be left from past quiescent hydrologic
conditions.

In summary, the comparison of data from our field site with the comprehensive national data set collected by the BGS suggest the existence of some
general vertical patterns of arsenic mobilization and transport. Results from a
variety of research groups [48,66,68,71] suggest that arsenic mobilization is
linked to reductive processes driven by organic carbon, but we still do not
fully understand what causes the spatial patterns of dissolved arsenic found
within the aquifers. Understanding the spatial and temporal patterns of
arsenic mobilization will require better hydrogeologic models that describe
how groundwater flow carries solutes into the subsurface, and flushes solutes
from the subsurface.

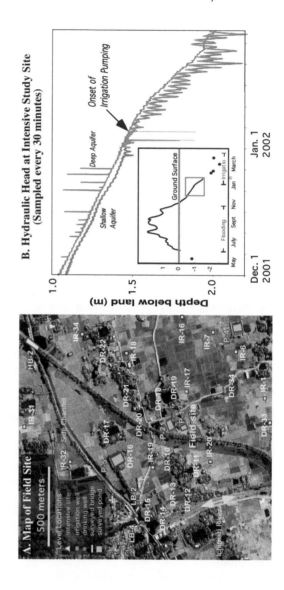

B. Hydraulic Head at Intensive Study Site
(Sampled every 30 minutes)

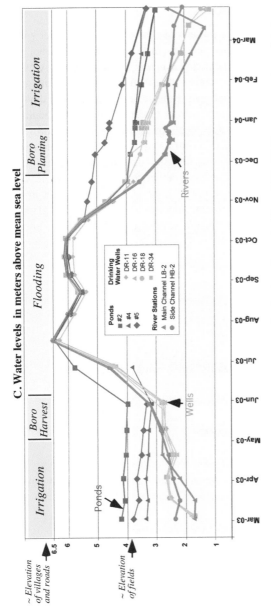

Figure 6 (A) Map of the water level recording locations for the data shown in (B) and (C). (B) Pressure transducer data from the intensive field area during 2001–2002, with the data from December and January expanded to show the effects of irrigation pumping. (C) Hand measured water levels in drinking water wells, ponds, and rivers at locations shown in (A).

6. CONCLUSIONS AND OUTLOOK

Researchers have developed predictive models of arsenic behavior under some controlled conditions (e.g., sorption to pure goethite or ferrihydrite) but we cannot yet accurately predict arsenic behavior *a priori* in many complex environments where hydrologic processes create sharp spatial gradients in geochemical conditions and drive transience over time-scales ranging from days to centuries. These problems will continue to attract research interest, particularly on large-scale problems, such as the high levels of arsenic in the groundwater of Bangladesh. In technically developed countries more efficient methods will be devised to lower arsenic concentrations to new standards and the sequestration of waste arsenic will pose important questions about the long-term stability of adsorption media. The resources necessary to deliver safe drinking water in the developing countries are simply not present, and countries such as Bangladesh will likely continue to receive higher arsenic exposures unless serious efforts are undertaken.

ABBREVIATIONS

ARD	acid rock drainage
AVS	acid volatile sulfide
BGS	British Geological Survey
CAD	Canadian dollars
DPHE	Department of Public Health Engineering (Bangladesh)
Eh	redox potential (volts)
IAEA	International Atomic Energy Agency
IP	ionization potential
MCL	maximum contaminant level
ppb	parts per billion
ppm	parts per million
SMOW	standard mean ocean water, that standard for oxygen-18 measurements
TRI	toxic release inventory
TU	tritium unit
USEPA	United States Environmental Protection Agency

REFERENCES

1. Yamauchi H, Fowler BA. In: Nriagu JO, ed. Arsenic in the Environment, Part II. Human Health and Ecosystem Effects. New York: Wiley, 1994:35–54.
2. Smith AH, Lopipero P, Bates M, Steinmaus C. Science 2002; 2145–2146.
3. Yu W, Harvey CM, Harvey CF. Water Resour Res 2003; 39:1146.
4. National Research Council, Arsenic in Drinking Water, 1999.
5. Cullen WR, Reimer KJ. Chem Rev 1989; 89:713–764.

6. Nriagu JO. Arsenic in the Environment. In: Advances in Environmental Science and Technology. Vol. 26. Wiley Interscience, 1994.
7. Jain CK, Ali I. Water Res 2000; 34:4304–4312.
8. Matschullat J. Sci Total Environ 2000; 249:297–312.
9. Mandal BK, Suzuki KT. Talanta 2002; 58:201–235.
10. Smedley PL, Kinniburgh DG. Appl Geochem 2002; 17:517–568.
11. Nicholas DR, Ramamoorthy S, Palace V, Spring S, Moore JN, Rosenzweig RF. Biodegradation 2003; 14:123–137.
12. Han FX, Yi Su, Monts DL, Plodinec MJ, Banin A, Triplett GE. Naturwissenschaften 2003; 90:395–401 (DOI: 10.1007/s00114-003-0451-2).
13. Welch AH, Stollenwerk KG, eds. Arsenic in Ground Water. Norwell, MA: Kluwer, 2003:475.
14. Bissen M, Frimmel FH. Acta Hydrochim Hydrobiol 2003; 31:9–18.
15. Boyle RW, Jonasson IR. J Geochem Explor 1973; 2:251–296.
16. Grosz AE, Grossman JN, Garrett R, Friske P, Smith DB, Darnley AG, Vowinkel E. Appl Geochem 2004; 19:257–260.
17. Williams M. Environ Geol 2001; 40:267–278.
18. Langmuir D. Aqueous Environmental Chemistry. Englewood Cliffs, NJ: Prentice Hall, 1997.
19. Nordstrom DK, Archer DG. Chapter 1. In: Welch AH, Stollenwerk KG, eds. Arsenic in Ground Water. Norwell, MA: Kluwer, 2003.
20. McBride BC, Wolfe RS. Biochemistry 1971; 10:4312–4317.
21. Mackenzie FT, Lantzy RJ, Paterson V. J Int Assoc Math Geol 1979; 11:99–142.
22. Guilbert JM, Park CF. The Geology of Ore Deposits. New York: W. H. Freeman, 1986.
23. Glanzman RK. U.S. EPA Workshop on Managing Arsenic Risks to the Environment: Characterization of Waste, Chemistry, Treatment and Disposal, Denver, CO, May 1–3, 2001.
24. Nordstrom DK. U.S. EPA Workshop on Managing Arsenic Risks to the Environment: Characterization of Waste, Chemistry, Treatment and Disposal, Denver, CO, May 1–3, 2001.
25. Fleet M, MacLean P, Barbier J. In: Keays R, Ramsay W, Groves DI, eds. The Geology of Gold Deposits: The Perspectives in 1988. Vol. 6. Econ Geol Monogr 1989:356–362.
26. Morse JW, Luther GW. Geochim Cosmochim Acta 1999; 63:3373–3378.
27. Savage KS, Tingle TN, O'Day PA, Waychunas GA, Bird DK. Appl Geochem 2000; 15:1219–1244.
28. Berner RA. Am J Sci 1987; 287:177–198.
29. Rickard D, Schoonen MA, Luther GW. Chapter 9. In: Vairavamurthy MA, Schoonen MAA, eds. Geochemical Transformations of Sedimentary Sulfur. Vol. 612. ACS Sym Ser 1995:168–193.
30. Pokrovski GS, Kara S, Roux J. Geochim Cosmochim Acta 2002; 66:2361–2378.
31. King RJ. Geol Today 2002; 18:72–75.
32. Craw D, Falconer D, Youngson JH. Chem Geo 2003; 199:71–82.
33. Rittle KA, Drever JI, Colberg PJS. Geomicrobiol J 1995; 13:1–11.
34. Newman DK, Beveridge TJ, Morel FMM. Appl Environ Microbiol 1997; 63:2022–2028.
35. Huber G, Sacher M, Vollman A, Huber H, Rose D. Syst Appl Microbiol 2000; 23:305–314.

36. Stollenwerk KG. Chapter 3. In: Welch AH, Stollenwerk KG, eds. Arsenic in Ground Water. Norwell, MA: Kluwer, 2003.
37. Waychunas GA, Rea BA, Fuller CC, Davis JA. Geochim Cosmochim Acta 1993; 57:2251–2269.
38. Fuller CC, Davis JA, Waychunas GA. Geochim Cosmochim Acta 1993; 57:2271.
39. Ford RG. Environ Sci Technol 2002; 36:2459–2463.
40. Dixit S, Hering J. Environ Sci Technol 2003; 37:4182–4189.
41. Manning BA, Goldberg S. Environ Sci Technol 1997; 31:2005–2011.
42. Su C, Puls RW. Environ Sci Technol 2001; 35:4562–4568.
43. Pierce ML, Moore CB. Water Res 1982; 16:1247–1253.
44. Redman AD, Macalady DL, Ahmann D. Environ Sci Technol 2002; 36:2889–2896.
45. Swedlund PJ, Webster JG. Water Res 1999; 33:3413–3422.
46. Jain A, Loeppert RH. J Environ Qual 2000; 29:1422–1430.
47. Kim M-J, Nriagu J, Haack S. Environ Sci Technol 2000; 34:3094–3100.
48. Appelo CAJ, Vanderweiden MJJ, Tournassat C, Charlet L. Environ Sci Technol 2002; 36:3096–3103.
49. Meng X, Bang S, Korfiatis GP. Water Res 2000; 34:1255–1261.
50. Arai Y, Sparks DL, Davis JA. Environ Sci Technol 2004; 38:817–824.
51. Postma D, Jakobsen R. Geochim Cosmochim Acta 1996; 60:3169–3175.
52. Wu X, Delbove F. Econ Geol 1989; 64:420–423.
53. Jambor JL, Blowes DW, Ritchie AIM, eds. Environmental Aspects of Mine Wastes. Short-Course Vol. 31. Ottawa: Canadian Mineralogical Association, 2003:415.
54. McGuire MM, Edwards KJ, Banfield JF, Hamers RJ. Geochim Cosmochim Acta 2001; 65:1243–1258.
55. Nesbitt HW, Muir IJ, Pratt AR. Geochim Cosmochim Acta 1995; 59:1773–1786.
56. Nordstrom DK, Alpers CN. Proc Natl Acad Sci USA 1999; 96:3455–3462.
57. Thompson NA, Schultz SR. U.S. EPA Workshop on Managing Arsenic Risks to the Environment: Characterization of Waste, Chemistry, Treatment and Disposal Denver, CO, May 1–3, 2001.
58. McCreadie H, Blowes DW, Ptacek CJ, Jambor JL. Environ Sci Technol 2000; 34:3159–3166.
59. Martin AJ, Pedersen TF. Environ Sci Technol 2002; 36:1516–1523.
60. Langmuir D, Mahoney J, MacDonald A, Rowson J. Geochim Cosmochim Acta 1999; 63:3379–3394.
61. Donahue R, Hendry MJ, Landine P. Appl Geochem 2000; 15:1097–1119.
62. Moldovan BJ, Jiang T, Hendry MJ. Environ Sci Technol 2003; 37:873–879.
63. Paktunc D, Foster A, Heald S, Laflamme G. Geochim Cosmochim Acta 2004; 68:969–983 (doi: 10.1016/j.gca.2003.07.013).
64. Richmond WR, Loan M, Morton J, Parkinson GM. Environ Sci Technol 2004; 38:2368–2372.
65. Bothe J, Brown PW. Environ Sci Technol 1999; 33:3806–3811.
66. British Geological Survey (BGS) and Department of Public Health Engineering (DPHE, Bangladesh). In: Kinniburgh DG, Smedley PL, eds. Arsenic Contamination of Groundwater in Bangladesh, Vol. 1–4, British Geologic Survey Report WC/00/19, (http://www.bgs.ac.uk/arsenic/Bangladesh), 2001.
67. Kinniburgh DG, Smedley PL, Davies J, Milne CJ, Gaus I, Trafford JM, Burden S, Ihtishamul Huq SM, Ahmad N, Ahmad MK. The scale and causes of the groundwater arsenic problem in Bangladesh. In: Welch AH, Stollenwerk KG, eds. Arsenic in

Ground Water: Geochemistry and Occurrence. Boston, MA: Kluwer Academic Publishers, 2003: 211–257.

68. Nickson R, McArthur J, Burgess W, Ahmed KM, Ravenscroft P, Rahman M. Nature 1998; 395:338.
69. Abedin MJ, Cotter-Howells J, Meharg AA. Plant Soil 2002; 240:311–319.
70. Maharg AA, Rahman M. Environ Sci Technol 2003; 37:229–234.
71. Harvey CF, Swartz C, Badruzzman ABM, Keon-Blute N, Yu W, Ali MA, Jay J, Beckie R, Niedan V, Brabander D, Oates P, Ahsfaque K, Islam S, Hemond H, Ahmed MF. Science 2002; 298:1602–1606.
72. Chowdhury TR, Basu GK, Mandal BK, Biswas BK, Samanta G, Chowdhury UK, Chand CR, Lodh D, Roy SL, Saha KC, Roy S, Kabir S, Quamruzzaman Q, Chakraborti D. Nature 1999; 401:545.
73. Swartz CH, Keon NE, Badruzzman B, Ali A, Brabander D, Jay J, Islam S, Hemond HF, Harvey CF. Geochem Acta 2004; in press.
74. Ali MA. In: Fate of Arsenic in the Environment, Ahmed MF, ed., Arsenic Contamination: Bangladesh Perspective. ITN-Bangladesh, June, 2003.
75. Shankar B, Halls A, Barr J. Int J Water 2004 (accepted).
76. Aggarwal PK, Basu AR, Poreda RJ. Isotope Hydrology of Groundwater in Bangladesh: Implications for Characterization and Mitigation of Arsenic in Groundwater, IAEA-TC Project Report: BGD/8/016, IAEA, Vienna, 2002.
77. Aggarwal PK, Basu AR, Kulkarni KM. Science 2003; 300:584.
78. McArthur JM, Ravenscroft P, Safiullah S, Thirlwall MF. Water Resour Res 2001; 37:109–117.

7

Anthropogenic Impacts on the Biogeochemistry and Cycling of Antimony

William Shotyk, Michael Krachler, and Bin Chen

Institute of Environmental Geochemistry, University of Heidelberg, Im Neuenheimer Feld 236, D-69120 Heidelberg, Germany

1. INTRODUCTION

Antimony is a potentially toxic trace element whose environmental significance clearly outweighs the attention it has received to date. No biological function has yet been found for Sb, and its toxicity to animals is comparable to that of As and Pb. Like Pb, Sb is a cumulative poison [1]. In fact, there are so many similarities between the anthropogenic geochemical cycles of Sb and Pb that the environmental importance of Sb in many ways is comparable to that of Pb [2]. However, unlike the geochemical cycle of Pb which has been exhaustively studied and is comparatively well understood [3], far less is known about the environmental fate and significance of Sb [4,5] and the impact of human activities on the geochemical cycle of Sb is poorly known. The summary presented here is intended to illustrate the environmental significance of Sb, and to identify some of the most significant gaps in our knowledge about the geochemical cycle of this element.

2. CHEMISTRY AND BEHAVIOR IN THE ENVIRONMENT

Antimony ($Z = 51$, atomic weight 121.75) lies between As and Bi in the periodic table, with the electronic configuration [Kr] $4d^{10}5s^25p^3$. Antimony commonly exhibits four formal oxidation states: (V), (III), (0) and ($-$III) with the following radii: Sb^{5+} 62 pm, Sb^{3+} 89 pm, Sb^0 182 pm [6] and Sb^{3-} 245 pm [7]. Elemental Sb is a blue-white, brittle metalloid which melts at 630.5°C and has a density of 6.69 g cm^{-3}. Antimony readily loses either three or five electrons, yielding Sb^{3+} and Sb^{5+}; while the former may be stable in aqueous solutions [8], the latter is certainly not. Upon hydrolysis, $[Sb(OH)_3]^0$ and $[Sb(OH)_6]^-$, respectively, are formed; these are the predominant aqueous species within the Eh and pH range of most natural waters (Fig. 1). While Sb(V) also forms polynuclear complexes [9], these are only significant in solutions containing elevated Sb concentrations (i.e., $>10^{-2}$ M). The Sb(III) species, $[Sb(OH)_3]^0$, has no electrical

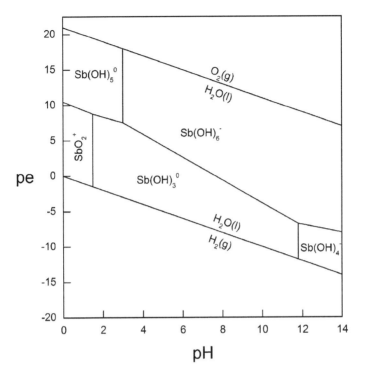

Figure 1 pe–pH diagram for Sb–H_2O system, at 25°C. The outlined regions indicate the conditions under which the given Sb solution species is predominant [9].

charge which allows it to pass more readily across cell membranes; this might help to explain why Sb(III) tends to be more toxic than Sb(V).

In zones of active bacterial sulfate reduction such as anoxic sediments, sulfide complexes may dominate the speciation of Sb(III), with the pH-dependence of $[H_2Sb_2S_4]^0$, $[HSb_2S_4]^-$, and $[Sb_2S_4]^{2-}$ shown in Fig. 2. All of the known Sb solids are fairly soluble. At any given value of Eh and pH, the concentration of total dissolved Sb in equilibrium with Sb_2O_3 is ~$10^{-4.2}$ M; the solubility of Sb_2O_5 is approximately four orders of magnitude higher (Fig. 3). As a consequence, the behavior and fate of Sb in soils and sediments may be independent of precipitation/dissolution reactions, and controlled by adsorption/desorption reactions.

Given the redox potential of most soils, therefore, Sb(V) should be the most important oxidation state and the migration of Sb is likely to be controlled by the adsorption of $[Sb(OH)_6]^-$. The most important sorbent phases are probably those with a net positive charge in the pH range of soils and waters especially the oxides and hydroxides of Al and Fe [10]. In fact, Goldschmidt showed long ago that Sb is markedly enriched in iron oxide ores (tens to hundreds of $\mu g\ g^{-1}$) indicating the importance of adsorption processes for the natural geochemical cycle of Sb [11]. Similarly, Boyle and Jonasson [12] found up to 580 $\mu g\ g^{-1}$ Sb in lacustrine Fe

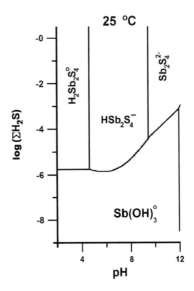

Figure 2   Plot of log $M_{(H_2S+HS^-)}$ vs. pH, showing the field of predominance of Sb-bearing species [156].

and Mn oxide formations and report Sb concentrations as high as 2440 µg g^{-1} in limonite. Given the association of Sb with oxides of Fe, and Mn, ferrous metallurgy could be an important source of Sb to the environment.

A summary of the most important solid, liquid, and gaseous Sb species in the environment is given in Table 1. In addition to chemical and physical processes affecting inorganic Sb species, stibine (SbH$_3$) can be produced under strongly reducing conditions and in geothermal systems [13]: this gas is both volatile and toxic. The fungus *Scopulariopsis brevicaulis* has been found to methylate Sb from Sb$_2$O$_3$ and Sb$_2$O$_5$ under oxic conditions, producing trimethylstibine, Sb(CH$_3$)$_3$ [14]. Subsequent studies have found that Sb(CH$_3$)$_3$ is emitted from anoxic soils, sediments, and landfills [15].

Although Sb has 40 isotopes, the only two stable isotopes are [121]Sb (57.213% of natural abundance) and [123]Sb (42.787%). Analyses of the isotopic composition of Sb in a range of geological materials using multicollector ICP

Table 1   Summary of the Most Important Solid, Aqueous, and Gaseous Sb Species in the Environment

Solid	Sb, Sb$_2$S$_3$, Sb$_2$O$_3$, Sb$_2$O$_5$
Aqueous (oxic)	[Sb(OH)$_6$]$^-$, [Sb(OH$_5$)]^0
Aqueous (sub-oxic)	[Sb(OH)$_3$]^0
Aqueous (sulfidic)	[H$_2$Sb$_2$S$_4$]^0, [HSb$_2$S$_4$]$^-$, [Sb$_2$S$_4$]$^{2-}$
Gas	[SbH$_3$]^0, [Sb(CH$_3$)$_3$]^0

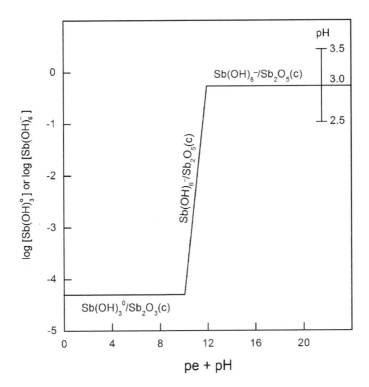

Figure 3 The effects of redox potential (pe + pH) on the solubility of $Sb_2O_3(c)$ and $Sb_2O_5(c)$ and the nature of dominant aqueous species [9].

mass spectrometry (MC-ICP-MS) has recently suggested that there is a significant fraction of Sb isotopes in nature [16]. Using gas chromatography coupled to MC-ICP-MS, Wehmeier et al. [17] showed that biological methylation of Sb may lead to considerable differences in the isotopic composition of trimethylstibine. In the future, therefore, the isotopic composition of Sb may become a useful way to "fingerprint" the predominant sources, pathways, and fate of Sb in the environment. One of the fallout radionuclides created by the detonation of atomic bombs and the nuclear reactor accident at Chernobyl is [125]Sb; with a half-life of 2.73 years, the brief occurrence of this species has been very helpful for studying the fate of Sb in lakes [18]. Two other artificial nuclides, [122]Sb ($t_{1/2} = 2.75$ days) and [124]Sb ($t_{1/2} = 60.2$ days) are used in neutron activation analyses [19].

3. ABUNDANCE AND OCCURRENCE IN THE ENVIRONMENT

The abundance of Sb in the Earth's crust is ~ 0.3 $\mu g\ g^{-1}$, with no significant difference between the upper and lower continental crust [20]. Although there is no marked preference for intrusive mafic or silicic rocks [21], volcanic rocks show a systematic increase in Sb concentrations from mafic to felsic types, from <1 to ~ 8 $\mu g\ g^{-1}$ [22]. The Chondritic meteorites contain comparable concentrations as crustal rocks, with slightly higher Sb concentrations found in iron meteorites [21]; Sb is most abundant in meteorites containing sulfide phases which attests to its strongly chalcophilic character [22]. In rock forming minerals, both Sb(III) and Sb(V) can substitute for Fe(II) and Fe(III). Antimony is found in seven silicate minerals, but all are rare [22]. According to Bowen [23], the concentration of Sb in soils is typically 1 $\mu g\ g^{-1}$. In both seawater and freshwaters, a typical concentration of Sb is 0.2 $\mu g\ L^{-1}$ [23].

There are 113 known Sb-bearing sulfide, antimonide, arsenite and telluride minerals, and 35 oxide, hydroxide, and halide minerals [22]. While Sb certainly does occur as the native metal, most often associated with silver ores, the main Sb ore mineral is stibnite, Sb_2S_3, which is dark gray. Antimony, however, is found in all sulfide minerals where it is often highly enriched [12]. For example, orpiment (As_2S_3) has been found to contain up to 1.5% by weight Sb, galena (PbS) up to 8000 $\mu g\ g^{-1}$, sphalerite (ZnS) up to 3800 $\mu g\ g^{-1}$, arsenopyrite (FeAsS) 1160 $\mu g\ g^{-1}$, pyrite (FeS) 1000 $\mu g\ g^{-1}$, and chalcopyrite ($CuFeS_2$) 180 $\mu g\ g^{-1}$ [12]. The greatest reported concentrations of Sb associated with occurrences of native metals are 100 $\mu g\ g^{-1}$ with Au, 1000 with Cu, 2000 with Bi, 5000 with Ag, and 90,000 $\mu g\ g^{-1}$ Sb with As [12]. Anglesite ($PbSO_4$) which is a valuable lead ore, has been found to contain up to 1000 $\mu g\ g^{-1}$ Sb. Pitchblende, a massive form of uraninite (UO_2) found in hydrothermal, sulfide-bearing veins, has been reported to contain up to 800 $\mu g\ g^{-1}$ Sb. Gold tellurides have been found to contain up to 8% by weight Sb, and bismuthinite (Bi_2S_3) up to 8.8% [12]. Not surprisingly then, Sb is recovered as a by-product of both primary and secondary Pb and Cu refining, as well as from silver and gold ores [12]. The melting point of Sb_2S_3 is only 548°C and that of Sb_2O_3 656°C [24]. The history and intensity of Sb environmental contamination, therefore, must be to a large extent linked to the recovery of metals from sulfide minerals, especially the production and refining of As, Pb, Zn, Cu, and Bi, but also Ag, Au, and U.

While the abundance of Sb in sulfides reflects its chalcophilic character [25], Sb(V) is less chalcophilic than Sb(III) [24] allowing Sb to occur in many other forms. Other minerals bearing Sb as a major constituent include the antimonides, sulfosalts, oxides, as well as a few antimonates and antimonites. A complete list of Sb-bearing minerals is given by Wedepohl [26]. With respect to commercial production of Sb, after stibnite, other important Sb minerals are kermesite, Sb_2S_2O; senarmontite, Sb_2O_3 (cubic); valentinite, Sb_2O_3 (orthorhombic); cervantite, Sb_2O_4; jamesonite, $Pb_4FeSb_6S_{14}$; boulangerite, $5PbS \cdot 2Sb_2S_3$, as well as the sulfantimonides of Cu, Ag, and Ni. Senarmontite and valentinite were found to be the dominant forms of particulate Sb released to the atmosphere by a metallurgical factory in Belgium [27].

According to the United States Geological Survey (USGS), global Sb production (2003) amounts to 142,000 tonnes per year, with most produced by China

(88%), South Africa (4%), Russia (3%), Tajikistan (2%), and Bolivia (2%). The important mining districts in China are in the provinces of Hunan, Kwangsi, Kweichow, Kwantung, and Szechuan [28]. Other producers of Sb in the past include the US (where Sb was recovered as a by-product of silver mining), Italy, Peru, Mexico, and France [6]. In the past, several former states (Czechoslovakia, Yugoslavia) were also important Sb producers. In Canada, Sb is recovered from Pb–Zn smelting and refining at Trail, BC, and Bathurst, NB as well as from secondary Pb refiners in Toronto and Montreal which recycle lead acid batteries.

Antimony is also commonly enriched in coal [29]. Up to 3000 μg g^{-1} Sb has been found in the ash of some German coals, but the "worldwide average" concentration of Sb in coals given by Valkovic [30] is 3 ppm. Considering the amount of mineral matter in coals (typically below 10%) and the abundance of Sb in the Earth's crust (0.3 μg g^{-1}), Sb certainly has a tendency to be enriched in coals. Again, the history and intensity of Sb emissions to the environment must also have a link to the combustion of fossil fuels. Antimony is also enriched in black shales, with the famous Kupferschiefer containing up to 90 μg g^{-1} [12].

4. BIOLOGICAL CHEMISTRY AND TOXICITY

In general, only limited information is available to accurately assess the impact of Sb on human health. Similar to other elements, the toxicological and physiological behavior of antimony depends on its oxidation state, the presence of potential ligands as well as on the solubility of the Sb compound [5,31]. Elemental Sb is more toxic than its salts and generally trivalent Sb compounds exert a 10 times higher toxicity than pentavalent Sb species. Sb(III) shows a high affinity for red blood cells and sulfhydryl groups of cell constituents, while erythrocytes are almost impermeable to pentavalent Sb.

Potassium antimony tartrate [Sb(III)] was found to be an inhibitor of gluthathione-S-transferase (GST) from human erythrocytes, whereas sodium stibogluconate [Sb(V)] had no effect on GST [32]. Based on these inhibition characteristics and the aforementioned preferential accumulation of Sb(III) in mammalian erythrocytes, Poon and Chu [32] deduced that in the case of high Sb intake, for example, during therapeutic treatment of leishmaniasis, Sb(III) concentrations in erythrocytes may be high enough to depress GST activity, which might compromise the ability of erythrocytes to detoxify electrophilic xenobiotics. Sb(III) oxide by inhalation has shown to cause lung cancer in rats [33]. The International Agency for Research on Cancer (IARC) has assigned antimony trioxide to the group of substances which are suspected of being carcinogenic in humans [31]. There is evidence that in mammals antimony, unlike arsenic, is not detoxified via methylation, but it still remains unclear what mechanism is responsible for antimony's genotoxicity [33]. Compared to As(III), Sb(III) proved to be five times less cytotoxic and one order of magnitude less potent in induction of micronuclei in human lymphocytes *in vitro* [34]. It was also proposed that Sb(III) and As(III) cause DNA damage by inhibition of enzymes involved in DNA repair [34].

5. TOTAL ANTIMONY DETERMINATION

5.1. Analytical Methods

The concentrations of total Sb found in unpolluted environmental samples are often very low ranging from 0.2 μg L^{-1} in natural waters to 1.0 μg g^{-1} in soil [23] with even lower values found in human tissue [35]. Methodologies for the environmental analysis of total antimony including speciation analyses of Sb have recently been reviewed by Nash et al. [36]. Therefore, this section mainly focuses on recent developments of analytical techniques for the determinations of total Sb. During the last 7 years, over 70 publications about the determination of total antimony in a variety of samples employing different analytical approaches have been reported. The most commonly documented detection systems for the quantification of Sb include: atomic absorption spectrometry (AAS), inductively coupled plasma-mass spectrometry (ICP-MS), atomic fluorescence spectrometry (AFS) and inductively coupled plasma-optical emission spectrometry (ICP-OES). To improve detection limits needed for environmental samples which often contain very low Sb concentrations, the coupling of hydride generation (HG) with AAS, ICP-MS, AFS, and ICP-OES techniques has been widely reported.

Even though AAS is not the most sensitive detection system, it is the most frequently documented technique for the determination of total Sb. Graphite furnace atomic absorption spectrometry (GFAAS) has been used to determine Sb in natural water samples [37–40], soils [41] and plant samples [42,43]. The detection limits (LOD) of GFAAS are in the low μg L^{-1} level, while employing online pre-concentration methods or separation techniques, LOD of GFAAS could be improved by 1–2 orders of magnitude. Thus, LOD down to 30 ng L^{-1} in tap water and snow water samples with selective sorption of Sb species on Polyorgs-31 [37], and 50 ng L^{-1} in river and rain water using tantalum wire for Sb adsorption [40] have been reported.

Even lower LOD, however, could be obtained when GFAAS or FAAS (flame AAS) is coupled to HG. The advantages of HG-AAS include: (a) separation of the analyte from the matrix, reducing matrix interferences during the measurement; (b) selective pre-concentration of the analyte; (c) selective production and analysis of analyte; (d) high sample introduction efficiency. The lowest LOD for Sb by HG-AAS so far reported, amounts to 10 ng L^{-1} for plant samples [44] and 21 ng L^{-1} for biological tissues [45], respectively. A new method, derivative hydride generation atomic absorption spectrometry (DHGAAS), provides a LOD of 20 ng L^{-1} in water samples [46].

In the continuous search for reliable and more powerful techniques for the detection of Sb at ultra low levels (ng L^{-1} or lower), ICP-MS is well known for its high sensitivity and ability to rapidly simultaneously determine many elements. Richter et al. [47] reported a LOD of 5 ng L^{-1} for cloud water while Delves et al. [48] measured Sb in several biological tissues with a LOD of 27 ng L^{-1}. Using double focusing ICP-MS, Barbante et al. [49] lowered the LOD down to 0.3 ng L^{-1}. More recently, Krachler et al. [50] adopted strict clean room procedures, using ICP-sector field mass spectrometry (ICP-SMS) for the determination of 19 elements in polar ice samples and obtained a LOD of 0.07 ng L^{-1} for Sb, which is by far the most sensitive method that has ever been reported. Coupling with hydride generation or pre-concentration methods, however, conventional ICP-MS can provide LOD as low as 2.5 ng L^{-1} [51] and 0.7 ng L^{-1} [52] in natural water samples. However, possibly because less expensive alternatives exist, fewer than 10 publications employing ICP-MS for Sb analysis were published during the seven past years. So far, reported LOD have not substantially improved and ICP-MS has still not proved to be sufficiently robust for the determination of Sb in complex matrices [53].

More recently, AFS coupled to hydride generation (HG-AFS) received increasing attention and has been widely utilized in geochemistry [54–58], environmental monitoring [59–62], food industry [63–65], agriculture [43,66], medicine [67–70], and biology [71–73]. This technique provides the lowest LOD for hydride forming elements (Sb, As, Se, etc.) and are comparable to those of ICP-MS. Many authors have demonstrated the advantages of HG-AFS such as wide dynamic range, less interferences and high sensitivity for the determination of Sb. Cava-Montesinos et al., for example, obtained a LOD of 3 ng L^{-1} for the determination of Sb in milk samples [74]. However, remarkably different LOD have also been observed, ranging from 80 ng L^{-1} by Sayago et al. [75] and 800 ng L^{-1} by De Gregori et al. [43] of Sb(III), respectively, during antimony speciation analysis. These differences are partly caused by the application of different atomizers and light sources, but also the HG procedure itself can affect the sensitivity of the analytical method and requires careful optimization. In order to investigate the anthropogenic impact of Sb in ancient peat samples, Chen et al. [54] have discussed and optimized a HG-AFS procedure with a detection limit of 8 ng L^{-1} using the same HG-AFS system as in Refs. [43,75].

ICP-OES enables rapid multi-element analysis with large dynamic calibration ranges, but its detection capability can nowadays hardly meet the LOD required for Sb analysis of environmental samples. Vaisanen et al. measured Sb in lead pellets by ICP-OES with a LOD of 0.4 mg L^{-1} [76]. Coupling ICP-OES with HG, Rigby and Brindle [77] measured Sb in synthetic solutions with a LOD of 2.0 μg L^{-1}, Feng et al. [78] determined Sb in marine sediment with a LOD of 0.8 μg L^{-1}. As mentioned earlier, Sb concentrations in soils are typically ~1.0 μg g^{-1}, and in natural waters ~0.2 μg L^{-1}. Therefore, ICP-OES cannot be helpful for measuring Sb in most environmental samples.

In their review, Nash et al. have also summarized the less common analytical techniques for the determination of total Sb in aqueous samples [36]. Since

1997, a few other techniques have been employed including: anodic stripping voltammetry (ASV), cathodic stripping voltammetry (CSV), differential pulse polarography (DPP), X-ray fluorescence spectrometry (XRF) [79], hydride generation coupled with microwave induced plasma-optical emission spectrometry (MIP-OES) [80,81] and neutron activation analysis (NAA). Because of their insufficient sensitivity, they have now almost completely been replaced by more popular techniques such as AAS, ICP-MS, and AFS. Chen et al. [54], for example, have also elucidated in their work that NAA is not appropriate for the determination of Sb at the low ng g^{-1} level in ancient peat samples. While NAA avoids the need to digest the samples, it requires relatively large amounts of sample (8–10 g) which is, in many cases, too wasteful to be used for valuable, archival samples such as ancient peats [54,82]. In an advanced study, Pacquette et al. investigated the combination of HG with laser induced fluorescence (HG-LIF) and obtained a high sensitivity for Sb determination $\sim$0.3 ng L^{-1} [83].

5.2. Sample Preparation Procedures

For the determination of Sb in solid samples, different digestion approaches to dissolve the sample matrix are required and these vary with the kind of sample studied. Depending on the composition of the sample matrix, different acid mixtures and digestion systems are necessary. For the determination of Sb in metals, HNO$_3$ alone with open vessel digestion on hot-plate are often utilized [76,80,84]. To release Sb quantitatively from biological samples which are of high organic content, a microwave-assisted digestion procedure using HNO$_3$–H$_2$O$_2$ mixtures are widely applied and reported to be effective [74,85,86]; when using an open vessel digestion procedure, stronger mixtures of oxidizing acids (HNO$_3$–H$_2$SO$_4$–HClO$_4$) are often required [45]. However, the digestion solutions of plant and geological samples such as soils, sediments, and peat samples only treated with HNO$_3$ always contained a precipitate of silicates, which was found to be a serious interference for Sb determination using HG-AAS [44]. When silicates are not specifically attacked, low recoveries or high relative standard deviations for Sb are found in reference materials containing high concentration of silicates. Krachler et al. [87] developed three different digestion procedures for dissolving peat materials which comprise very resistant natural polymers of both organic (e.g., lignin) and inorganic (e.g., silicate) content. To decompose both fractions and release Sb quantitatively, hydrofluoric acid (HF) or tetrafluoroboric acid (HBF$_4$) together with oxidizing acids is needed. In fact, HBF$_4$ is preferable to HF because both HBF$_4$ and HF have a nearly identical potential for dissolving silicates, and HBF$_4$ is less dangerous than HF.

When using HG techniques, antimony should be in the most reactive state, i.e., Sb(III). Thus the quantitative pre-reduction of Sb(V) to Sb(III) is necessary because Sb(V) is the dominant species under oxidizing conditions, i.e., in the digestion solution. It was also found that low concentrations of Sb(III) can be rapidly converted to Sb(V) in high purity water [88]. Experiments [54] revealed that freshly prepared Sb(III) standard solutions provided nearly the same sensitivity as Sb(V) which have been pre-reduced by L-cysteine, whereas a 10 μg L^{-1} Sb(III) standard solution which was prepared 5 days earlier showed a similar low sensitivity as Sb(V). Thus it is important to realize that a pre-reduction of Sb(III) standard solutions is also necessary in order to obtain the correct calibration results and sufficient sensitivity, and the quantitative pre-reduction of Sb(V) to Sb(III) increased the sensitivity by seven times when the HG-AFS technique was used. Pre-reduction reagents including potassium iodide/ascorbic acid, L-cysteine and thiourea are frequently reported to be effective. The most common pre-reduction involves the addition of KI/ascorbic acid reagent, for which the Sb pre-reduction is immediate [89].

However, it should be noted that KI/ascorbic acid requires daily preparation in relatively high acidic media (>2 mol L^{-1}). The concentration of KI which is effective for pre-reduction is normally $>5\%$ [38,44–46,82,90]. Thiourea is often used when simultaneous hydride generation methods are undertaken. Thiourea serves not only as a pre-reductant, but also a masking agent in eliminating the interference of copper [80]. Compared to KI/ascorbic acid and thiourea, L-cysteine is a stronger chelating reagent which reacts with interfering ions, especially Fe, Ni, Co, and Cu [78], and reduces Sb(V) at much lower acid concentrations (0.01 mol L^{-1}) and more quickly, making it more suitable for water sample analysis. L-Cysteine is reported to enhance Sb(III) stability in solution and the efficiency of hydride generation at low acid concentrations, as well as to increase the stability of the plasma and the adsorption intensity. L-Cysteine is thus considered the most beneficial reagent for Sb analysis using HG techniques.

6. ANCIENT AND MODERN USES OF ANTIMONY

The first recorded use of stibnite (Sb$_2$S$_3$) as a pigment to make mascara is mentioned in an Egyptian papyrus dating from 1600 BC. Known to the ancients as "white lead", there has been some question as to whether Sb was really known at that time as a metal, or whether it was only familiar as the sulfide. Part of a 5000 year old Mesopotamian vase now in the Louvre is made of almost pure Sb [1]. Antimony oxide, combined with lead oxide or lead carbonate, was used to glaze decorative bricks in Babylon during the 7th and 6th centuries BC. Without doubt, Sb has been used directly and indirectly by man for several thousand years, often in medicine. Discorides, for example (1st century AD), recommended stibnite for skin complaints and burns. During the middle ages, pills made of metallic Sb were used as a laxative; these could be recovered, reused, and were often kept in families for generations. Starting in the 16th century, Sb was valued as an emetic because it caused the patient to vomit immediately; the Sb was usually consumed after dissolving the metal in wine. The importance of Sb in medicine and pharmacy during the 17th century is reflected in "The Triumphal Chariot of Antimony" which was published in 1685 by Johann Thölde who wrote under the pen name Basil Valentine. Even at this time, however, it was well known that Sb could serve both as a medicine and as a poison: "To write on Antimony, there is needed profound meditation, a large mind, a wide knowledge of its preparation, and of its true soul, in which consists all its usefulness. If you are familiar with these, you can truly tell what is good and medicinal or what is bad and poisonous in it" [91].

While Sb has been used to provoke vomiting and as a laxative, the medical dose is dangerously near to the toxic dose [6]. Mozart, for example, may have been accidentally poisoned with Sb: he had been prescribed with antimony tartrate, and the symptoms of his death were identical to those of Sb poisoning. There are many famous instances of persons having been deliberately poisoned with Sb [1].

Today, Sb is used in a diverse array of processes and industrial and commercial products. Highly purified (>99.999%) metallic Sb is used in semiconductor technology to prepare the intermetallic compounds In, Al, Ga antimonide for diodes and infrared detectors. Antimony is also used as an alloying ingredient to harden Pb and other metals, and to improve corrosion resistance. Alloys of Sb are used in the manufacture of batteries, cable sheathing, bearings, castings, sheets and pipes, plumbing solder, and antifriction materials. "Hard lead" containing 3–9% Sb is used in the grids, terminals, and lead-oxide paste of automotive and standby batteries. When lead acid batteries are recycled, much of the Sb can be recovered. Antimonial lead linings are used to protect pipes, valves, pumps, and sheets used in the chemical industry. Antimonial lead is also used for solders and for making bearings. Casting alloys contain up to 13% Sb. Printing type was traditionally made from an alloy consisting typically of 60% Pb, 30% Sb and 10% Sn. When leaded pewter became unpopular, it was replaced with an alloy consisting of 89% Sn, 7% Sb, and 2% of each of Cu and Bi [1]. Other important Sb alloys include Britannia metal (Pb–Sb–Cu), Queen's metal (Sb–Sn–Cu–Zn) and Sterline (Cu–Sb–Zn–Fe). Antimony trisulfide is used in the manufacture of safety matches [92]. Automotive brake linings may contain several weight percent Sb which is added to improve their heat-resistance.

Non-metallic Sb products include paint pigments, ceramic enamels, plastics, glass, pottery, ammunition primers, and fireworks. Some Sb compounds are used in the vulcanising of rubber; other organic Sb compounds may be bactericides or fungicides [5]. The single greatest use of Sb today (approximately two-thirds of total Sb production) is as a flame retardant: antimony trioxide (Sb_2O_3) is added to plastics such as polyvinyl chloride (PVC), especially in car components, televisions, electrical insulation, furnishing fabrics, and other synthetic fibres, and cot mattresses; these applications account for approximately two-thirds of all antimony consumed today. Antimony trioxide is also the catalyst used to manufacture polyethylene terepthalate (PET); this material may contain a few hundred parts per million of Sb. However, because Sb_2O_3 is a suspected carcinogen, it must be handled with caution. Potassium antimony tartrate is widely used in pharmaceutical industry as an emetic and sodium stibogluconate (pentosam) is the treatment of choice for a number of tropical parasites, including the one which causes leishmaniasis [93].

7. ATMOSPHERIC EMISSIONS TO THE ENVIRONMENT

According to Ref. [94], the most important anthropogenic sources of Sb to the global atmosphere today are fossil fuel combustion, primary and secondary non-ferrous metals refining (Cu, Pb, and Zn), waste incineration, and incineration of sewage sludge (Table 2). With the growing use of Sb in plastics and other non-metallic products, the fate of Sb in industrial and consumer waste streams is a growing concern. In fact, emissions of Sb from incinerating municipal solid waste exceed those of As, Cd, Cr, Hg, Ni, Se, and Sn.

In Japan, Nakamura et al. [95] found that a typical concentration of Sb in household waste was $\sim$7.6 g T^{-1} with some individual components breaking down as follows: paper, plastic, leather and rubber, a few μg g^{-1} Sb up to $\sim$25 μg g^{-1} Sb; cassette tape, 38 μg g^{-1} Sb; polyester clothing, stuffed toys, and bedding, $\sim$120–160 μg g^{-1} Sb; other textiles 250 μg g^{-1} Sb; matchboxes 1600 μg g^{-1} Sb; curtains, 2100 μg g^{-1} Sb. The glass of a cathode ray tube of a television set was found to contain 0.2% Sb, and the plastic cover of that set 3.1% Sb. However, during combustion, $\sim$54% of the Sb was found in the bottom ash fraction, 45% in the fly ash collection by electrostatic precipitation, and <1% was found in the exhaust gas stream. Paoletti et al. [96] investigated the thermodynamic aspects of Sb during combustion, and showed that most of the Sb should be volatilized as Sb oxides and chlorides. In their experimental studies at a research waste incineration facility, however, they found almost no gaseous Sb in the flue gas stream. In fact, approximately one-half of the Sb was found in the grate ash fraction which they suggested might be due to the formation of calcium antimonate, $Ca_3(SbO_4)_2$.

7.1. Antimony in the Atmosphere

Antimony concentrations in the air are variable, ranging from <1 pg m^{-3} in Antarctica to >50 ng m^{-3} in Paris in the early 1970s [97]. Hinkley et al. [98], have estimated that volcanoes contribute $\sim$5 tonnes of Sb to the atmosphere each year. We have estimated that the natural background rate of total atmospheric Sb deposition in the pre-anthropogenic past ($\sim$6000–9000 years ago) was 90–154 tonnes per year [2]. Thus, volcanic emissions appear to account for only 3–5% of global emissions of Sb to the atmosphere. Because there is no quantitative data about global emissions of trimethyl antimony from the biosphere, we assume that 95–97% of the Sb to the atmosphere was originally supplied by soil dust. The correlation between Sb and Sc in peat samples dating from pre-anthropogenic times [2] supports this assumption. If most of the Sb in the atmosphere is supplied by soil dust derived from the weathering of crustal

Table 2 Predominant Anthropogenic Sources of Sb to the Global Atmosphere (tonnes per year)

Source	Asia	North America	Europe	South America	Africa	Australia Oceania	Total (by source)
Fossil fuels, stationary	308	226	124	1	31	40	730
Primary copper production	159	25	19	87	26	3	319
Municipal waste	98	58	78			1	235
Lead production	86	14	15	8	7	4	134
Zinc production	39	15	29	6	2	4	95
Sewage sludge		34	3				37
Secondary copper production		2	3				5
Total (by region)	690	374	271	102	66	52	1555

Source: Adapted from Ref. [93].

rocks, it can be helpful to compare the ratio of Sb to a conservative, lithogenic element such as Al, Sc, or Ti. Specifically, the Sb/Sc ratio is normalized to the corresponding value in crustal rocks [20]. Of the long list of elements studied by Rahn [99], Sb showed one of the greatest ranges in EF, from a low of approximately 3.8 (Saharan dust aerosols) to a high of approximately 24,000 (urban particulates in Ghent, Belgium). The lowest Sb EF values reported [99] are consistent with the pre-anthropogenic values (4.8 $\pm$ 1.4) found in peat samples dating from 6000 to 9000 years ago [2]. For the discussion presented here, therefore, we assume that the "natural background" Sb/Sc ratio is approximately five times the value in the Earth's crust [20].

Compared to the assumed background value, measurements of Sb/Sc (or Sb/Al) in aerosols from urban, rural, and even remote areas show that Sb is highly enriched. The data presented by Bogen [97] for the city of Heidelberg, Germany, for example, revealed an Sb EF of 496 which is comparable to the value reported for Chicago, USA, in 1970 (Sb EF = 516) [19]. Studying aerosols in Britain (Lake Windermere district) at about the same time, Peirson et al. [100] reported a crustal EF for Sb of 1350. Collecting aerosols over the North Atlantic, Duce et al. [101] reported Sb EF = 2300; Buat-Menard and Chesselet [102] reported Sb EF = 250 for marine aerosols over the Tropical North Atlantic. In fine aerosol fractions (<2.5 μm) collected in remote northern Norway [103] reported Sb EF on the order of 1000. In the same size fraction, Güllü et al. [104] reported Sb EF values on the order of 100 in aerosols from the Mediterranean. Even the air in Antarctica is enriched in Sb. According to Cunningham and Zoller [105] Sb EF values range from 125 (summer) to 2840 (winter); similar enrichment factors for the Antarctic had been published [106]. In contrast, Mroz and Zoller [107] found that volcanic emissions from Iceland had Sb EF values of "only" 97–115. More recently, Gomez et al. [108] found that the EF for Sb in volcanic ashes in Argentina are on the order of 2–6, relative to crustal abundance. Volcanic emissions, therefore, cannot explain the very high enrichment factor values (1000–10,000) found in urban, rural, and even some remote locations.

Heinrichs and Brumsack [109] showed that "urban particulates" in Germany today contain Sb EF values on the order of several hundred. In contrast, fly ash particles from lignite and anthracite combustion had Sb EF values on the order of 1000, and particulates from waste combustion >1000; particles emitted from automobile brake linings had Sb EF >10,000. Based on their studies, they estimated that 75% of the Sb in urban particulates today is due to vehicular traffic.

7.2. Antimony in Soils

We know of no published studies of the environmental biogeochemistry of Sb in uncontaminated soil. The behavior of Sb during chemical weathering of soils, therefore, is essentially unknown. Steinnes [110] has shown that there is a strong S–N gradient with respect to Sb concentrations in the surface layers of soils of Norway: from 2.4 ± 0.3 µg g^{-1} in southern Norway, compared with 0.22 ± 0.04 µg g^{-1} in the northern part of the country. The regional differences have been attributed to long-range atmospheric transport of contaminants.

7.2.1. Soils Contaminated by Lead Smelting

Antimony contamination of soils impacted by lead smelters have been reported by several investigators [111–114]. Studying soils (0–2 cm) in the vicinity of the lead smelter at Kellogg Valley, Idaho, which has operated for approximately one century, Ragaini et al. [111] found Sb concentrations in the range 5–260 µg g^{-1}. Using the Sc concentrations and normalizing to the abundance of Sb in crustal rocks, they calculated an average Sb EF in the soils of 1150. Ambient air collected at the same site showed an average EF for Sb of 46,000. Samples of grass from this area contained up to 110 µg g^{-1} [111].

In England, Li and Thornton [112] found up to 4.0–42.5 µg g^{-1} Sb in the top 0–15 cm of soils in the Derbyshire mining area, up to 50.0 µg g^{-1} Sb in the Shipham mining area, and up to 154 µg g^{-1} Sb near a Pb smelting site at Derbyshire; samples of slag contained up to 1230 µg g^{-1} Sb. In the vicinity of two secondary lead smelters in California, Kimbrough and Suffet [113] reported up to 13,000 µg g^{-1} Sb in solids (drosses, slags, foundary wastes) around Site 1, and up to 210 µg g^{-1} in soils around Site 2. In the soils near a Pb smelter in the Czech Republic, Sb concentrations up to 980 µg g^{-1} were found [114].

7.2.2. Soils Contaminated by Lead Ammunition at Shooting Ranges

Antimony is highly enriched in soils from shooting ranges because the lead alloy used in making ammunition contains 1–2% by weight Sb. It has been estimated that 100,000 tonnes of bullets accumulate in shooting ranges each year, representing an addition of one to two thousand tonnes of Sb. Wersin et al. [115] reported up to 17,460 µg g^{-1} Sb in soils from shooting ranges, and up to 4586 µg L^{-1} total dissolved Sb in leachates (ratio solid/water of 0.1). They suggest further that Sb in these soils is rapidly oxidized to Sb(V) which is mobilized as [Sb(OH)$_6$]$^{-}$. Lintschinger et al. [116], also found that while most of the Sb which can be extracted from contaminated soils is Sb(V), the total yield of extractable Sb is small due to the affinity of Sb(V) for Fe and Al oxides. However, using only water as an extractant, they [116] were able to extract $\sim$1% of total Sb, giving solution concentrations of 80–180 µg L^{-1} and indicating a marked solubility.

Fuentes et al. [117] examined the availability and redox state speciation of Sb in soils from a region of Chile receiving anthropogenic Sb from a Cu smelter. In their soils which contained between 1.8 and 13 µg g^{-1} total Sb, they were able to extract between 0.1% and 4.0% of the total Sb using water alone; most of the Sb extracted was Sb(V) whereas Sb(III) was below the limit of detection of 17 ng L^{-1}. However, contradictory conclusions have been reported by Scheinost et al. [118] who have used EXAFS to study the solid state speciation of Sb in shooting range soils: they suggested that Sb(III) concentrations are limited by the solubility of the corresponding hydroxide, and that Sb(V) is adsorbed to Fe and Mn oxides; they conclude that the mobility of Sb in shooting range soils is fairly limited. In a small country such as Switzerland with approximately 2000 shooting ranges, this is now an urgent research theme.

7.2.3. Soils Contaminated by Copper Smelting

Crecelius et al. [119] investigated As and Sb contamination in the vicinity of a Cu smelter near Tacoma, WA. This facility was known to release 300 tonnes of particulate material to the atmosphere each year, with the stack dust containing ~2% Sb. Soils within 5 km of the smelter contained up to 204 $\mu g\ g^{-1}$ Sb in the top 3 cm, compared to soils north of Seattle containing 3–5 $\mu g\ g^{-1}$. In Korea, soils (0–15 cm) in the vicinity of the Dalsung Cu–W mine contain up to 161 $\mu g\ g^{-1}$ Sb [120]; elevated Sb concentrations were also found in alluvial soils in the same region (up to 3.7 $\mu g\ g^{-1}$ Sb) and in garden soils (up to 12.6 $\mu g\ g^{-1}$). In Chile, De Gregori et al. [66] also determined Sb concentrations in soils impacted by a Cu smelter: in the top 20 cm of agricultural soils, they found Sb concentrations in the range 0.4–6.7 $\mu g\ g^{-1}$.

7.2.4. Dust in Urban Areas

In a study of atmospheric Sb in the city of Munich, Germany, Dietl et al. [121] found that Sb was strongly affected by vehicle traffic. Most of the Sb was associated with particles <10 μm. Moreover, 81% of this element was found in the thoracic particle fraction (i.e., 10–2.5 μm) and 17% in the respirable fraction (i.e., <2.5 μm). The rates of atmospheric Sb deposition which they reported for Munich [1.2–13.1 $\mu g\ m^{-2}\ day^{-1}$] are ~1250–13,700 times greater than the "natural background" rates of atmospheric Sb deposition in central Europe recorded by peat samples dating from pre-anthropogenic times [2].

In a study of a diverse group of trace elements in the city of Cologne, Germany, Weckwerth [122] found that Sb was the single most highly enriched element, and attributed 99% of it to the abrasion of automobile brake linings. In plant (*Nerium oleander*) and dust samples collected in Messina (pop. 230,000), Siciliy, Italy, Dongarra et al. [123] reported 0.1–6.3 $\mu g\ g^{-1}$ Sb in plant samples, and 9.1–72.8 $\mu g\ g^{-1}$ Sb in dust samples. The dust samples were fractionated according to particle size, with the Sb EF increasing from 36 in the 500–250 μm fraction, to 234 in the 10–2.5 μm fraction. Antimony is now sometimes used as a tracer of vehicle traffic [124].

7.3. Antimony in Sediments

In the vicinity of the Cu smelter near Tacoma, WA, marine sediments contain up to 12,500 $\mu g\ g^{-1}$ Sb, compared to "background" values which are typically below 1 $\mu g\ g^{-1}$ Sb [125]. Sediments from storm drains near a secondary lead smelter were found to contain up to 2300 μg g^{-1} Sb [126] which indicates that industrial facilities with significant atmospheric Sb emissions also may have aquatic discharge problems. Floodplain soils of the Bitterfield industrial area of the former East German contain up to 6.8 μg g^{-1} Sb, most of which originated in coal fly ash [127]. Sediments of the Malter Reservoir, Ore Mountains, Germany, contain up to 6.5 μg g^{-1} Sb, representing an EF, relative to the abundance of Sb in shale, of 25 [128]; in contrast to the other elements, Sb showed no significant temporal change in concentration.

7.4. Bioavailability of Antimony to Plants

Conflicting results have been published regarding the availability of Sb to plants growing in soils contaminated with Sb. One fundamental problem involves trying to distinguish between Sb which has been taken up by plants from soils via the soil solution and root system, vs. Sb contamination of plants by local atmospheric soil dust. A second, and perhaps more complex problem, are the implications of Sb contaminated plants for ecosystem, animal, and human health. Ainsworth et al. [129] examined Sb contamination of air, soils, and plants near an Sb smelter in northeast England where Sb has been produced since 1846 and soil Sb concentrations (0–5 cm) are up to 1489 μg g^{-1}. Antimony concentrations in grass samples in the vicinity of the smelter were certainly contaminated with Sb, with concentrations ranging from 8 to 336 μg g^{-1}; similarly high Sb concentrations were found in plants potted in contaminated soil and exposed at varying distances from the smelter for 1 month, and up to 1200 μg g^{-1} Sb was found in bags of *Sphagnum* moss which had been installed to collect atmospheric particles. All of these concentrations are far higher than the concentrations in plants from the control site where Sb was <1 μg g^{-1}. However, grass plants growing in pots of contaminated soil in a growth chamber yielded much lower Sb concentrations, with the maximum value of 2.2 μg g^{-1}. Ainsworth et al. [129] concluded that the plants growing near the smelter were certainly contaminated with Sb, but that most of this was due to local atmospheric dust.

Similarly observations have been made by others. Li and Thornton [112,130], for example, have indicated that the concentrations of Sb in washed samples of herbage from former mine sites are not significantly different from the control sites; however, the concentrations which they measured in plant samples approached the lower limits of detection (0.04 μg g^{-1}) which they achieved for Sb using HG-ICP-AES. As a consequence, an unambiguous interpretation of their findings is difficult. Perhaps more importantly, these authors pointed out that concentrations of most trace elements in unwashed herbage was generally higher than in washed samples, suggesting that herbage samples contaminated by metalliferous dust could be an important exposure pathway for grazing livestock, depending on the bioavailability of Sb to the animal. In their study of crop plants growing near the Cu–W mine in Korea, Jung et al. [120] found that Sb concentrations were below the lower limit of detection (0.01 μg g^{-1} Sb) both in the case of the mining site and the control site. In settling ponds adjacent to an ore mining and processing plant in Siberia, the belowground parts of wetland plants such as cattail (*Typha latifolia*), sylvan bulrush (*Scirpus sylvaticus*), and reed (*Phragmites australis*) contained up to 1300 μg g^{-1} Sb and the aboveground parts of these same plants averaged 15, 19, and 15 μg g^{-1} Sb, respectively [131]. Again, however, it is not clear how much of the Sb contamination by plants is due to biological uptake, and how much to aerial contamination by local dust.

In soils from Germany which have been contaminated by mining since the 15th century and contain up to 486 μg g^{-1} total Sb, Hammel et al. [132] were able to extract 0.02–0.29 μg g^{-1} Sb (0.06–0.59% of total soil Sb) using 1 M NH$_4$NO$_3$ for 2 h. However, the concentrations of Sb in fruits and grains of plants grown on these soils contained no more Sb than is typically found in plants growing on uncontaminated soils and was generally below the lower limit of detection (0.02 μg g^{-1}). In contrast, however, the leaves of endive, kale, parsley, and spinach contained more Sb, with the maximum reported concentration (2.2 μg g^{-1} Sb) in the leaves of endive (*Cichorium endiva*). In soil artificially contaminated with different forms of Sb and aged for 6 months, the authors reported up to 6.8% of total Sb could be extracted using NH$_4$NO$_3$ and that leaves of spinach (*Spinacia oleracea*) grown under these conditions contained up to 399 μg g^{-1} Sb.

In soils from a former Sb mine in southern Tuscany containing up to 15,112 μg g^{-1} Sb, Baroni et al. [133] reported up to 1367 μg g^{-1} Sb in the leaves of *Achillea ageratum*. Hundreds of parts per million of Sb were also found in the leaves of *Plantago lanceolata* and *Silene vulgaris*. In the case of these soils, however, concentrations of "soluble" Sb (i.e., extracted using only water) and "extractable" Sb (using 0.43 M acetic acid) were very high. One fundamental difference between these two studies is the Mediterranean climate at the Italian site which may have created an entirely different weathering regime for the Sb-bearing sulfide minerals compared to the German site.

In soils from Chile which have been contaminated by a nearby Cu smelter, total concentrations of Sb in the soils were much lower (maximum 6.7 μg g^{-1} Sb) than in these other studies, but even here, alfalfa plants were found to contain up to 1.7 μg g^{-1} Sb. Using 0.05 M EDTA, between 0.4% and 8.0% of the total Sb in the soils could be extracted. However, the authors concluded that the transfer of Sb to alfalfa plants could not be unambiguously understood because the concentrations involved were so small [66]. Methylated Sb species were found in moss samples contaminated with Sb (up to 190 μg g^{-1}) in a gold mining area of northern Canada [134], but these species represented a very small percentage of total Sb.

7.5. Antimony in Natural Waters

This topic has been thoroughly reviewed by Filella et al. [35,135] and is not discussed further here, with the following exception: Millen [136] has recently described a fascinating account of naturally elevated concentrations of Sb in spring waters in the vicinity of Schwaz, Tyrol, Austria, containing up to 3 mg L^{-1} total dissolved Sb. These anomalous concentrations originate via dissolution of Sb-bearing ores (10–20% Sb) by oxic waters at pH 7.5–8.2. Even though the Sb concentrations are far in excess of the maximum suggested limit for drinking water in Europe (0.005 mg L^{-1}), local inhabitants apparently have been drinking this water for centuries, with no apparent health effects.

8. ARCHIVES OF ATMOSPHERIC ANTIMONY DEPOSITION

8.1. Polar and Alpine Snow and Ice

Data on Sb concentrations in polar and alpine snow and ice are very scarce. There are no published studies to date reporting temporal trends for Sb. Because of the extremely low concentrations of Sb in this matrix, ICP-sector field-mass spectrometry (high resolution ICP-MS) emerged as the method of choice for determining Sb at extremely low concentration characteristic of polar snow and ice. Detection limits for Sb as low as 0.07 pg g^{-1} have been recently reported [50]. Antimony concentrations determined in 120 ice samples from a core from the Canadian High Arctic ranged from 0.47 to 108 pg g^{-1} with a median concentration of 31 pg g^{-1}. Considering the Sb/Sc ratio of 0.044 in the upper continental crust (UCC), Sb in these polar ice samples is enriched by at least 100 times, with some samples being enriched in Sb by as much as 10,000 times. The reasons for the general enrichment of Sb relative to Sc (100×) and the anomalous enrichments (10,000×) remains to be elucidated [50]. In 68 samples of a 2.7 m deep snow pit at Summit (central Greenland), Sb concentrations ranged from 0.2 to 4.4 pg g^{-1} [137]. Somewhat higher Sb concentrations (1.6, 2.5, 5.1, 111 pg g^{-1}) were found in four ice samples from the glacier Sajama in Bolivia at an altitude of 6542 m [138]. Antimony concentrations in 366 Eastern Alpine surface snow samples range from 1.7 to 6200 pg g^{-1} with a median of 31 pg g^{-1} [49]. Lower Sb concentrations were found in snow and ice samples from the Mont Blanc (median: 11 pg g^{-1}, range: 0.2–109 pg g^{-1}, $n = 74$) and Monte Rosa (median: 19 pg g^{-1}, range: 2–689 pg g^{-1}, $n = 25$) [49].

8.2. Biomonitors

Atmospheric Sb deposition using 495 moss samples (*Hylocomium splendens*) collected in Norway indicated that Sb is transported over long distances [139]. Actual concentrations of Sb in moss samples range from <0.01 to 0.64 μg g^{-1}, with a median of 0.09 μg g^{-1}. In comparison to Sb data generated in 1977, the 1990 results showed a decline to ~30–40% of the previous Sb levels. Medians from a very similar survey in Norway carried out in the years 1985 and 1977, respectively, revealed a median of 0.16 and 0.17 μg g^{-1} [140]. Already that study (1985) indicated a 50% decline in Sb concentrations in the south of Norway compared to 1977.

Another moss species, *Pleurozium schreberi*, was utilized to assess atmospheric Sb deposition in the Netherlands in 1992 [141]. Employing Monte-Carlo Assisted Factor Analysis, the metallurgical industry was identified to account for almost 100% of the Sb found in the moss samples (median: 0.47 μg g^{-1}, range: 0.21–1.9 μg g^{-1}, $N = 66$).

Temporal and spatial trends of Sb have been established by the use of an Environmental Specimen Bank (ESB) [142]. An ESB provides the unique opportunity to analyze samples retrospectively, as sub-samples of all specimens that have been collected over the years are stored in the vapour phase above liquid nitrogen to preserve their chemical composition. Various representative plant samples (spruce shoots, poplar leaves) from the Federal ESB of Germany collected annually between 1985 and 1998 revealed no Sb concentration trend over time [142]. However, spruce shoots from a semi-natural area (National Park of Berchtesgaden, Germany) showed Sb concentrations ($\sim$22 ng g^{-1}, range: 17–29 ng g^{-1}) that were approximately four times lower than those in corresponding samples from an urban-industrialized area [142]. Similarly, elder leaves were used to demonstrate the influence of car traffic on Sb concentrations along motorways in Germany [142]. In these cases, the Sb contamination is mainly from wear on brake linings, as well as from rubber tires which contain Sb from the vulcanization process.

Concentrations of Sb in poplar leaves from Chile and Argentina (4 ng g^{-1}) are the lowest values found in plant samples and these low Sb concentrations are assumed to represent natural background levels of Sb [142].

8.3. Peat Bogs

Analyses of Sb in cores from a Swiss bog which has been accumulating peat for almost 15,000 years revealed "natural background" concentrations of Sb (8 ± 3 ng g^{-1}) and Sc (76 ± 16 ng g^{-1}) in peat samples between six and nine thousand years old. These values represent a useful reference level for comparison with modern plants, especially bryophytes which today are used extensively in monitoring programs. Moreover, these values allow Sb concentrations in modern peat samples to be separated into "natural" and "anthropogenic" components. This has been done using two replicate peat cores collected from the Swiss bog in 1991 (core 2F) and in 1993 (core 2K); these results are shown in Fig. 4 along with the distribution of natural and anthropogenic Pb.

The small differences between the replicate cores are due to differential compression during sample collection as well as the thickness and imprecision of the sectioning of the peat cores (they were cut by hand using a bread knife into 3 cm sections). Despite these limitations, the profiles show a number of salient features. First, the radiocarbon age dates of the deeper parts of the core indicate that there was already significant atmospheric Sb contamination during the Roman period. Second, the inventory of anthropogenic Sb clearly dwarfs the inventory of natural Sb. Third, the occurrence of anthropogenic Sb pre-dates the

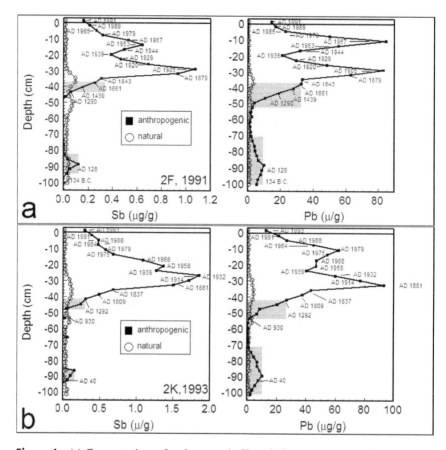

Figure 4 (a) Concentrations of anthropogenic Sb and Pb concentrations (■) and natural concentrations (O) in the 2F peat profile from Etang de la Gruère, Switzerland [2]. The age dates of samples since the middle of the 19th century were obtained using ^{210}Pb, and the older dates using ^{14}C (decay counting of bulk samples); the radiocarbon dates were calibrated and expressed here as calendar years. The shaded areas denote significant concentrations of anthropogenic Sb during the Medieval Period as well as the Roman Period. (b) Concentrations of anthropogenic Sb and Pb concentrations (■) and natural concentrations (O) in the 2K peat profile from Etang de la Gruère, Switzerland which was collected 2 years after the 2F core [Shotyk, unpublished data].

Industrial Revolution, with Sb contamination also found in peat samples dating from the Medieval period. Fourth, the distribution of anthropogenic Sb resembles that of anthropogenic Pb; this is logical, considering the chemical and mineralogical association of these two metals. Fifth, the maximum concentration of anthropogenic Sb was already seen in the early part of the 20th century (i.e., the early 1900s). Even though the concentrations of anthropogenic Sb have gone into

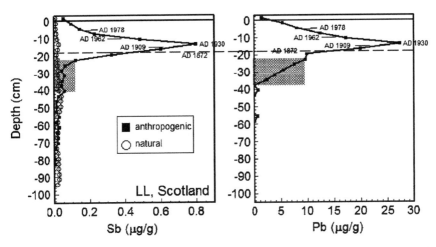

Figure 5 Concentrations of anthropogenic Sb and Pb concentrations (■) and natural concentrations (○) in the peat profile from Loch Laxford, northwest Scotland [2]. The horizontal dashed line indicates the point below which unsupported ^{210}Pb could no longer be quantified. The greatest concentrations of anthropogenic Sb, therefore, are found in peats which are more than ∼150 years old. However, peat samples which are older than 150 years also are significantly enriched in Sb (shaded area).

decline since that time, they are still very much greater than the natural concentrations of lithogenic Sb.

For comparison with the Swiss peat core, a peat core from one of the most remote corners of northwest Scotland shows a similar story (Fig. 5): in recent peat samples, anthropogenic Sb concentrations are much higher than the natural values, and correlate with anthropogenic Pb. Again, the maximum concentrations of Sb (and Pb) date from the early part of the 20th century. A similar chronology of atmospheric Sb deposition was found in a Scottish peat core collected near Glasgow [143]. Here, the maximum rate of Sb deposition (∼1.5 mg m^{-2} year^{-1}) is approximately 4000 times greater than the pre-anthropogenic rate recorded by the Swiss bog [2].

While the possible importance of post-depositional migration of Sb in peat cores has not yet been systematically studied, the correlation with Pb (which is known to be immobile in peat bogs) suggests that peat cores from ombrotrophic (i.e., rain-fed) bogs could be very helpful for reconstructing the change rates and sources of atmospheric Sb deposition.

9. ANTIMONY AND HUMAN HEALTH

Occupational exposure to Sb compounds is known to cause irritation to the respiratory tract and may lead to pneumoconiosis [31]. Dusts containing Sb or its compounds can induce dermatitis, keratitis, conjunctivitis, suppuration of the nasal septum and gastritis [31]. However, many other health threatening effects of Sb compounds are presently known [31]. Anthropogenic activities lead to pronounced exposure to Sb of particular sub-groups of the population. In Japan, about 20,000 tonnes of Sb are employed in different industrial processes every year, while "only" 100 tonnes of arsenic, a well known potentially toxic element, is used per year [144].

Among several industrial uses of Sb compounds, antimony trioxide (Sb_2O_3) is largely employed in the production of glassware and ceramics. Moreover, Sb_2O_3 is added to molten glass as a clarifying agent and is employed as a pigment in dyes and paints as well as in the textile industry. Several Sb compounds are used as additives to metal coatings and to rubber, whereas others are added to textiles as flame retardants. Lately, some concern has been raised upon the exposure of young infants to antimony trihydride generated from Sb_2O_3 added to polyvinyl chloride cot mattress covers as fire retardants potentially leading to the sudden infant death syndrome (SIDS) [145,146]. However, no evidence to support a causal role for antimony in SIDS could be established [145,146].

Lead battery workers are exposed to antimony trioxide (Sb_2O_3) and stibine (SbH_3) resulting in elevated Sb concentrations in urine [147]. Stibine is a highly toxic gas that can cause both serious injury to the central nervous system and hemolysis [31]. This exposure to Sb compounds leads to at least 10-times increased concentrations of Sb in urine and whole blood (Table 3). These elevated Sb load, however, is still thought to lie within a tolerable range for industrial hygiene [147]. Both Sb_2O_3 and SbH_3 exhibit very similar characteristics with regard to pulmonary absorption and renal elimination. Cumulative effects are not to be expected under regular conditions. The half-life of renal elimination amounts to $\sim$4 days [147].

In that context, the importance of chemical speciation of Sb compounds should be addressed. As already mentioned in Section 4, the toxicity of Sb strongly depends on its oxidation state and the presence of complex-forming ligands. For the elucidation of the pathways of incorporated Sb into the body and for a fair risk assessment, the Sb compounds found in different compartments of the human body need to be specified.

Even though huge improvements in the speciation of Sb in environmental and biological matrices have been made during recent years, distinct drawbacks still hamper the application of speciation procedures of Sb to real world samples [148]. One major problem with speciation of Sb in real samples consists of the generally low extraction yield for Sb. Quite frequently, only a few percent of total Sb could be released from the matrix. Thus, most of the Sb species information is retained within the sample and speciation procedures are not that useful in gaining further knowledge about the fate of Sb in environmental samples [148,149]. Another problem is encountered with the availability of standard compounds which are needed for identification and quantification of the Sb species. Besides the two inorganic forms of antimony, Sb(III) and Sb(V),

Table 3 Concentrations ($\mu g\ L^{-1}$) of Sb in Various Human Body Fluids

Matrix	Comments	Concentration	Mean	N^a	LOD	Reference
Urine	Preterm infants	0.03–1.7	0.28	26	0.02	[48]
	Term infants	<0.02–3.0	0.07	132	0.02	
	Infants <1 year	<0.02–0.90	0.05	210	0.02	
	Unexposed subjects	0.19–1.1	0.79 ± 0.07	360	n. r.	[153]
	Healthy adults	0.19–1.8	—	—	n. r.	[154]
	Unexposed subjects	<0.12	—	2	0.12	[151]
	Exposed-battery production	5.1–8.3	—	2	0.12	
Serum	Infants <1 year	0.07–0.76	0.18	220	0.02	[48]
	Unexposed subjects	0.01–1.7	0.5 ± 0.1	22	n. r.	[153]
	Healthy adults	~0.8	—	—	n. r.	[154]
Whole blood	Infants <1 year	0.08–0.88	0.26	15	0.02	[48]
	Starter battery production					[147]
	Casters	0.5–3.4	2.6[b]	7	n. r.	
	Formation workers	0.5–17.9	10.1[b]	14	n. r.	
	Male	<0.5–7.54	0.60[b]	63	0.5	[155]
	Female	<0.5–3.58	0.47[b]	73	0.5	
	Geogenically exposed	<0.5–7.54	0.57[b]	89	0.5	
	Reference	<0.5–2.27	0.48[b]	47	0.5	
	Unexposed subjects	0.03–3.5	2.16 ± 0.45	27	n. r.	[152]
	Healthy adults	0.3–3	—	—	n. r.	[154]

[a]Number of samples analyzed.
[b]Median.
Note: n. r., not reported.

only a trimethylated Sb(V) compound, namely trimethylantimony dichloride ($TMSbCl_2$), trimethylantimony dihydroxide [$TMSb(OH)_2$], or trimethylantimony oxide (TMSbO), are currently available to the scientific community. Attempts to synthesize soluble mono- and dimethylated Sb compounds which can be used for speciation studies always failed so far because these compounds tend to polymerize upon dissolution or cannot even be synthesized as a monomer, or they are not stable at ambient air [148].

Generally, Sb species are separated by chromatography (HPLC or GC) coupled online to an element-specific detector such as a hydride generation-atomic absorption spectrometer (HG-AAS), HG-atomic fluorescence spectrometer (HG-AFS), inductively coupled plasma-optical emission spectrometer or ICP-mass spectrometer (ICP-MS). Using HPLC-HG-AAS, detection limits of $0.4~\mu g~L^{-1}$ for $TMSbCl_2$, $0.7~\mu g~L^{-1}$ for Sb(III) and $1.0~\mu g~L^{-1}$ for Sb(V) were reported [150]. Replacing HG-AAS by ICP-MS, detection limits can be lowered by at least 100-times [88,151].

So far only one study is reported regarding speciation of Sb in human urine of occupationally exposed and unexposed persons [151]. In the two urine samples of non-exposed subjects, the total Sb concentration was below the method detection limit of $0.12~\mu g~Sb~L^{-1}$ [151]. Concentrations for Sb(V) and Sb(III) were also below the method detection limits, except for $TMSbCl_2$, which was found in ultratrace amounts in one sample. Total Sb urine concentrations of two persons working in the lead battery producing industry were about 100 times higher than for non-exposed persons. These workers are exposed to antimony trioxide and stibine (SbH_3). Surprisingly Sb(V), but not Sb(III), followed by $TMSbCl_2$ were the predominant Sb species determined in the two urine samples. Sb(III), if detectable, was only present at very low concentrations. As Sb(III) is known to exert a 10-times higher toxicity than Sb(V), it seems that the human body can detoxify Sb(III) by oxidation to Sb(V), and possibly biomethylates Sb(III) and/or Sb(V) to $TMSbCl_2$. Additionally, an unknown Sb species was detected in the urine of one exposed person at a concentration of $0.13~\mu g~L^{-1}$.

Overall, huge efforts still need to be undertaken to understand the meaning and fate of Sb and its compounds in biological and environmental systems.

10. SUMMARY AND CONCLUSIONS

Antimony is a potentially toxic trace element with no known biological function. Antimony is commonly enriched in coals, and fossil fuel combustion appears to be the largest single source of anthropogenic Sb to the global atmosphere. Abundant in sulfide minerals, its emission to the atmosphere from anthropogenic activities is linked to the mining and metallurgy of non-ferrous metals, especially Pb, Cu, and Zn. In particular, the geochemical and mineralogical association of Sb with Pb minerals implies that, like Pb, Sb has been emitted to the environment for thousands of years because of Pb mining, smelting, and refining. In the US alone, there are more than 400 former secondary lead smelting operations

[152] and worldwide there are 133 Pb–Zn smelters in operation today. Antimony is used in creating and improving dozens of industrial and commercial materials including various alloys, ceramics, glasses, plastics, and synthetic fabrics, making waste incineration another important source of Sb to the environment.

Enrichments of Sb in atmospheric aerosols, plants, soils, sediments, as well as alpine and polar snow and ice suggest that Sb contamination is extensive, but there are very few quantitative studies of the geographic extent, intensity, and chronology of this contamination. There is an urgent need to quantify the extent of human impacts and how these have changed with time. The decreasing inventories of anthropogenic Sb with time in peat cores from Switzerland and Scotland suggest that the atmospheric Sb flux may be declining, but there have been too few studies to make any general conclusions. In fact, some studies of sediments and biomonitors in central Europe show little decline in Sb concentrations during the past decades. There is an obvious need for reliable data from well dated archives such as polar snow and ice, peat bogs, and sediments. The air concentrations, extent of enrichment, particle size distribution, and rate of deposition of Sb in urban areas is cause for concern. The natural processes which controlled the Sb flux to the atmosphere in the pre-anthropogenic past are poorly understood.

The cumulative amount of anthropogenic Sb in soils has not yet been quantified. The long-term fate of Sb in soils, including weathering and mobilization, has only started to be investigated. However, the limited data available suggests that, in some locations at least, anthropogenic Sb in soils may be more mobile than anthropogenic Pb. Further study of this problem is needed, as well as the chemical speciation of Sb in soil–water–plant–sediment systems, and the implications which this has for human and ecosystem health.

ACKNOWLEDGMENT

Our sincere thanks to the Alexander von Humboldt Foundation for a postdoctoral fellowship to Bin Chen.

ABBREVIATIONS

AAS	atomic absorption spectrometry
AFS	atomic fluorescence spectrometry
ASV	anodic stripping voltammetry
CSV	cathodic stripping voltammetry
DHGAAS	derivative hydride generation atomic absorption spectrometry
DPP	differential pulse polarography
EDTA	ethylenediamine-N,N,N',N'-tetraacetate
EF	enrichment factor
Eh	standard potential

ESB	environmental specimen bank
EXAFS	extended absorption fine structure spectroscopy
FAAS	flame atomic absorption spectrometry
GC	gas chromatography
GFAAS	graphite furnace atomic absorption spectrometry
GST	glutathione *S*-transferase
HG	hydride generation
HG-AAS	hydride generation-atomic absorption spectrometry
HG-AFS	hydride generation-atomic fluorescence spectrometry
HG-ICP-AES	hydride generation-inductively coupled plasma-atomic emission spectrometry
HG-LIF	hydride generation-laser induced fluorescence spectrometry
HPLC	high pressure liquid chromatography
IARC	International Agency for Research on Cancer
ICP-MS	inductively coupled plasma-mass spectrometry
ICP-OES	inductively coupled plasma-optical emission spectrometry
ICP-SMS	inductively coupled plasma-sector field mass spectrometry
INAA	instrumental neutron activation analysis
LOD	limit of detection
MC-ICP-MS	multi collector-inductively coupled plasma-mass spectrometry
MIP-OES	microwave induced plasma-optical emission spectrometry
NAA	neutron activation analysis
pe	redox intensity
PET	polyethylene terepthalate
PVC	polyvinyl chloride
SIDS	sudden infant death syndrome
TMSb	trimethylantimony
USGS	United States Geological Survey
UCC	Upper Continental Crust
XRF	X-ray fluorescence spectrometry

REFERENCES

1. Emsley J. Nature's Building Blocks. An A–Z Guide to the Elements, Oxford: Oxford University Press, 2001.
2. Shotyk W, Krachler M, Chen B. Global Biogeochem Cycles 2004; 18:1016 (doi: 1010.1029/2003GB002113).
3. Hutchinson TC, Meema K, eds. Lead, Mercury and Arsenic in the Environment, Published on behalf of the Scientific Committee on Problems of the Environment of the International Council of Scientific Unions. Chichester: John Wiley and Sons, 1987:360.
4. Bodek I, Lyman WJ, Reehl WF, Rosenblatt DH, eds. Environmental Inorganic Chemistry. Properties, Processes, and Estimation Methods, New York and Oxford: Pergamon Press, 1988.

5. Fowler BA, Goering PL, Merian E. Antimony. In: Merian E, ed. Metals and Their Compounds in the Environment. VCH: Weinheim, 1991:743–750.

6. Emsley J. The Elements, 3rd ed., Oxford: Clarendon Press, 1998.

7. Ruben S. Handbook of the Elements. Indianapolis, USA: Howard. Sams W, 1968.

8. Cotton FA, Wilkinson G. Advanced Inorganic Chemistry, 4th ed., New York: John Wiley and Sons, 1980.

9. Rai D, Zachara JM. Chemical Attenuation Rates, Coefficients, and Constants in Leachate Migration. Volume 1. A Critical Review, Electric Power Research Institute, EPRI EA-3356, Volume 1, Research Project 2198-1. Final Report, February, 1984, pp. 4–1 to 4–5. Battelle Pacific Northwest Laboratories, Richland, WA, USA, 1984.

10. Brannon JM, Patrick WH. Environ Pollut 1985; 9:107–126.

11. Goldschmidt VM. Chem Soc J 1937; 655–673.

12. Boyle RW, Jonasson IR. J Geochem Explor 1984; 20:223–302.

13. Hirner AV, Feldmann J, Krupp E, Gruemping R, Goguel R, Cullen WR. Org Geochem 1998; 29:1765–1778.

14. Jenkins RO, Craig PJ, Goessler W, Miller D, Ostah N, Irgolic KJ. Environ Sci Technol 1998; 32:882–885.

15. Haas K, Feldmann J, Wennrich R, Stark HJ. Fresenius J Anal Chem 2001; 370:587–596.

16. Rouxel O, Ludden J, Fouquet Y. Chem Geol 2003; 200:25–40.

17. Wehmeier S, Ellam R, Feldmann J. J Anal At Spectrom 2003; 18:1001–1007.

18. Lindner G, Kaminski S, Greiner I, Wunderer M, Behrschmidt J, Schroeder G, Kress S. Verh Int Verein Limnol 1993; 25:238–241.

19. Dams R, Robbins JA, Rahn KA, Winchester JW. Anal Chem 1970; 42:861–867.

20. Wedepohl KH. Geochim Cosmochim Acta 1995; 59:1217–1232.

21. Onishi H. Handbook of Geochemistry, Vol. II/4, Berlin: Springer-Verlag, 1969:51-A-1–51-O-1.

22. Lueth VW. Antimony: Element and Geochemistry, The Encyclopedia of Geochemistry. Dordrecht: Kluwer Academic Publishers, 1999:15–16.

23. Bowen HJM. Environmental Chemistry of the Elements. London: Academic Press, 1979.

24. Roesler HJ, Lange H. Geochemical Tables. NY: Elsevier, 1972:468.

25. Goldschmidt VM. In: Muir A, ed. Geochemistry. Oxford: Clarendon Press, 1954.

26. Wedepohl KH. Handbook of Geochemistry, Vols I and II, Berlin, Heidelberg, New York: Springer, 1969.

27. Bloch P, Vanderborght B, Adams F. J Environ Anal Chem 1983; 14:257–274.

28. Gornitz V. The Encyclopedia of Geochemistry and Environmental Sciences. Pennsylvania: Dowden, Hutchinson, Ross, Stroudsburg, 1972.

29. Goldschmidt VM. Ind Eng Chem 1935; 27:1100–1102.

30. Valkovic V. Trace Elements in Coal. CRC Press, 1983.

31. Deutsche Forschungsgemeinschaft (DFG), Analyses of Hazardous Substances in Biological Materials, 1994:51.

32. Poon R, Chu I. J Biochem Mol Toxicol 2000; 14:169–176.

33. Gebel T. Chem-Biol Interact 1997; 107:131–144.

34. Schaumloffel N, Gebel T. Mutagenesis 1998; 13:281–286.

35. Filella M, Belzile N, Chen YW. Earth-Sci Rev 2002; 57:125–176.

36. Nash MJ, Maskall JE, Hill SJ. J Environ Monit 2000; 2:97–109.

37. Garbos S, Bulska E, Hulanicki A, Shcherbinina NI, Sedykh EM. Anal Chim Acta 1997; 342:167–174.
38. Kubota T, Kawakami A, Sagara T, Ookubo N, Okutani T. Talanta 2001; 53:1117–1126.
39. Cabon JY. Anal Bioanal Chem 2002; 374:1282–1289.
40. Amin N, Kaneco S, Nomura K, Suzuki T, Ohta K. Microchim Acta 2003; 141:87–91.
41. Lopez I, Garcia, Sanchez Merlos M, Hernandez Cordoba M. Spectrochim Acta, Part B-At Spectro 1997; 52:437–443.
42. Canuto MH, Siebald HGL, de Lima GM, Silva JBB. J Anal At Spectrom 2003; 18:1404–1406.
43. De Gregori I, Pinochet H, Fuentes E, Potin-Gautier M. J Anal At Spectrom 2001; 16:172–178.
44. Krachler M, Burow M, Emons H. Analyst 1999; 124:777–782.
45. Krachler M, Burow M, Emons H. Analyst 1999; 124:923–926.
46. Sun HW, Ha J, Sun JM, Zhang DQ, Yang LL. Anal Bioanal Chem 2002; 374:526–529.
47. Richter RC, Swami K, Chace S, Husain L. Fresenius J Anal Chem 1998; 361:168–173.
48. Delves CE, Sieniawska CE, Fell GS, Lyon TDB, Dezateux C, Cullen A, Variend S, Bonham JR, Chantler SM. Analyst 1997; 122:1323–1329.
49. Barbante C, Cozzi G, Capodaglio G, van de Velde K, Ferrari C, Boutron C, Cescon P. J Anal At Spectrom 1999; 14:1433–1438.
50. Krachler M, Zheng J, Fisher D, Shotyk W. J Anal At Spectrom 2004; 19:1017–1019.
51. Chwastowska J, Zmijewska W, Sterlinska E. J Radioanal Nucl Chem 1995; 196:3–9.
52. Garbos S, Bulska E, Hulanicki A, Fijalek Z, Soltyk K. Spectrochim Acta Part B-At Spectro 2000; 55:795–802.
53. Enger J, Marunkov A, Chekalin N, Axner O. J Anal At Spectrom 1995; 10:539–549.
54. Chen B, Krachler M, Shotyk W. J Anal At Spectrom 2003; 18:1256–1262.
55. Tao S, Huang J, Ma W, Chen L, Chen D. At Spectro 1998; 19:180–185.
56. De Gregori I, Lobos MG, Pinochet H. Water Res 2002; 36:115–122.
57. Guo XW. Lab Robotics Autom 2000; 12:67–73.
58. Li GW, Li L, Zhong GP, Hou XD. Spectro Lett 2003; 36:275–285.
59. Caballo-Lopez A, de Castro MDL. Anal Chem 2003; 75:2011–2017.
60. Yan XP, Yin XB, Jiang DQ, He XW. Anal Chem 2003; 75:1726–1732.
61. Aucelio RQ, Rubin VN, Smith BW, Winefordner JD. J Anal At Spectrom 1998; 13:49–54.
62. Morrison MA, Weber JH. Environ Sci Technol 1997; 31:3325–3329.
63. Bohari Y, Lobos G, Pinochet H, Pannier F, Astruc A, Potin-Gautier M. J Environ Monit 2002; 4:596–602.
64. Ebdon L, Foulkes ME, Le Roux S, Munoz-Olivas R. Analyst 2002; 127:1108–1114.
65. Chen LC, Yang FM, Xu J, Hu Y, Hu QH, Zhang YL, Pan GX. J Agric Food Chem 2002; 50:5128–5130.
66. De Gregori I, Fuentes E, Rojas M, Pinochet H, Potin-Gautier M. J Environ Monit 2003; 5:287–295.
67. Suo YR, Li TC. Spectro Spect Anal 2002; 22:850–852.
68. Yang LL, Gao LR, Zhang DQ. Anal Sci 2003; 19:897–902.
69. Yin XB, He XW, Jiang Y, Yan XP. Chem Analit 2003; 48:45–53.

70. Sun HW, Suo R, Lu YK. Anal Sci 2003; 19:1045–1049.
71. Lu YK, Sun HW, Yuan CG, Yan XP. Anal Chem 2002; 74:1525–1529.
72. Gomez-Ariza JL, de la Torre MAC, Giraldez I, Sanchez-Rodas D, Velasco A, Morales E. Appl Organomet Chem 2002; 16:265–270.
73. Chen CY, Zhao JJ, Zhang PQ, Chai ZF. Anal Bioanal Chem 2002; 372:426–430.
74. Cava-Montesinos P, Cervera ML, Pastor A, de la Guardia M, Talanta 2003; 60:787–799.
75. Sayago A, Beltran R, Gomez-Ariza JL. J Anal At Spectrom 2000; 15:423–428.
76. Vaisanen A, Suontamo R, Rintala J. J Anal At Spectrom 2002; 17:274–276.
77. Rigby C, Brindle ID. J Anal At Spectrom 1999; 14:253–258.
78. Feng YL, Chen HW, Chen HY, Tian LC. Fresenius J Anal Chem 1998; 361:155–157.
79. Haffer E, Schmidt D, Freimann P, Gerwinski W. Spectrochim Acta Part B-At Spectro 1997; 52:935–944.
80. Nakahara T, Li YM. J Anal At Spectrom 1998; 13:401–405.
81. Gong ZB, Chan WF, Wang XR, Lee FSC. Anal Chim Acta 2001; 450:207–214.
82. Krachler M, Shotyk W, Emons H. Anal Chim Acta 2001; 432:303–310.
83. Pacquette HL, Elwood SA, Ezer M, Simeonsson JB. J Anal At Spectrom 2001; 16:152–158.
84. Feng XJ, Fu B. Anal Chim Acta 1998; 371:109–113.
85. Keenan F, Cooke C, Cooke M, Pennock C. Anal Chim Acta 1997; 354:1–6.
86. Rojas I, Murillo M, Carrion N, Chirinos J. Anal Bioanal Chem 2003; 376:110–117.
87. Krachler M, Emons H, Barbante C, Cozzi G, Cescon P, Shotyk W. Anal Chim Acta 2002; 458:387–396.
88. Krachler M, Emons H. Anal Chim Acta 2001; 429:125–133.
89. Rahman L, Corns WT, Bryce DW, Stockwell PB. Talanta 2000; 52:833–843.
90. Flores EMD, dos Santos EP, Barin JS, Zanella R, Dressler VL, Bittencourt CF. J Anal At Spectrom 2002; 17:819–823.
91. Valentine B. The Triumphal Chariot of Antimony. English version of the original Latin (Published in 1685), translated by Waite AE edited by Bouleur J. Edmonds, WA, USA: Alchemical Press, Holmes Publishing Group. 1992.
92. Stwertka A. Guide to the Elements, revised ed. New York: Oxford University Press, 1998.
93. Ulrich N. Antimony, 81, Chemical & Engineering News, September 8, 2003:126.
94. Pacyna JM, Pacyna EG. Environ Rev 2001; 9:269–298.
95. Nakamura K, Kinoshita S, Takatsuki H. Waste Manage 1996; 16:509–517.
96. Paoletti F, Seifert T, Vehlow J, Sirini P. Waste Manage Res 2000; 18:141–150.
97. Bogen J. Atmos Environ 1973; 7:1117–1125.
98. Hinkley TK, Lamothe PJ, Wilson SA, Finnegan DL, Gerlach TM. Earth Planet Sci Lett 1999; 170:315–325.
99. Rahn KA. The Chemical Composition of the Atmospheric Aerosol, Technical Report, Graduate School of Oceanography, Kingston, R.I. USA: University of Rhode Island, 1976:265.
100. Peirson DH, Cawse PA, Salmon L, Cambray RS. Nature 1973; 241:252–256.
101. Duce RA, Hoffman GL, Zoller WH. Science 1975; 187:59–61.
102. Buat-Menard P, Chesselet R. Earth Planet Sci Lett 1979; 42:399–411.
103. Maenhaut W, Cornille P, Pacyna JM, Vitols V. Atmos Environ 1989; 23:2551–2569.
104. Güllü GH, Oelmez I, Tuncel G. The Impact of Desert Dust Across the Mediterranean. Dordrecht, The Netherlands: Kluwer Academic Publishers, 1996:339–347.

105. Cunningham WC, Zoller WH. J Aerosol Sci 1981; 12:367–384.
106. Zoller WH, Gladney ES, Duce RA. Science 1974; 183:198–200.
107. Mroz EJ, Zoller WH. Science 1975; 190:461–464.
108. Gomez D, Smichowski P, Polla G, Ledesma A, Resnizky S, Rosa S. J Environ Monit 2002; 4:972–977.
109. Heinrichs H, Brumsack HJ. Geochemie und Umwelt. Berlin: Springer, 1997:442.
110. Steinnes E. Water Air Soil Pollut 1997; 100:405–413.
111. Ragaini RC, Ralston HR, Roberts N. Environ Sci Technol 1977; 8:773–781.
112. Li XD, Thornton I. Environ Geochem Health 1993; 15:135–144.
113. Kimbrough DE, Suffet I. Environ Sci Technol 1995; 29:2217–2221.
114. Rieuwerts J, Farago M. Appl Geochem 1996; 11:17–23.
115. Wersin P, Johnson CA, Furrer G. Geochim Cosmochim Acta 2002; 66:A829–A829.
116. Lintschinger J, Michalke B, Schulte-Hostede S, Schramel P. Int J Environ Anal Chem 1998; 72:11–25.
117. Fuentes E, Pinochet H, De Gregori I, Potin-Gautier M. Spectrochim Acta Part B-At Spectro 2003; 58:1279–1289.
118. Scheinost AC, Rossberg A, Marcus M, Pfister S, Kretzschmar R. Physica Scripta 2004; in press.
119. Crecelius EA, Johnson CJ, Hofer GC. Water Air Soil Pollut 1974; 3:337–342.
120. Jung MC, Thornton I, Chon HT. Sci Total Environ 2002; 295:81–89.
121. Dietl C, Reifenhauser W, Peichl L. Sci Total Environ 1997; 205:235–244.
122. Weckwerth G. Atmos Environ 2001; 35:5525–5536.
123. Dongarra G, Sabatino G, Triscari M, Varrica D. J Environ Monit 2003; 5:766–773.
124. Cal-Prieto MJ, Carlosena A, Andrade JM, Martinez ML, Muniategui S, Lopez-Mahia P, Prada D. Water Air Soil Pollut 2001; 129:333–348.
125. Crecelius EA, Bothner MH, Carpenter R. Environ Sci Technol 1975; 9:325–333.
126. Kimbrough DE, Carder NH. Environ Pollut 1999; 106:293–298.
127. Schulze D, Kruger A, Kupsch H, Segebade C, Gawlik D. Sci Total Environ 1997; 206:227–248.
128. Muller J, Ruppert H, Muramatsu Y, Schneider J. Environ Geol 2000; 39:1341–1351.
129. Ainsworth N, Cooke JA, Johnson MS. Environ Pollut 1990; 65:65–77.
130. Li XD, Thornton I. Appl Geochem 1993; 8:51–56.
131. Hozhina EI, Khramov AA, Gerasimov PA, Kumarkov AA. J Geochem Explor, 2001; 74:153–162.
132. Hammel W, Debus R, Steubing L. Chemosphere 2000; 41:1791–1798.
133. Baroni F, Boscagli A, Protano G, Riccobono F. Environ Pollut 2000; 109:347–352.
134. Koch I, Wang LX, Feldmann J, Andrewes P, Reimer KJ, Cullen WR. Int J Environ Anal Chem 2000; 77:111–131.
135. Filella M, Belzile N, Chen YW. Earth-Sci Rev 2002; 59:265–285.
136. Millen BMJ. Mitt Oesterr Geol Ges 2001; 94:139–156.
137. Barbante C, Boutron C, Morel C, Ferrari C, Jaffrezo LJ, Cozzi G, Gaspari V, Cescon P. J Environ Monit 2003; 5:328–335.
138. Ferrari CP, Clotteau T, Thompson LG, Barbante C, Cozzi G, Cescon P, Hong SM, Maurice-Bourgoin L, Francou B, Boutron CF. Atmos Environ 2001; 35:5809–5815.
139. Berg T, Roeyset O, Steinnes E, Vadset M. Environ Pollut 1995; 88:67–77.
140. Steinnes E, Hanssen JE, Rambaek JP, Vogt NB. Water Air Soil Pollut 1994; 72:121–140.
141. Kuik P, Wolterbeek HT. Water Air Soil Pollut 1995; 84:323–346.

142. Krachler M, Burow M, Emons H. J Environ Monit 1999; 1:477–481.
143. MacKenzie AB, Logan EM, Cook GT, Pulford ID. Sci Total Environ 1998; 222:157–166.
144. Zheng J, Ohata M, Furuta N. Analyst 2000; 125:1025–1028.
145. Jenkins RO, Craig PJ, Goessler W, Irgolic KJ. Human Exp Toxicol 1998; 17:138–139.
146. Cullen A, Kiberd B, Devaney D, Gillan J, Kelchan P, Matthews TG, Mayne P, Murphy N, O'Regan M, Shannon W, Thornton L. Arch Dis Childhood 2000; 82:244–247.
147. Kentner M, Leinemann M, Schaller K-H, Weltle D. Int Arch Occup Environ Health 1995; 67:119–123.
148. Krachler M, Emons H, Zheng J. Trac-Trends Anal Chem 2001; 20:79–90.
149. Krachler M, Emons H. Fresenius J Anal Chem 2000; 368:702–707.
150. Krachler M, Emons H. J Anal At Spectrom 2000; 15:281–285.
151. Krachler M, Emons H. J Anal At Spectrom 2001; 16:20–25.
152. Eckel WP, Rabinowitz MB, Foster GD. Environ Pollut 2002; 117:273–279.
153. Minoia C, Sabbion E, Apostoli P, Pietra R, Pozzoli L, Gallorini M, Nicolaou G, Alessio L, Capodaglio E. Sci Total Environ 1990; 95:89–105.
154. Caroli S, Alimonti A, Coni E, Petrucci F. Crit Rev Anal Chem 1994; 24:363–398.
155. Gebel T, Claussen K, Dunkelberg H. Int Arch Occup Environ Health 1998; 71:221–224.
156. Zotov AV, Shikina ND, Akinfiev NN. Geochim Cosmochim Acta 2003; 67:1821–1836.

8

Microbial Transformations of Radionuclides: Fundamental Mechanisms and Biogeochemical Implications

Jon R. Lloyd and Joanna C. Renshaw

*School of Earth, Atmospheric and Environmental Sciences and
The Williamson Research Centre for Molecular Environmental Sciences,
The University of Manchester, Manchester M13 9PL, UK*

1. INTRODUCTION

The release of radionuclides from nuclear sites and their subsequent mobility in the environment is a subject of intense public concern. Natural sources of radioactivity include U (present in the Earth's crust at a concentration of 1.8 ppm), Th, Ra isotopes, and radon [1], while significant quantities of natural and artificial/manmade radionuclides were also released as a consequence of nuclear weapons testing in the 1950s and 1960s, and via accidental release, e.g., from Chernobyl in 1986. The major burden of anthropogenic environmental radioactivity, however, is from the controlled discharge of process effluents produced by industrial activities allied to the generation of nuclear power.

The inventory of radionuclides generated during the last 60 years of operating fission reactors is long and includes ^{237}Np, Pu isotopes, Am, ^{3}H, ^{14}C, ^{85}Kr, ^{90}Sr, ^{99}Tc, ^{129}I, and ^{137}Cs. Wastes containing some or all of these radionuclides are produced at the many steps in the nuclear fuel cycle, and vary considerably from low level, high-volume radioactive effluents produced during uranium mining to the intensely radioactive plant, fuel and liquid wastes produced from reactor operation and fuel reprocessing. All wastes pose a potential threat to the environment and require (i) treatment prior to release, and (ii) a much deeper understanding of the biological and chemical factors controlling the mobility of radionuclides in the environment should they be dispersed by accident or as part of a controlled/monitored release, e.g., in effluents.

The aim of this chapter is to give an overview of what is known about the interactions of microorganisms with key radionuclides, and where appropriate to discuss how such interactions can impact on the mobility of radionuclides in the environment. In addition, as there is intense interest in harnessing these natural processes for *in situ* and *ex situ* remediation of radioactive waste, the biotechnological applications of radionuclide–microbe interactions are also discussed.

2. SOURCES AND NATURE OF RADIOACTIVE WASTE

The majority of radioactive waste, especially that containing transuranic elements, has been generated by nuclear power and nuclear weapons programs [2]. Fallout from nuclear weapons tests account for most of the total activity released into the environment [3], estimated to be $>14.8 \times 10^{15}$ Bq ($\sim$3800 kg) ^{239}Pu and ^{240}Pu [4]. Contamination from fuel reprocessing is much less than from weapons testing [2,4,5], but this mode of release can result in localized high levels of contamination [4,6]. The smallest contribution comes from accidental release [2,4], for example, the much publicized Chernobyl accident is thought to have released 62×10^{12} Bq ^{239}Pu and ^{240}Pu [4].

The scale of the legacy of nuclear activities is perhaps best illustrated by the inventories at the US Department of Energy's 120 sites, which contain 1.7 trillion gallons of contaminated ground water and 40 million cubic metres of contaminated soil and debris. More than 50% of the sites are contaminated with radioactive waste, with the priority radionuclides being ^{137}Cs, ^{239}Pu, ^{90}Sr, ^{99}Tc, and ^{238}U/^{235}U, in addition to toxic heavy metals including chromium, lead, and mercury [7]. These problems are not exclusive to the US, however, and sites in the former USSR that pose concern include Kyshtym (high level waste tank explosion), Chelyabinsk (leaks and high level waste disposal to surface waters) and, as previously mentioned, Chernobyl (reactor explosion). In the UK, there are also nuclear sites with contaminated soil and groundwater that need monitoring or remediation, including Sellafield, Aldermaston, and Dounreay.

When looking at the environmental impact of the nuclear fuel cycle (Fig. 1) in more detail, the predominant waste-producing operations are defined as mining and milling of uranium ore, reprocessing of spent nuclear fuel and the decommissioning of facilities [8]. Focusing on the first stage in the cycle, ore bodies contain $<1\%$ U, and the wastes (mine tailings and contaminated groundwater) are high volume, e.g., tens of millions of tonnes of debris, with relatively low but significant activity from the daughter elements of both ^{235}U and ^{238}U. Dominating the inventory of this type of waste are ^{230}Th and ^{226}Ra as well as Po and Pb. Recent studies have emphasized poor stability of the solid waste if

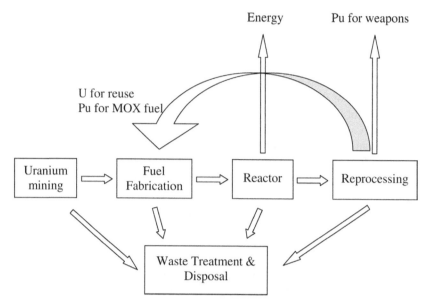

Figure 1 The nuclear fuel cycle.

conditions are not maintained to inhibit microbial metabolism. Soluble Ra is often co-precipitated with $BaSO_4$ by the addition of $BaCl_2$ to the sulfate-rich tailing effluents. The resulting $Ba_nRa_{2-n}SO_4$ can be utilized by sulfate-reducing bacteria, releasing H_2S, Ba^{2+}, and Ra^{2+} [9]. Dissimilatory iron-reducing bacteria, e.g., *Shewanella putrefaciens*, have also been shown to release dissolved ^{226}Ra from the tailings where Ra was precipitated using $Fe(OH)_3$ [10].

During reprocessing [11], the cladding is first removed, then the fuel repro-cessed. Fission reactions in the fuel material during nuclear power generation leads to the production of medium-weight elements including radioiodine, noble gases, and rare earth elements. Neutron capture results in the formation of the high mass transuranic elements and their decay products including various isotopes of Pu, Am, and Np. These products, together with residual U, dominate the waste inventory from nuclear power-generating facilities. In the PUREX process, the spent fuel is dissolved in nitric acid and clarified to remove any solid precipitates. After conditioning (for acidity and uranium content), the U and Pu are selectively co-extracted into an organic phase containing tri-*n*-butyl phosphate (TBP) in an organic diluent. The residual, highly radio-active, waste is currently further processed prior to calcination and vitrification for long term storage. Historically this waste has not always been treated this way, e.g., the Hanford site tanks with aqueous fuel reprocessing waste. Many nations are also seeking methods of removal of long-lived transuranics from this waste (isotopes of Np, Am, Cm, and Cf) for transmutation. The U and Pu are then separated into separate product streams through several further solvent extraction operations and control of the Pu oxidation states (and hence extractability). The final oxide products can then be recycled as U-oxide or U/Pu mixed oxide (MOX) fuel.

At the end of their useful life, nuclear reactors, processing buildings, and all associated equipment need to be decontaminated or disposed of as waste. Structural materials, including fuel cladding and reactor walls can also be activated by neutrons and gamma rays during nuclear power generation. Subsequent decontamination of surfaces during the decommissioning of reactors can be achieved using chelating agents which form selective and strong complexes with radionuclides. The aminopolycarboxylic chelating agents, ethylenediaminetetraacetic acid (EDTA), diethylenetriaminepentaacetic acid (DTPA), and nitrilotriacetic acid (NTA) are frequently used and form stronger complexes with actinides than soil humic acids, preventing immobilization of the radionuclides in soil [12]. Indeed, EDTA-enhanced nuclide migration has been implicated at both the Oak Ridge National Laboratory and Maxey Flats radioactive waste burial sites [12]. EDTA is biodegraded slowly in the environment [13] by two possible pathways [14], the first via the stepwise removal of acetate groups to leave ethylenediamine, or the second via NTA–aldehyde with the removal of an iminodiacetate, a glycine and then an ammonium group. There has been little work done on the biodegradation of EDTA, NTA, or DTPA chelated to uranium or transuranic elements. This is in contrast to studies on metal–citrate biodegradation described in Section 4.3. Citric acid, a naturally-occurring organic complexing agent, has also found use in the decontamination of nuclear reactors [15].

It is clear that there are a wide variety of waste types produced by the nuclear industry, from high level waste destined for vitrification and storage in deep repositories, through intermediate waste that may be encapsulated in cement before disposal, to the very large volumes of low level waste that will be stored in drums and buried in near-surface repositories, or in some cases released into the marine environment. In many examples, deficiencies in past disposal strategies have left us with considerable legacy problems that must be dealt with. Indeed, the emerging market for the clean up of radioactive contamination in the US may already be worth as much as $1 trillion, while current estimated costs of decontamination and safe disposal of radioactive waste in the UK have been estimated at £48 billion in a recent UK Government White paper [16].

Unfortunately, existing chemical techniques are not always economical or cost effective for the remediation of water or land contaminated with metals and radionuclides. Current strategies for land contaminated with metals include the use of "dig and dump" approaches that only move the problem to another site, and are expensive and impractical for large volumes of soil or sediment. Likewise, soil washing, which removes the smallest particles that bind most of the metals, is useful but can be prohibitively expensive for some sites. "Pump and treat" technologies rely on the removal of metals from the site in an aqueous phase which is treated *ex situ* (e.g., above land). These approaches can cut down on excavation costs but are still expensive, and metal removal can be inefficient. A potentially economical alternative is to develop biotechnological approaches based on natural biogeochemical transformations, that could be used in the sediment or soil (*in situ*) to either extract the metals or stabilize them in forms that are immobile or non-toxic.

In addition to developing biotechnologies to treat contaminated land, there is also considerable interest in more effective techniques that can be used to treat water contaminated with metals from a range of industrial processes. Problems inherent in currently used chemical approaches include a lack of specificity associated with some ion exchange resins, or the generation of large quantities of poorly settling sludge through treatment with alkali or flocculating agents. Again, by harnessing radionuclide–microbe interactions, it may be possible to obviate these problems. In order to understand the fundamental mechanisms underpinning the biogeochemical cycling of radionuclides, we must first consider the often complicated and unique chemical characteristics of these substrates.

3. A BRIEF INTRODUCTION TO THE CHEMISTRY OF RADIONUCLIDES

A wide range of isotopes are present in nuclear waste. For example, thermal fission of U-235 yields more than 100 fission products ranging from ^{72}Zn to ^{161}Tb but many of the isotopes present are either stable or have short half-lives [17]. The most important radionuclides from an environmental perspective are those that have significant activity, long half-lives, are produced in relatively large quantities and/or are bioavailable. The key radionuclides that will be considered here in most detail are given in Table 1, with their half-lives, major decay mode, and oxidation states.

The actinides U, Np, Pu, and Am all have isotopes with long half-lives and are all present in the environment in significant quantities [18]. Of these, only U occurs naturally in significant amounts and is present in the Earth's crust at a concentration of 1.8 ppm. The transuranic elements Np, Pu, and Am are all produced in nuclear reactors. The other radionuclides listed are all fission products present in nuclear waste, with the exception of ^{60}Co. ^{60}Co is generated by neutron activation and emits beta and very penetrating gamma radiation. It is widely used in industrial applications requiring a radioactive source, for example, in checking welded structures for faults, and is also found in the steel cladding surrounding spent nuclear fuel. All of these fission products are present in the environment in significant quantities as a result of nuclear weapons testing, nuclear accidents and nuclear fuel reprocessing and are bioavailable [17,19,20]. For example, both ^{99}Tc and ^{129}I can be accumulated in the thyroid gland, whilst ^{90}Sr is accumulated in bone [17]. Both ^{99}Tc and ^{129}I have very long half-lives (2.15×10^5 years and 1.57×10^7 years, respectively). ^{90}Sr and ^{137}Cs have much shorter half-lives

Table 1 Characteristics of Key Radionuclides of Environmental Importance

Element	Isotope	Half-life (y)	Oxidation states[a]	Major decay mode
Cobalt	Co-60	5.271	+2, +3	Beta, gamma
Strontium	Sr-90	29.1	+2	Beta
Technetium	Tc-99	2.15×10^5	+4, +7	Beta
Iodine	I-129	1.57×10^7	−1, 0, +1	Beta, gamma
Cesium	Cs-137	30.17	+1	Gamma
Uranium	U-238	4.47×10^9	+3, +4, +5, **+6**	Alpha
Neptunium	Np-237	2.14×10^6	+3, +4, **+5**, +6, +7	Alpha
Plutonium	Pu-238	87.7	+3, **+4**, +5, +6, +7	Alpha
	Pu-239	2.41×10^4		Alpha
	Pu-240	6.55×10^5		Alpha
	Pu-241	14.4		Beta
Americium	Am-241	432.7	(+2), **+3**, +4, +5, +6, +7	Alpha

[a]For the actinides, the dominant oxidation state in oxic environments is in bold.

(29.1 years and 30.17 years, respectively) but are environmentally mobile and therefore are also significant contaminants.

The environmental fate of radionuclides will depend on a number of environmental factors, but the key property of the radionuclide is the oxidation state (Table 1). The mid actinides (U, Np, Pu, and Am) exhibit variable oxidation states [21] leading to very complicated chemical behavior, especially with respect to aqueous solution chemistry where key actinides show a strong tendency for complex formation [22]. These complexes have high coordination numbers and variable geometries [21,23]. The multiplicity of oxidation states available to the actinides means that it can be difficult to predict what species will be present and how these species will interact with microorganisms, and behave in the environment.

The oxidation state of uranium can vary from III to VI [24]. However, only the IV and VI oxidation states are stable in environmental conditions, with the VI oxidation state being the most stable form of U, as the linear uranyl cation $[UO_2]^{2+}$ [5,19]. The III oxidation state is readily oxidized, whilst the V oxidation state tends to disproportionate:

$$4H^+ + 2UO_2^+ \longrightarrow U^{4+} + UO_2^{2+} + 2H_2O$$

Both neptunium and plutonium can be found in oxidation states from III to VII [24]. However, Np only exists in the IV to VI states in the environment, with V being the most stable form [19,25,26]. The most stable oxidation state of Pu is IV, but III, V, and VI can all be environmentally stable [19,23,27]. Both Pu and Np can exist in the VII oxidation state in strongly alkaline solutions [28–31]. Americium can exist in oxidation states ranging from II to VII, although II is only found in the solid state and VII is not stable [24]. The most stable, and only environmentally important, oxidation state of Am is III [22,25].

All of these actinides can exist in the IV, V, and VI oxidation states. In the IV oxidation state, the actinide ions have a high charge and act as acids in solution [32]. Therefore, they are prone to hydrolysis and polymerization. In particular, the IV cations of U and Pu can form polymeric species containing oxo and hydroxo bridges, leading to very complicated solution chemistry [22,33–36]. The V and VI oxidation states are dominated by actinyl ions, $[AnO_2]^{n+}$ ($n = 1$ for V and 2 for VI; [37]). The $[AnO_2]^{n+}$ ion is very stable and essentially linear, with further coordination (4–6 ligands) around the equatorial plane [18].

The redox chemistry of the fission products is less complicated than that of the actinide elements. Both Cs and Sr only exist in one oxidation state, I for Cs and II for Sr. Therefore, their environmental behavior is not directly dependent on any redox chemistry. However, their oxidation states do account for their high bioavailablity. Radiocesium is released in large quantities during the controlled discharge of low level liquid wastes (nuclear fuel reprocessing streams and reactor cooling waters). In common with other alkali metals and unlike most other radionuclides of environmental significance, Cs is a very weak Lewis acid with a low tendency to interact with ligands [38]. It forms electrostatic (ionic) rather than covalent bonds with oxygen-donor ligands, binding only weakly to organic and inorganic ligands. Cs^+ is analogous to K^+ and can be taken up into cells via transport pathways for K^+ [13]. Sr, in contrast to Cs, but in common with other alkaline earth metals, e.g., Ca^{2+}, Ba^{2+}, and Ra^{2+}, readily forms complexes and insoluble precipitates. Radium is also of interest as it is present in dilute wastewaters from the uranium mining and milling industries, and in an unknown number of sites resulting from the historically widespread use of luminous paints containing Ra. The concentration of Ra^{2+} in wastewaters is typically in the picogram per liter range, compared to other cations such as Ca^{2+}, Mg^{2+}, and Na^+, which are typically in the order of several milligrams per liter [39]. Macaskie [12] calculated that for a biological treatment strategy to meet the current low permissible discharge limits, over 99% of the Ra present must be removed against this high concentration of contaminating cations.

There are two environmentally stable oxidation states for Tc, VII as the pertechnetate anion $[TcO_4]^-$, and IV [13]. $[TcO_4]^-$ is the most stable chemical form of Tc under aerobic conditions [17] and is an analogue of sulfate, enabling it to be taken up into the food chain [40,41]. Under anaerobic conditions, Tc(IV) is stable and tends to exist as the insoluble dioxide [42]. Iodine can exist in $-I$, 0, I, III, and V oxidation states. The most important oxidation states in the environment are $-I$, 0 and V, as I^-, I_2 and IO_3^-, respectively [43]. In aqueous systems, the anionic I^- and IO_3^- species are dominant [43]. Cobalt can exist in both II and III oxidation states in aqueous solution. Co(II) is the dominant oxidation state in seawater, but complexes of Co(III) can also be found [44]. For example, a number of US Department of Energy sites are contaminated with ^{60}Co(III)–EDTA complexes [13].

3.1. Environmental Chemistry

Oxidation state is the most important factor controlling the speciation and therefore the environmental behavior of the actinides. The most obvious influence of oxidation state on environmental transport is on the complexes and compounds formed, whether they are soluble, and therefore mobile, or insoluble. For example, in the V and VI oxidation states, as dioxo cations, actinide ions tend to be soluble whilst the III and IV cations are much more insoluble [42]. Therefore, U and Np, which are most stable as the dioxo cations, will tend to be more mobile in the environment than Pu and Am, which are most stable in the IV and III oxidation states, respectively. I_2, with iodine in the 0 oxidation state, is volatile and can react with organic compounds, while the other environmental oxidation states of iodine are more stable [42]. More detail on the role the microorganisms play in controlling radionuclide oxidation state is given in Sections 4 and 5.

The distribution of radionuclides between solid and soluble phases is controlled by four processes: precipitation, complexation, sorption, and colloid formation [18], dependent on the oxidation state of the radionuclides and environmental factors such as pH and Eh, presence and concentration of ligands, and the nature of organic matter, and mineral surfaces present. Microorganisms can also cause dramatic changes in the mineralogy of sediments, and so are capable of indirect interactions with radionuclides, in addition to direct interactions with the cell, and its components (examples discussed in Sections 4.4 and 4.5).

As previously stated, the pH and Eh can effect both the oxidation state and speciation in the environment. Natural waters usually have a pH range of pH 5–7 and a wide range of redox potentials (-300 to $+500$ mV) [23]. For radionuclides that can exist in several oxidation states, pH and Eh can stabilize particular states. For example, the IV oxidation state of Tc is stabilized in conditions of neutral pH and where Eh is $< +220$ mV [45]. However, radionuclides such as Cs and Sr are less directly affected by pH and Eh, as they exist only as Cs^+ and Sr^{2+}, which are soluble across a wide pH range [42].

Complexation by ligands will also have a significant affect on the environmental behavior of radionuclides. Ligand complexation can either increase the solubility and mobility of an ion or can lead to precipitation. With hydroxide, $[UO_2]^{2+}$ forms insoluble precipitates at neutral pH values, but $[UO_2]^{2+}$ carbonato species are soluble [22]. Np(V) is usually soluble, but at high carbonate concentrations, it forms insoluble species [46]. In some cases ligands, including those of microbial origin, e.g., carbonate (Section 4.4), can stabilize particular oxidation states. The most environmentally stable oxidation state of Pu is IV, but the stability of Pu(V) is increased by carbonate complexation [23]. The two most important ligands in many environments are probably hydroxide and carbonate

[23], but other inorganic and organic ligands may also be present, including phosphate, sulfate, fluoride, nitrate, and carboxylic acids. Of particular importance are humic and fulvic acids, which are abundant in natural waters [47]. They contain carboxylic, hydroxy, and phenolic groups and so can bond strongly to metals, forming stable complexes [25].

Formation of colloids or sorption to minerals will also significantly affect the mobility of radionuclides in the environment. Colloids containing radionuclides can form either through condensation of particular radionuclide species by a hydrolytic or precipitation process or through sorption onto colloids of other material, for example, iron oxyhydroxides or organic matter [18,25]. Whilst sorption to mineral surfaces will obviously reduce radionuclide mobility, sorption to or formation of colloids can either retard or enhance mobility, depending on the size and charge of the colloid. Larger colloids will travel more rapidly than smaller colloids, relative to the groundwater. If the colloid has the same charge as the surrounding solid phases, electrostatic repulsion will increase the relative velocity further. Colloids with the opposite charge may be sorbed onto the solid phases [18]. The role of microbial products (e.g., extracellular polysaccharides) in promoting colloid formation is an interesting, but generally poorly studied area, especially in the context of radionuclide migration. In general, cations of radionuclides tend to sorb onto minerals and colloidal material whilst neutral and anionic species do not sorb and so remain more mobile [46]. Both Sr^{2+} and Cs^+ strongly sorb to clay minerals [42]; ^{60}Co is found to associate with Fe/Mn oxides [46]. Tc(IV) reacts readily with mineral surfaces such as iron oxyhydroxides and sulfides and with humic acids [42]. However, $[TcO_4]^-$ is only poorly sorbed by solid phases and so is more mobile than Tc(IV) [48].

Actinides are also reported to sorb to colloids and minerals. Both U(VI) and Am(III) are reported to be strongly sorbed to aquatic colloids, rich in humic and fulvic acids [49], whilst Am(III) and Pu(IV) are reported to sorb strongly to minerals such as hematite [23]. However, Np(V) is only negligibly sorbed to the colloids and minerals [23]. The low sorption [50] is probably a result of the relatively low charge density on the metal center [23].

As can be seen from these few examples, the environmental chemistry of radionuclides can be very complicated, particularly when multiple oxidation states can exist and there are a number of environmental factors that will affect mobility. Microbes will also have a very significant impact on the behavior of the radionuclide, either directly, by affecting the oxidation state or speciation of the radionuclide, or indirectly, by altering the local environment, for example, through changes in geochemistry or mineralogy. The potential biological mechanisms underpinning such radionuclide–microbe interactions is the focus of Section 4.

4. RADIONUCLIDE–MICROBE INTERACTIONS: FUNDAMENTAL MECHANISMS

From the preceding sections, it is apparent that the environmental fate of a radionuclide is governed by the interplay between the background matrix of the radioactive material, the often complex chemistry of the radionuclide in question and a broad range of chemical factors associated with the environment that has been impacted by radioactive material. In addition, microbial activity will have a profound effect on the solubility of radionuclides via the following mechanisms: biosorption, bioaccumulation, biotransformations, biomineralization, and microbially-enhanced chemisorption of heavy metals (MECHM).

4.1. Biosorption

The term biosorption is used to describe the metabolism-independent sorption of heavy metals and radionuclides to biomass. It encompasses both adsorption; the accumulation of substances at a surface or interface and absorption; the almost uniform penetration of atoms or molecules of one phase forming a solution with a second phase [51]. Both living and dead biomasses are capable of biosorption and ligands involved in metal binding include carboxyl, amine, hydroxyl, phosphate, and sulfhydryl groups.

Dead biomass often sorbs more metal than its live counterpart [52,53], presumably due to an increase in accessible metal-binding sites. Thus, dead biomass may be better suited to treatment of highly toxic radioactive wastes. Biosorption is generally rapid and unaffected over modest temperature ranges and, in many cases, can be described by isotherm models such as the Langmuir, Freundlich, and Brunauer–Emmett–Teller (BET) isotherms [54–56]. Ultimately, however, the amount of residual metal remaining in solution at equilibrium is governed by the stability constant of the metal–ligand complex [12], and the only way to change the equilibrium position is to transform the metal from a poorly-sorbing species to one which has a higher ligand-binding affinity, e.g., by a change of metal valence (Section 2.3), or by modifying the binding ligand (e.g., through genetic engineering) to one which has a greater binding affinity for the given radionuclide. For a more recent and extensive review of radionuclide biosorption, the reader is directed to Ref. [13].

4.2. Metabolism-Dependent Bioaccumulation

Energy-dependent metal uptake has been demonstrated for most physiologically important metal ions, and some radionuclides enter the cell as chemical "surrogates" using these transport systems. Once in the cell, toxic metals (and potentially radionuclides), may be sequestered by cysteine-rich metallothioneins [57,58] or, in the case of fungi, compartmentalized into the vacuole [51,59]. In this context, it should be emphasized that the uptake of higher mass radionuclides, e.g., the actinides, into microbial cells has been reported sporadically and remains poorly characterized [52,60–62]. Indeed, a recent review proposed that intracellular accumulation of uranium in such studies was due to increased membrane permeability caused by uranium toxicity, and was not driven by metabolism-dependent transport mechanisms [63]. Cs, however, is actively taken up into microbial biomass due to its close similarity to the K^+ ion, using broad-specificity alkali metal uptake transporters (Section 5.2).

Factors that inhibit cellular energy metabolism can prevent bioaccumulation of metals and may, therefore, limit microbial uptake in toxic/highly radioactive waste. For example, monovalent cation transport, e.g., for K^+ and analogs, is linked to the plasma membrane-bound H^+-ATPase via the membrane potential, and is affected by increasing metal concentrations that deenergize the cell membrane [64]. Other factors that can reduce active metal and radionuclide uptake by some microorganisms include the absence of substrate, anaerobiosis, incubation at low temperatures and the presence of respiratory inhibitors such as cyanide [64].

4.3. Enzymatically-Catalyzed Biotransformations (Bioreduction, Biomethylation, Biodegradation)

Microorganisms can catalyze the direct transformation of toxic metals and metalloids to less soluble or more volatile forms via two enzymatic mechanisms. Bioreduction can result in precipitation of the metal/metalloid [65], while biomethylation can yield highly volatile derivatives (e.g., for Se, Te, and Hg; [66]). Although the mechanisms are distinct, the end result is the same: a decrease in the concentration of soluble metals in contaminated water. While the microbial reduction of radionuclides including U(VI), Pu(IV), Np(V), and Tc(VII) has been demonstrated [67–70], biomethylation of radionuclides has received little attention, with the exception of ^{129}I. The high instability of alkylated actinides [71] may, in part, explain this observation. Enzymatic biodegradation of metal-chelates can also control the solubility of key radionuclides.

4.3.1. Bioreduction of Radionuclides

In the absence of oxygen, specialist microorganisms are able to respire through the reduction of alternative electron acceptors. The environmental relevance of nitrate, sulfate or carbon dioxide reduction has long been recognized, but more recent studies have shown that high valence metals and radionuclides [e.g., Fe(III), Mn(IV), and U(VI)] can also function as alternative electron acceptors during anaerobic respiration by specialist organisms [72].

Fe(III) is the most abundant electron acceptor in many sedimentary environments [73], and respiration using Fe(III) has been studied in most detail [73]. As Fe(III) can be a surrogate for U(VI) in biological systems, outcompeting U(VI) in biosorption experiments [63], it is not unreasonable to expect the reduction of U(VI) (and possibly other actinides) to proceed by mechanisms similar to those of Fe(III) reduction (see Ref. [72] for a review of mechanisms of Fe(III) reduction). Indeed, as far as the authors are aware, all Fe(III)-reducing bacteria that have been tested are also able to reduce U(VI). These factors, in combination with indirect effects of Fe(III) reduction on the solubility of Tc and some actinides (Section 5), suggest that the biogeochemical cycles of Fe(III), and several important radionuclides are linked.

4.3.2. Biodegradation of Associated Organic Compounds

Citrate has found use as a chelating agent in decontamination operations forming highly soluble radionuclide–citrate complexes, which can be degraded by microorganisms resulting in subsequent re-precipitation of radionuclide [15]. It should also be noted that citrate, and other organic acids with strong radionuclide-complexing capabilities, can also be produced by microbial activity (see the reviews [74,75]). Early work suggested that the type of complex formed between the metal and citric acid plays an important role in determining its biodegradability, with binuclear uranium–citrate complexes recalcitrant to microbial degradation [15] but subsequent photodegradation was possible for these problematic complexes. This opened up the way for a multi-stage process for the biodegradation of mixed metal–citrate complexes via microbiological and photochemical steps [15].

Cultures of *Pseudomonas aeruginosa* and *P. putida* were also able to grow using a range of metal–citrate complexes as carbon, with metal precipitation promoted by the addition of inorganic phosphate [76]. A unique aspect of this study was the use of *P. putida* to treat Ni–citrate waste generated by cleaning a bioinorganic ion exchange column previously used to remove Ni. EDTA, another industrial chelating agent, also forms very strong complexes with di- and trivalent metals that are degraded by a mixed microbial population [77] and by bacterial strain DSM 9103 [78]. However, complexes between radionuclides and these aminopolycarboxylic chelating agents may be more recalcitrant to microbial degradation.

4.4. Biomineralization via Microbially-Generated Ligands

Metals and radionuclides can precipitate with enzymatically-generated ligands, e.g., phosphate, sulfide, or carbonate. The concentration of residual free metal at equilibrium is governed by the solubility product of the metal complex (e.g., 10^{-20} to 10^{-30} for the sulfides and phosphates, higher for the carbonates). Most of the metal or radionuclide should be removed from solution if an excess of ligand is supplied. This is difficult to achieve using chemical precipitation methods in dilute solutions; the efficiency of microbial biomineralization processes is due to the high concentrations of ligand in juxtaposition to the cell surface, which can also provide nucleation foci for the rapid onset of metal precipitation; effectively the metals are concentrated "uphill" against a concentration gradient.

4.4.1. Phosphates

The best-documented organism for metal phosphate biomineralization, a *Citrobacter* sp., grows well on cheaply available substrates and viable cells are not required for metal uptake since this relies on the activity of a single enzyme (phosphatase) catalyzing hydrolytic cleavage of a supplied organic phosphate donor [79].

It should be noted that the *Citrobacter* species has now been reclassified as a *Serratia* sp. on the basis of molecular methods, the presence of the *phoN* phosphatase gene and the production of pink pigment under some conditions [80]. The role of a phosphatase in metal accumulation was confirmed by the finding that the expression of the cloned *phoN* gene in *Escherichia coli* conferred the ability to bioprecipitate uranyl phosphate [81].

4.4.2. Sulfides

Sulfide precipitation, catalyzed by mixed cultures of sulfate-reducing bacteria, was utilized first to treat water co-contaminated by sulfate and zinc [82] and more recently, soil leachate contaminated with sulfate alongside metal and radionuclides [83]. Ethanol was used as the electron donor for the reduction of sulfate to sulfide in both these *ex situ* treatment processes. The ubiquitous distribution of sulfate-reducing bacteria in acid, neutral, and alkali conditions [84], suggests that they have the potential to play a critical role in controlling the migration of radionuclides in a range of environments.

4.4.3. Carbonates and Hydroxides

Ralstonia eutropha (formerly *Alcaligenes eutrophus*) is able to precipitate metals (and potentially some radionuclides) via plasmid-borne resistance mechanisms. Here, proton influx countercurrent (antiport) to metal efflux results in localized alkalinization at the cell surface [85,86]. Metal hydroxides and carbonates are formed at these high local pH values, the latter from carbon dioxide formed through respiration. Although this system has not been tested against radionuclides, microbial respiration has also been shown to play a role in the formation of extracellular strontium carbonate by *P. fluorescens* [87]. Microbial activity was also implicated in the deposition of a strontium calcite phase at a groundwater discharge zone [88]. The deposition of the mineral was thought to be the result of carbonate precipitation by epilithic cyanobacteria (driven by HCO_3^- / OH^- exchange during photosynthesis), with a Sr/Ca ratio that promoted deposition of $SrCO_3$ in calcite (up to 1% Sr). Although Sr carbonates are insoluble, this is not true of uranium–carbonate complexes; in fact carbonate has been used as a lixiviant to extract uranium from contaminated soils [89,90]. Furthermore, recent studies have shown that microbial activity (CO_2 production) can enhance the dissolution of uranyl hydroxide and uranyl hydroxophosphato species, via the formation of soluble uranyl carbonate species [91].

4.5. Microbially-Enhanced Chemisorption of Radionuclides

MECHM is a generic term to describe a class of reactions whereby microbial cells first precipitate a biomineral of one metal or radionuclide ("priming deposit"). The priming deposit then acts as a nucleation focus, or "host crystal" for the subsequent deposition of the metal of interest ("target" metal), acting to promote and accelerate target metal precipitation reactions [92–94]. Examples of priming deposit include hydrogen uranyl phosphate (HUP), laid down by phosphatase activity, and consisting of sheets of uranyl phosphate ions separated by water molecules creating a regular network of hydrogen bonds [95,96]. The overall outcome of this highly organized crystal lattice is a high mobility of protons in the interlamellar space [97] which gives rise to an ion-exchange intercalative property that can accommodate other metals and radionuclides [96]. Other hexavalent actinides ($[AnO_2]^{2+}$) can also substitute directly for $[UO_2]^{2+}$ in the backbone of the lattice, displacing the uranyl ion to give a hybrid crystal [98].

Another well studied example of MECHM is when H_2S produced by sulfate-reducing bacteria reacts with iron to form FeS, with cell-bound FeS acting as a sorbent for "target" metals, including radionuclides [99]. Insoluble Fe(III) oxide minerals formed by, for example, Fe(II)-oxidizing bacteria [66], can also sorb considerable quantities of target radionuclides (e.g., U(VI) [13]). Finally, Fe(II)-bearing minerals (e.g., magnetite) produced by iron-reducing bacteria, are able to abiotically reduce and precipitate high valence metals and radionuclides, e.g., Tc(VII) and Cr(VI) [100,101]. Tetravalent uranium formed by the action of Fe(III)-reducing bacteria is also able to abiotically reduce and precipitate Tc(VII) [102].

5. THE BIOGEOCHEMISTRY OF KEY RADIONUCLIDES

It is clear from previous sections that microorganisms have the potential to control the solubility of radionuclides via many mechanisms, either directly through interactions with the microbial cell, or via indirect transformations driven by local changes in the chemistry of the environment attributed to microbial activity. Over the last decade there have been significant advances in our understanding of the biogeochemical controls on key radionuclides, facilitated by advances at the interface between biology and chemistry. Contributing factors include improvements in techniques available in specialist areas as diverse as microbiology, molecular biology (including genomics), spectroscopy, surface science, and radiochemistry. However, there is in many cases still a very significant gap between knowledge of radionuclide–microbe interactions obtained from laboratory studies using pure cultures of microorganisms under well-defined conditions, and a detailed understanding of the biogeochemical cycles of key radionuclides in complex microbial communities against a far more complex geochemical/mineralogical background associated with most terrestrial environments.

This knowledge gap is most apparent for key actinides, such as plutonium. Building on previous sections on nuclear waste, the chemistry of radionuclides and the principal mechanisms by which microorganisms interact with metals, this section details the current understanding of biogeochemical cycles for key radionuclides. It encompasses results from both ends of the research spectrum, including pure culture/biochemical studies through to field investigations where appropriate. Where it is possible to tie together these sometimes disparate strands of research, we have tried to propose appropriate conceptual biogeochemical models for key radionuclides. Where knowledge gaps do exist they are highlighted.

5.1. Actinides

The interactions of actinides with microorganisms have been discussed in detail in recent reviews [102–106], including several that focus on the application of microbial systems for the bioremediation of actinide containing sediments and water [12,13,15,107,108]. As described in Section 3, the oxidation state is a key factor that governs the mobility of actinides with the environment, and it is therefore important to give a more detailed overview of the potential impact of microorganisms on the oxidation state of the most environmentally important actinides.

Figure 2 shows the expected oxidation states of key actinides as a function of redox potential at pH 7.0. It is apparent from this figure that the standard redox potential of ferrihydrite/Fe^{2+} ($\sim$0 V [109]) is more electronegative than the potentials for Pu(V)/Pu(IV), Np(V)/Np(IV), and U(VI)/U(IV). Thus, Fe(III)-reducing bacteria have the metabolic potential to reduce these radionuclides enzymatically, or via Fe(II) produced from the reduction of Fe(III) oxides. This is significant because the tetravalent actinides are potentially less soluble due to their high ligand-complexing abilities [13], and are immobilized in sediments containing active biomass [110]. Thus, although it is possible for Fe(III)-reducing bacteria to reduce and precipitate actinides in one step, e.g., the reduction of the highly soluble uranyl cation ($[UO_2]^{2+}$) to form insoluble uraninite (UO_2) [69,111] (see below), some transformations do not result in direct formation of an insoluble mineral phase but in the formation of a cation, that is, potentially less mobile in the environment. A good example here is the reduction of Np(V) to Np(IV) by S. putrefaciens [68] (see below). Finally, it should also be noted, from Fig. 2, that Th(IV), and Am(III) are stable across most Eh values encountered in radionuclide-contaminated waters, but are removed easily through biosorption or biomineralization by a range of ligands including biogenic phosphate [13].

Although it is apparent from several laboratory studies on actinide–microbe interactions, that microbial processes have the potential to control radionuclide mobility, there is little information on the relative importance of these microbial processes in the environment. For example, most studies on the behavior of Pu and Np have focused on physical factors, including mixing of sediments and tidal effects, or on correlating distribution in the environment with historical discharge records. Indeed, for Pu and some other radionuclides,

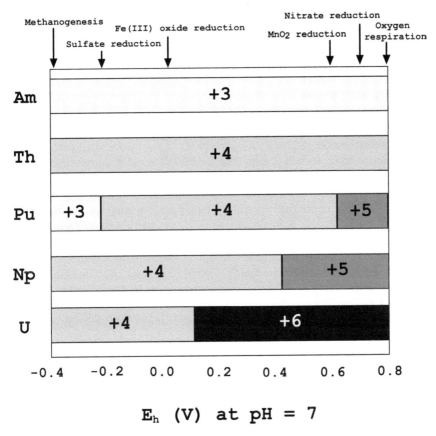

Figure 2   The potential impact of microbial processes on the oxidation state of key actinides. Adapted from Ref. [105].

there remains controversy over the factors that are most important in influencing environmental behavior. The few papers that have investigated (i) variations (seasonal or vertical) in actinide distribution and (ii) the correlation with diagenetic constituents such as Fe, Mn, or dissolved organic carbon (DOC) for evidence of coupling to microbially driven processes, have in many cases been inconclusive or contradictory.

5.1.1. Americium

Americium is stable in the trivalent oxidation state across a wide range of environmental conditions, making it one of the simplest actinides to discuss in the context of metal–radionuclide interactions. There is some evidence that microbes can play an important role in controlling Am(III) solubility from studies on the seasonal variations in actinide porewater concentrations in a salt marsh in west Cumbria [112]. The summer minimum in soluble Am(III) (and also Pu; see below) correlated with maximal microbial biomass levels, although the mechanism of removal from solution was not investigated.

However, early work has shown efficient sorption of [241]Am onto various algal biomasses, which was three orders of magnitude more greater than for Pu, which was, in turn, more than tenfold greater than for Np [113]. Uptake by *E. coli* and a marine bacterium challenged with 0.24 nM Am has also been studied [114] with the production of a higher-affinity ligand implicated as the cells reached stationary phase. Biosorption of [241]Am by *Rhizopus arrhizus* was also reported to be 40-fold greater than for Pu(IV) at pH 2, comparable at pH 4 and only 30% of that of Pu(IV) at pH 7 [115]. Removal of 90% [241]Am from an initial solution of 299 nCi/mL was reported for *Candida utilis* [99].

In addition to removal of Am by biosorption, a study of enzymatically-mediated metal phosphate biodeposition of [241]Am has also been reported [93], with 100% removal from solution. Finally, while discussing trivalent actinides, of which Am is the best studied example, it should be noted that the microbial uptake of an analogous actinide, Cm(III) has been little-studied. Cm(III) in HCl carrier (100 ng/mL initial concentration, in HCl, pH 2) was partially removed by sterilized sediments but almost completely removed by natural sediments after 4 months [110]. The mechanism of removal was not investigated.

5.1.2. Thorium

The oxidation state of thorium is also stable across a wide range of environmental conditions, this time in the tetravalent form. In aqueous solution, An(IV) species are only available as the An^{4+} cation at low pH values (pH 2.0), where protonation of biomass ligands will also occur, effectively reducing the scope for biosorption. Despite this, several groups have reported biosorption of Th(IV) onto fungal biomasses [116,117]. It is difficult to develop accurate models for the biosorption of Th(IV) under these conditions, because if the localized pH is above that of the bulk solution then the sorbed species will not be An(IV) but a species carrying 1–4 hydroxyl groups. Indeed, at pH 4.0 Th(IV) hydrolyzed, forming colloidal $Th(OH)_4$, leading to increased uptake by *R. arrhizus* [118].

In a more recent study on Th(IV) biosorption by *Mycobacterium smegmatis* at pH 1, loadings of $\sim$4% of the biomass dry weight was reported, comparable to values for $[UO_2]^{2+}$ with that strain [119]. Thorium removal as an insoluble biogenic phosphate was, as also reported by Yong and Macaskie [120,121], promoted by the incorporation of NH_4^+, to form thorium ammonium phosphate. Further, co-challenge of the cells with La and Th improved Th removal (to 90% from a 300 μM solution) [121], with analysis of the recovered precipitate showing a hybrid crystal of Th/La phosphate, with molar proportions of Th/La of 1 : 1 by proton-induced X-ray emission analysis.

5.1.3. Neptunium

Neptunium-237 is an alpha emitting radionuclide produced in tonne quantities in nuclear reactors. It will eventually predominate in wastes after 10^4 to 10^7 years due to decay of its grandparent ^{241}Pu and parent ^{241}Am. It has a long half-life (2.13×10^6 years), high biological toxicity and is very mobile as the neptunyl cation ($[Np^{(V)}O_2]^+$) which sorbs relatively poorly to surfaces. Despite the importance of this radionuclide, there have been few attempts to study the interaction of microorganisms with Np.

Np solubility did not share a seasonal cycle linked to microbial activity, unlike Pu and Am in a west Cumbrian salt marsh [112]. Fisher et al. [113] also reported negligible Np uptake by marine algae and in the only detailed study performed prior to 1993 low uptakes (10 mg/g dry weight) was reported with *P. aeruginosa*, *Streptomyces viridochromogenes*, *Scenedesmus obliquus* and *Micrococcus luteus* [122]. Even the metal phosphate-accumulating *Citrobacter* sp. (now *Serratia*) which accumulated 100% of ^{241}Am and 50% of $^{238/239}$Pu (discussed in Section 5.1.5) was largely ineffective against Np [123]. However, more recent studies using *P. fluorescens* have suggested that appreciable quantities of Np can sorb to biomass (85% of 4.75 μM Np(V) removed from solution at pH 7) [50].

Some anaerobic bacteria are also able to reduce Np(V) to the less soluble Np(IV) [50,68]. For example, *S. putrefaciens* previously reported to reduce other radionuclides including Tc(VII) [67] and U(VI) [65,69] reduced Np(V), and co-removal of the reduced species was accomplished by the concerted use of *S. putrefaciens* and the *Citrobacter* sp. to reduce and simultaneously precipitate Np as neptunium phosphate in the presence of NH_4^+ to promote Np(IV) precipitation (see above) [68]. The importance of such transformations in the environment remains to be investigated.

5.1.4. Uranium

In contrast to the other actinides discussed in this section, the biological removal of $[UO_2]^{2+}$ is well-documented (e.g., reviewed in Refs. [12,13]) and removal of the other hexavalent actinide species [e.g., Pu(VI)] would also be predicted from these data. Here it is worth stressing that although radiolytic oxidation of Pu(IV) to Pu(VI) has been reported in saline solution [124], microbial oxidation of Pu(IV) or Np(V) has not been identified, although bio-oxidation of other metals including Fe(II), Mn(II), and As(III) is well-documented [66].

5.1.4.1. Biosorption and MECHM: Most of the studies on uranium uptake by microorganisms have focused on biosorption by various biomasses; the literature is large and the reader is referred to a very detailed discussion on this topic in Ref. [13]. There are also examples of microbially-generated mineral deposits that have been shown to sorb U(VI). For example, ferric (and manganese) oxides, which can be formed by a variety of microorganisms [66], form the basis of the commercial Enhanced Actinide Recovery Process at

British Nuclear Fields Ltd (BNFL; Sellafield, UK). The geochemical literature also provides good evidence for sorption of U(VI) to sulfide minerals [125] with concomitant partial reduction of U(VI) to U(IV) suggested using X-ray photoelectron spectroscopy, Auger electron spectroscopy and Fourier transform infrared (FTIR) analysis. Indeed, MECHM uptake of metals via biogenic FeS, precipitated by sulfate-reducing bacteria, is generally more efficient than "geochemical" FeS [99].

5.1.4.2. Biomineralization: The best documented biomineralization system for U(VI) is its precipitation using inorganic phosphate generated by the mobilization of cellular polyphosphate [126] or via the cleavage of a supplied organic phosphate donor ligand such as tributyl phosphate [127] or glycerol 2-phosphate [79] via phosphatase activity localized in the periplasmic space [60] and exocellularly [128]. Further studies on the latter system have suggested that the production of exocellular enzyme together with phosphate-containing extracellular polymer enables effective resistance to metal toxicity and promotes metal phosphate biomineralization [128]. Blockage by accumulated metal phosphate has never been seen; presumably the crystals are effectively compartmentalized away from the enzyme and sites of substrate access. Also, although the pH optimum of the enzyme is $\sim$5–7.5 [129,130], even acidic U mine waters have been successfully remediated at pH values of 3–4 using this system [131]. It is assumed that exocellular phosphate (and carboxyl) groups act as natural buffers for pH-stasis. Although this approach has been successful for lab-scale biotreatment of U(VI) contaminated water, it is not clear how important such transformations are in controlling uranium solubility in the environment.

5.1.4.3. Bioreduction: The first demonstration of dissimilatory U(VI) reduction was by Lovley and coworkers [69] who reported that the Fe(III)-reducing bacteria *Geobacter metallireducens* (previously designated strain GS-15) and *S. oneidensis* (formerly *Alteromonas putrefaciens* and then *S. putrefaciens*) can conserve energy for anaerobic growth via the reduction of U(VI). It should be noted, however, that the ability to reduce U(VI) enzymatically is not restricted to Fe(III)-reducing bacteria. Other organisms including a *Clostridium* sp. [15] and the sulfate-reducing bacteria *Desulfovibrio desulfuricans* [132] and *D. vulgaris* [133] also reduce uranium, but are unable to conserve energy for growth via this transformation. To date, *D. vulgaris* remains the only organism in which the enzyme system responsible for U(VI) reduction has been characterized in detail. Purified tetraheme cytochrome c_3 was shown to function as a U(VI) reductase *in vitro*, in combination with hydrogenase, its physiological electron donor [133]. *In vivo* studies using a cytochrome c_3 mutant of the close relative *D. desulfuricans* strain G20 confirmed a role for cytochrome c_3 in hydrogen-dependent U(VI) reduction, but suggested additional pathways from organic electron donors to U(VI), that bypassed the cytochrome [134].

More recent studies have identified a homologous cytochrome (PpcA), a triheme periplasmic cytochrome c_7 of the Fe(III)-reducing bacterium

G. sulfurreducens, that may also play a role in U(VI) reduction [135]. The protein was able to reduce U(VI) *in vitro*, while a *ppcA* deletion mutant supplied with acetate as an electron donor had lower activity against U(VI) [135]. Additional (if indirect) evidence linking the activity of this periplasmic protein with U(VI) reduction *in vitro* included the precipitation of the reduced product U(IV) in the periplasm, and the lack of impact of protease treatment of whole cells on the ability to reduce U(VI) [102]. This final result is particularly important, as it implies that U(VI) and Fe(III) are reduced by different mechanisms in *G. sulfurreducens*. U(VI) would seem to be reduced in the periplasm, while the reduction of insoluble Fe(III) oxides was inhibited dramatically by protease treatment, presumably due to removal of surface bound cytochromes required for reduction of the extracellular electron acceptor. The mechanism of U(VI) reduction by a *S. putrefaciens* strain has also been investigated [136]. A novel screening method was used to identify mutants that were unable to reduce U(VI). Evidence was presented to suggest that the mechanism of U(VI) reduction was distinct from those of Fe(III) and Mn(IV) reduction, but may share components of the nitrite reducing pathway [136].

Recent field trials on uranium mill tailings remedial action (UMTRA) sites in the US have demonstrated that microbial reduction of U(VI) could be a potentially useful approach to treat U(VI) contaminated water *in situ* (reviewed in detail in Ref. [108]). In these studies, uranium-contaminated aquifer sediments were collected from UMTRA sites in Colorado and New Mexico and incubated in the laboratory. Addition of acetate stimulated anaerobic conditions and the loss of soluble U(VI) from solution [137]. Loss of soluble U(VI) occurred in live sediments only, at the same time as Fe(II) production and prior to observed losses of sulfate [137]. *Geobacteraceae*, known Fe(III)- and U(VI)-reducing microorganisms, were greatly enriched (up to 40% of the bacterial community detected), supporting the hypothesis that these organisms play a role in U(VI) reduction [138]. In subsequent field studies at the Old Rifle UMTRA site, the loss of U(VI) from groundwater *in situ* was successfully stimulated through the addition of acetate via a gallery composed of 20 injection wells [139]. Removal of an average of 70% of initial concentrations of uranium was recorded in the 18 monitoring wells downstream from the injection wells, over ~50 days.

5.1.5. Plutonium

Studies on the biogeochemistry of plutonium are far more challenging than those discussed for the other actinides above, due to its high radiotoxicity and complex redox chemistry. Although the most stable oxidation state of Pu is (IV), Pu(III), (V), and (VI) can also be stable in the environment. Early studies showed a seasonal variation in Pu in surface waters of Lake Michigan [140] with >75% of total Pu (dissolved and suspended) lost from the epilimnion during summer months. The onset of Pu removal coincided with the end of a major plankton bloom in the lake, leading to a suggestion that Pu was scavenged by phytoplankton (mainly diatoms) and biogenic calcite, which settled to the bottom of the lake.

Although Sholkovitz et al. [141] agreed that there was a seasonal cycle for Pu, this author noted that the precise nature of the cycle is not clear from the earlier study. For example, the correlation between Pu and biogenic SiO_2 in the paper of Wahlgren et al. [140] was based on one data point and also the variation in Pu solubility varies from year to year at this site. Other studies on Pu in Gull Pond, Cape Cod, Massachusetts, suggested Pu that had sorbed to anoxic sediments at the bottom of the pond were released under anoxic conditions, either as the Fe(III)/Mn(IV) oxides in the sediments were reductively solubilized or as the sorbed Pu [presumably Pu(IV)] was reduced to a more soluble oxidation state [141].

More recent studies [142] on a wider range of samples, including pore water and sediments from seven sites in the north west Atlantic, have failed to show a relationship between pore water Pu and Fe/Mn that would be consistent with Pu solubilization due to the reductive dissolution of insoluble Fe/Mn oxides. Also, in the study of Malcom et al. [143] there was no consistent relationship between the Pu profile and any of the major diagenetic indicators (Fe^{2+}, Mn^{2+}, NO_3, PO_4^{3-}, SO_4^{2-}, or DOC).

Interestingly, a very recent study by Morris et al. [144] of pore waters from two sites in the River Esk Valley, North West England, noted changes in soluble Pu coincided with changes in pore water Fe(II) and Mn(II), with Pu minima in summer months coinciding with soluble Fe and Mn maxima. However, soluble divalent Fe/Mn concentrations were too low to suggest significant Fe(III) and Mn(IV) reduction, and Eh values indicated mildly oxidizing conditions, suggesting that the Fe, Mn, and Pu cycles were not related through direct, redox driven cycles, but that these results may reflect other microbial processes.

Several laboratory studies have attempted to clarify the potential role of microorganisms in controlling Pu solubility. Most studies on Pu(IV)–microbe interactions have used very low concentrations of the radionuclide, but one investigation utilized a stock solution of ^{239}Pu of 2 mg/mL as the carbonate, introduced to natural sediments [110]. All of the Pu was removed after 4 months, with only 34% of the removal attributed to biosorption using heat-killed controls, implicating additional biochemical mechanisms. On this note, laboratory studies have suggested that the reduction of Pu(IV) to Pu(III) can be achieved by Fe(III)-reducing bacteria, although the Pu(III) was reported to reoxidize spontaneously [70]. Although Pu(IV) reduction may lead to solublization of sediment-bound Pu(IV), it will yield a trivalent actinide that could react with a range of microbially produced ligands [13]. Efficient scavenging could then occur as for Am(III) as described above.

In addition to bioreduction of Pu(VI), biosorption of Pu(IV) has been observed by pure cultures of *R. arrhizus* at pH 4 and 9, but not by *Gibberella* spp. [115]. The capacity of the biomass at pH 9.4 (3 mg Pu/L solution) was 0.8 mg/g biomass. An alternative approach to the removal of tetravalent Pu [Pu(OH)$_4$] from solution employed *P. aeruginosa* immobilized on a plasma-treated polypropylene web [145]. This was intially used in a batch operation against nuclear plant wastewater containing 1.7 nCi of Pu, and up to 95% of the Pu was removed, but Pu removal (possibly as PuO_2) was particle size-dependent. Finally, a metal phosphate-accumulating *Citrobacter* sp. removed 50% of the $^{238/239}$Pu from a 60 nM solution, by phosphatase-containing cells at steady-state in a flow-through system; none was removed by the corresponding phosphatase-deficient mutant [93]. Efficiency was much higher (100% removal) with the addition of La^{3+}, leading to the formation of $LaPO_4$, which in turn promoted precipitation of Pu via the MECHM mechanism described [121,124].

At present, it is fair to conclude that although microorganisms have the potential to control the solubility of Pu in a range of environments, the precise mechanism of Pu–microbe interactions underpinning the biogeochemistry of this very complicated and challenging element remain unclear.

5.2. Alkali Metals

5.2.1. Cesium

Although the chemical toxicity of stable Cs is low, the long half-life of ^{137}Cs (30.1 years) in combination with high solubility has led to concern over its release. Indeed, recent reports have shown that higher than expected levels of ^{137}Cs persisted in a mobile form in the environment following the Chernobyl accident in 1986, with accumulation of the Cs by microorganisms playing a role in passing the radionuclide to higher trophic levels in the food chain [146]. Pure culture studies have shown that microbial biosorbents are inefficient for Cs uptake [12], but uptake of Cs^+ by actively metabolizing microorganisms is more efficient. Due to the similarity of the K^+ and Cs^+ cations, both are taken up by the same metabolism-dependent transport systems. Indeed, broad-specificity alkali earth metal uptake transporters have been reported in all microbial groups [146]. The various mechanisms of Cs^+ uptake in *E. coli*, cyanobacteria, eukaryotic algae, and fungi are discussed in detail elsewhere [13,147].

Most studies on microbial Cs^+ accumulation have used relatively simple, well defined laboratory solutions. Plato and Denovan [148], however, quantified ^{137}Cs uptake by *Chlorella pyrenoidosa* against a background of K^+ ions. Approximately 83–88% of the Cs^+ was removed if K^+ was present at concentrations between 50 and 375 µM. Removal was <20% if the concentration of K^+ was below or above these values, due to poor growth of the organism or K^+ competition. Other studies have shown, however, that the presence of excess K^+ does not inhibit Cs^+ uptake in all examples. For example, Cs^+ uptake by *Synechocystis* PCC 6803 was unaffected by equimolar concentrations of K^+ [149]. High concentrations of H^+ ions have also been shown to exert a deleterious effect on Cs^+ uptake; in the same study bioaccumulation was enhanced at alkali pH values (pH 10). Similar results have been obtained with yeast cells where Cs^+ uptake was pH-independent at high pH, but decreased as the pH of the growth medium dropped from pH 5.5–3.0 [150].

5.3. Alkali Earth Metals

5.3.1. Strontium

Two studies have demonstrated the biomineralization of strontium by actively metabolizing microorganisms. *Pseudomonas fluorescens* grew in a minimal medium containing strontium complexed to citrate, which was supplied as the sole carbon source [87]. Strontium was excluded from the cell and precipitated in the growth medium as crystalline strontium carbonate, as determined by X-ray fluorescence spectroscopy, X-ray diffraction spectrometry and FTIR spectroscopy. A threefold increase in the concentration of dissolved carbon dioxide in the growth medium implied a role for the gas in mineral deposition. Microbial activity was also implicated in the deposition of a strontium calcite phase at a groundwater discharge zone [88]. The authors interpreted the deposition of the mineral to be the result of carbonate precipitation by epilithic cyanobacteria (driven by HCO_3^-/OH^- exchange during photosynthesis), with a Sr/Ca ratio that promoted deposition of $SrCO_3$ in calcite (up to 1% Sr).

Sr accumulation by a *Citrobacter* species has also been studied [151]. Cells immobilized in polyacrylamide gel removed 47% of Sr from 5.6 L of flow over 8 days (60 mL columns; flow rate of 30 mL/h) as an insoluble phosphate mediated via phosphatase activity. This corresponded to a loading of 363 mg Sr/g biomass dry weight [12]. Sr removal was poor at neutral pH and was maximal only at pH values above the pH optimum of the phosphatase. At alkaline pH and using a series of sequential columns removal of Sr was increased to >90% [151]. Following on from work of Zajic and Chiu [152] who noted Sr uptake by *Penicillium* sp. to 75 mg/g biomass, Watson et al. [153] made a detailed study of Sr biosorption by various microorganisms. *Rhizopus* was the most effective organism tested overall but *M. luteus* gave comparable uptake after 2 days (compared to 14 days incubation with the former). Other studies by this group showed microbial Sr uptake to be instantaneous (maximum loading of 1.2 mg/g cells) with metal uptake identical after 4 h for both free cells and cells immobilized in gelatin [154]. Immobilized cells were used in both batch and flow-through contactors.

deRome and Gadd [56] also studied the biosorption of strontium, along with uranium and cesium by *R. arrhizus* and *Penicillium chrysogenum*. The uptake mechanism for the two organisms was different. In the former, uptake was metabolism-independent. With the latter, uptake was biphasic in the presence of glucose, surface biosorption being followed by energy-dependent influx. Removal of the surface bound metals was achieved by elution with carbonate/bicarbonate solutions or mineral acids. Biosorptive uptake of strontium by waste *Saccharomyces cerevisiae* (to 21% of the biomass dry weight) was also noted by Avery and Tobin [155].

5.3.2. Radium

The biosorption of ^{226}Ra has been studied in detail. Tsezos and Keller [156] used equilibrium biosorption isotherms to quantify the Ra uptake capacity of various types of biomass and activated carbon. Maximum uptake was at pH 7.0, with reduced and negligible uptake at pH 4.0 and 2.0, respectively. The best absorbent was biomass from a municipal-wastewater-activated return sludge (40,000 nCi/g) which bound significantly more Ra than commercially-available activated carbon (3600 nCi/g). With some biomass types, over 99% of the Ra was removed from solution. Interestingly, *R. arrhizus*, which accumulated substantial quantities of U [118], was a poor accumulator of Ra, indicating a different mechanism of uptake for U and Ra. A subsequent study [157] determined that radium biosorption was rapid, with equilibrium attained in 2 min (*P. chrysogenum*) and 5 min (activated sludge). The rate of Ra adsorption was not affected by changes in pH (between the range pH 4 and 9) or the presence of competing ions including Ca^{2+}, Ba^{2+}, Cu^{2+}, or Fe^{2+}. Biosorption was also similar in synthetic laboratory and actual waste solutions. Immobilized sludge from a municipal wastewater treatment plant was used for radium recovery from Elliot Lake uranium tailing streams [39,158].

In a more recent study, Dwivedy and Mathur [159] reported the use of an *Arthrobacter* sp. for the precipitation of Mn and ^{226}Ra from neutralized uranium mill effluents. Yeast extract (0.005%) and sucrose (0.005%) were added as growth substrates, and 95% of the Ra was co-precipitated with 92% of the Mn after treatment (Mn was precipitated as $MnO_2 \cdot nH_2O$). The concentrations of the two elements were within the limits set by the International Committee of Radiological Protection (Mn <0.5 mg/L, Ra <3.0 pCi/L). Anaerobic reduction of Mn(IV) is possible [160], however, and could potentially remobilize the ^{226}Ra. Liberation of Ra from stored tailings via the biological oxidation of pyrite (generating sulfuric acid) or microbial ferric iron reduction [10] is already documented.

5.4. Transition Metals

5.4.1. Yttrium (Group 3)

^{90}Sr decays to a short-lived and relatively innocuous isotope ^{90}Y (half-life of 64 h). Biological interactions with Y have been little studied, although it should mirror that of La, which is bioaccumulated by several mechanisms. Both elements are stable as the trivalent cation, and immobilized cells of *Citrobacter* sp. accumulated Y^{3+} very rapidly in a similar manner to that described for La^{3+} (K. M. Bonthrone and L. E. Macaskie, personal communication). Lear and Oppenheimer [161] showed that ^{90}Sr and ^{90}Y were removed from seawater by marine microorganisms, following the early work of Spooner [162] which suggested that accumulation of ^{90}Y is greater than that of ^{90}Sr by marine algae.

A more recent study quantified Y and Sc biosorption by *S. cerevisiae* and *Aspergillus terreus* [163] and concluded that Al, Fe(III), and Ti affected biosorption of both elements (particularly Y). Biosorption was maximal at high pH and Y could be eluted by acid washing of the biomass. Tzesos et al. [164] also noted that Y biosorption by two Pseudomonads was inhibited by the presence of U, suggesting that the latter exhibited a higher affinity for the binding loci on the biomass. In an earlier pilot-scale study, Tsezos et al. [165] reported that Y was also displaced from immobilized *R. arrhizus* biomass by incoming U in a mine effluent. In addition to competing cations, the deleterious effects of organic chelating agents on biosorption was emphasized by the study of Sun et al. [166] who noted reduced uptake of rare earth elements (La, Gd, and Y) when supplied as citrate, NTA, or EDTA complexes.

5.4.2. Zirconium (Group 4)

Zirconium isotopes arise from fission and activation reactions, and also by corrosion of fuel cladding, but Zr is generally considered to be biologically inert [167]. ^{95}Zr and ^{93}Zr (half-lives; 64 days and 1.5×10^6 years, respectively) are present in some nuclear wastes. In a typical low-active waste stream the concentration of ^{95}Zr is 2.2 μCi/L, while in unreprocessed fuel the activity of ^{93}Zr is $\sim$20% of that of ^{99}Tc (in Bq, see Ref. [12]).

There are few studies on Zr bioaccumulation. Garnham et al. [168] reported energy-independent, pH-dependent Freundlich-type biosorption of Zr by microalgae and cyanobacteria. Biosorption of ^{95}Zr was observed by *R. arrhizus*, maximal at pH 2 and reduced by $\sim$40% at pH values of 4 and above [115]. A direct comparison between Zr(IV) uptake and uptake of Pu(IV) was not possible because the challenge concentrations were not identical but the behavior of Zr(IV) should be contrasted with that of Pu(IV), where increased pH gave increased biosorption of Pu(OH)$_n$.

5.4.3. Technetium (Group 7)

Technetium-99, a fission product of ^{235}U, is a long-lived (half-life: 2.1×10^5 years) and a significant pollutant in waste streams. It is normally present in the heptavalent form as the pertechnetate anion ([TcO$_4$]$^-$), which is both highly soluble and mobile in the environment [169]. Despite its artificial nature, the biological activity of Tc(VII) is high, acting as a sulfate analogue [40], and assimilation by plants facilitates entry into the food chain [41]. These factors, when considered in combination, led Trabalka and Garten [170] to conclude that Tc may be the critical radionuclide in determining the long-term impact of the nuclear fuel cycle.

Uptake of the pertechnetate anion by biosorption is low [171]. Indeed, like Np(V), Tc(VII) has weak ligand complexing capabilities and is difficult to remove from solution using conventional "chemical" approaches. Several reduced forms of the radionuclide are insoluble, however, and metal-reducing microorganisms can reduce Tc(VII) and precipitate the radionuclide as a low valence oxide.

Although microbial metabolism was known to decrease the solubility of Tc from earlier studies [172,173], Lloyd and Macaskie were the first to unequivocally demonstrate direct enzymatic reduction of Tc(VII) by microorganisms [67]. In this study, a novel phosphorimager technique was used to confirm reduction of the radionuclide by *S. putrefaciens* and *G. metallireducens*, with similar activities subsequently detected in laboratory cultures of *Rhodobacter sphaeroides*, *Paracoccus denitrificans*, some Pseudomonads [102], *E. coli* [174] and a range of sulfate-reducing bacteria [175–177]. Other workers have used this technique to show that *Thiobacillus ferrooxidans* and *T. thiooxidans* [178] and the hyperthermophile *Pyrobaculum islandicum* [179] are also able to reduce Tc(VII). It should be stressed that Tc(VII) reduction has not been shown to support growth in any of these studies, and seems to be a fortuitous biochemical side reaction in the organisms studied to date.

Finally, X-ray absorption spectroscopy studies have recently identified insoluble Tc(IV) as the final oxidation state produced when Tc(VII) is reduced enzymatically by *G. sulfurreducens* [101], *E. coli* [J. R. Lloyd and V. A. Solé, unpublished results], and *S. putrefaciens* [180]. Recent studies have also shown that Tc(VII) can be reduced by indirect microbial processes, for example, via biogenic sulfide [177], Fe(II) [101] or U(IV) [102]. Tc(VII) reduction and precipitation by biogenic Fe(II) is particularly efficient, and may offer a potentially useful mechanism for the remediation of Tc-contaminated sediments containing active concentrations of Fe(III)-reducing bacteria [101].

The biochemical basis of Tc(VII) reduction has been best studied in *E. coli*, where reduction is catalyzed by hydrogenase 3 of the formate hydrogenlyase (FHL) complex [174]. The identification of a hydrogenase as the Tc(VII) reductase of *E. coli* opened up the way for a program to screen for organisms with potentially enhanced activities against Tc(VII). Several organisms documented to have naturally high activities of FHL or uptake hydrogenase were tested, resulting in the identification of several strains of sulfate-reducing bacteria that were able to couple the oxidation of formate or hydrogen to Tc(VII) reduction [175]. Rates of reduction in some strains were approximately 64-fold greater than those recorded in anaerobic cultures of *E. coli* [181]. *D. desulfuricans* [176] and related strains [175] were also able to utilize formate as an efficient electron donor for Tc(VII) reduction. This is consistent with the existence of a rudimentary FHL complex (consisting of a formate dehydrogenase coupled to a hydrogenase via a cytochrome) located in the periplasm of these strains [182]. Accordingly, the site of reduced Tc precipitation was identified as the periplasm in *D. desulfuricans* [176], and more recent studies have confirmed a role for a periplasmic Ni-Fe hydrogenase in Tc(VII) reduction by a relative in the δ subclass of the *Proteobacteria*, the sulfate-reducing bacterium *D. fructosovorans* [183].

5.4.4. Cobalt and Nickel (Groups 9 and 10)

5.4.4.1. Cobalt: Although [60]cobalt is considered relatively innocuous, by virtue of a relatively short half-life (5.27 years), radioactive Co(III) complexed by EDTA is noted as a contaminant at some Department of Energy sites [184]. Recent studies have shown that iron-reducing bacteria are able to reduce Co(III) when complexed with EDTA [185,186]. The Co(II) formed does not associate strongly with EDTA [it is over 25 orders of magnitude less thermodynamically stable than Co(III)EDTA], and absorbs to soils, offering potential for *in situ* immobilization of the metal in contaminated soils. However, Gorby et al. [186] have demonstrated that reduced Co(II)EDTA can transfer electrons abiotically to solid Mn(IV) oxides, effectively acting as an electron shuttle between the bacterial cell and metal oxide. Mn(IV) minerals could, therefore, play an important role in maintaining concentrations of Co(III)EDTA in the subsurface.

5.4.4.2. Nickel: Nickel-59 is produced as an activation product from fuel assemblies and is considered problematic due to its long half-life (7.6×10^4 years). Microbial interactions with this isotope have been relatively poorly studied, but the uptake of "cold" nickel has been studied by several authors. *Oscillatoria* sp. accumulated Ni from freshwater to a concentration factor of 9000 [187]. The initial concentration of the Ni was 0.12 ppm. *Zoogloea* sp. accumulated up to 50% of 500 ppm Ni, with an extracellular gelatinous matrix implicated as a major site of metal biosorption [188]. A strain lacking the matrix accumulated only 25% of the Ni supplied.

More recently, Tsezos et al. [164] noted poor Ni biosorption by two Pseudomonads, severely depressed by the presence of Ag and Y. The Co-accumulating alga *S. natans* accumulated Ni poorly [B. Volesky, personal communication] and Ni is generally held to be fairly recalcitrant, for reasons which are still not clear. Tsezos et al. [189] in a study of biosorption of Ni by six strains, considered the hydrolysis behavior, chemical coordination, stereochemical and redox characteristics and suggested that solubility considerations alone could not explain the observed low biosorption of Ni; stereochemical factors were considered the most likely.

Addition of phosphate to a solution of Ni^{2+} released by biodegradation of citrate gave poor Ni^{2+} removal as compared to Co^{2+} removal by two Pseudomonads [76]. Removal of Ni^{2+} by phosphate-liberating *Citrobacter* sp. was, similarly, poor [190], although Ni^{2+} was removed by a strain of *A. eutrophus* [86] well-established to remove heavy metals as the carbonates via localized alkalinization concomitant with metal efflux [191]. MECHM-mediated nickel removal into biogenic HUP ($HUO_2PO_4 \cdot 4H_2O$) laid down by *Citrobacter* sp. would also seem to be efficient [190].

5.5. Halogens

5.5.1. Iodine

The major source of iodine in soils is thought to be the deposition of atmospheric iodine that has been evaporated from the oceans. However, there is currently comparatively little known about the subsequent global iodine cycle, and so the biogeochemistry of ^{129}I remains obscure. This is due in part to its complicated chemistry and the trace (but significant) concentrations of this radionuclide in environments impacted by radioactive waste. Nevertheless, [129]I remains a key radionuclide when assessing the potential environmental impact of the nuclear fuel cycle due to its bioavailability, long half-life, and mobility. Focusing on the chemical species in the environment, iodide, iodine, iodate and periodate are all potentially important, while reaction with organic compounds can lead to organo-iodides. All species show very different sorption properties when contacted with mineral phases and microbial biomass.

A wide range of soil organisms were shown to convert the iodide ion (I^-), to volatile methyl iodide (CH_3I), and this activity was enhanced by the addition of glucose in some cases [192]. Radiotracer experiments using ^{125}I confirmed volatilization of radioiodine, from soils. Further evidence that these activities were attributed to microbial metabolism, was provided by inhibition of volatilization in the presence of bacterial-specific antibiotics. Enzymatic reduction of iodate (IO_3^-) to iodide has also been demonstrated for cultures of sulfate- and Fe(III)-reducing bacteria, with indirect mechanisms also possible with these organisms, catalyzed by soluble ferrous iron, sulfide, or iron monosulfide (FeS) [193]. Strains related to *Roseovarius tolerans* also produced free iodine and organic iodine from iodide [194].

In another recent study, accumulation of iodine in soils was explained by the effects of microorganisms or their products (e.g., enzymes), or both [195], while iodine was observed to be desorbed from the flooded soils under reducing conditions (low Eh) created by the microbial activities [196]. Evaporation of biogenerated methyl iodide from the soil–plant system, specifically from rice plants, was also noted to decrease the iodine concentrations in lowland soils in this latter study.

6. CONCLUSIONS

Driven by technological advances at the interface between biology and geology, there has been rapid progress in our knowledge of the biogeochemical cycles that control radionuclide solubility. This work has been fostered by several important funding initiatives aimed to promote a better understanding of natural processes that could be harnessed to clean up land and water contaminated by radioactive waste.

There is, however, still an apparent disconnection between many laboratory- and field-based studies in this new area, and this must be alleviated if we are to obtain a better understanding of the biogeochemical cycles for environmentally relevant radionuclides. This will be possible given the imminent availability of genome sequences and genetic systems for key subsurface microorganisms, alongside technical advances in, for example, the analysis of radionuclide chemistry at low concentrations.

ACKNOWLEDGMENTS

The authors thank the UK Natural Environment Research Council (NERC) and the Natural and Accelerated Bioremediation Research (NABIR) program of the US Department of Energy for financial support through grants NER/A/S/ 2001/00960 and DE-FG02-02ER63422, respectively.

ABBREVIATIONS

An	actinides
BET	Brunauer–Emmett–Teller
BNFL	British Nuclear Fuels Ltd
DOC	dissolved organic carbon
DTPA	diethylenetriaminepentaacetic acid
EDTA	ethylenediaminetetraacetic acid
Eh	redox potential
FHL	formate hydrogenlyase
FTIR	Fourier transform infrared
HUP	hydrogen uranyl phosphate
MECHM	microbially-enhanced chemisorption of heavy metals
MOX	mixed oxide
NTA	nitrilotriacetic acid
TBP	tri-*n*-butyl phosphate
UMTRA	uranium mill tailings remedial action

REFERENCES

1. Livens FR, Bryan ND, Collison D, Faulkner S, Fox D, Goodall PS, Lloyd JR, May I, Renshaw JC, Sarsfield M, Steele H, Thomson S, Volkovich VA. Eur J Inorg Chem 2004, in press.
2. Morris K, Raiswell R. In: Keith-Roach MJ, Livens FR, eds. Interactions of Micro-organisms with Radionuclides. Amsterdam: Elsevier Science Ltd, 2002:101–141.
3. Perkins RW, Thomas CW. In: Hanson WC, ed. Transuranic Elements in the Environment. US: Department of Energy, 1980:53–82.
4. Pentreath RJ. Radionuclides. In: Guary JC, Guegueniat P, Pentreath RJ, eds. A Tool for Oceanography. London: Elsevier Science, 1988:12–34.
5. Choppin GR, Stout BE. Sci Total Environ 1989; 83:203–216.
6. Sholkovitz ER. Earth-Sci Rev 1983; 19:95–161.
7. McCullough J, Hazen TC, Benson SM, Metting FB, Palmisano AC. Bioremediation of Metals and Radionuclides—What is it and How it Works. Berkeley, California: Lawrence Berkeley National Laboratory, 1999.
8. Hutson GV. In: Wilson PD, ed. The Nuclear Fuel Cycle—From Ore to Waste. Oxford: Oxford University Press, 1996:161–183.
9. McCready RGL, Krouse HR. Geomicrobiology J 1980; 2:55–62.
10. Landa ER, Phillips EJP, Lovley DR. Appl Geochem 1991; 6:647–652.
11. Denniss IS, Jeapes AP. In: Wilson PD, ed. The Nuclear Fuel Cycle—From Ore to Waste. Oxford: Oxford University Press, 1996:116–137.
12. Macaskie LE. Crit Rev Biotechnol 1991; 11:41–112.
13. Lloyd JR, Macaskie LE. In: Lovley DR, ed. Environmental Microbe-Metal Interactions. Washington DC: ASM Press, 2000:277–327.

14. Belly RT, Lauff JJ, Goodhue CT. Appl Microbiol 1975; 29:787–794.
15. Francis AJ. J Alloys Compd 1994; 213/214:226–231.
16. Anon, "Managing the nuclear legacy; a strategy for action", UK Government White Paper, 2002.
17. Wilson PD. The Nuclear Fuel Cycle—From Ore to Waste, Oxford: Oxford University Press, 1996.
18. Silva RJ, Nitsche H. Radiochim Acta 1995; 70/71:377–396.
19. Konhauser KO, Mortimer RJG, Morris K, Dunn V. In: Keith-Roach MJ, Livens, FR, ed. Interactions of Microorganisms with Radionuclides. Amsterdam: Elsevier, 2002:61–100.
20. Yoshihara K. Top Curr Chem 1996; 176:17–35.
21. Choppin GR. Radiochim Acta 1988; 42:82–83.
22. Clark DL, Hobart DE, Neu MP. Chem Rev 1995; 95:25–48.
23. Runde W. Los Alamos Sci 2000; 26:392–411.
24. Seaborg GT. Radiochim Acta 1993; 61:115–122.
25. Dozol M, Hagemann R. Pure Appl Chem 1993; 65:1081–1102.
26. Morris K, Butterworth JC, Livens FR. Estuarine, Coastal Shelf Sci 2000; 51:613–625.
27. Choppin GR, Wong PJ. Aquat Geochem 1998; 4:77–103.
28. Bolvin H, Wahlgren U, Moll H, Reich T, Geipel G, Fanghanel T, Grenthe I. J Phys Chem A 2001; 105:11441–11445.
29. Gelis AV, Vanysek P, Jensen MP, Nach KL. Radiochim Acta 2001; 89:565–571.
30. Garnov AY, Yusov AB, Shilov VP, Krot NN. Radiochemistry 1998; 40:219–223.
31. Williams CW, Blaudeau JP, Sullivan JC, Antonio MR, Bursten B, Soderholm L. J Am Chem Soc 2001; 123:4346–4347.
32. Ahrland S. In: Katz JJ, Seaborg GT, Morss LR, eds. The Chemistry of the Actinide Elements. Vol. 1. London, UK: Chapman and Hall, 1986:1483–1495.
33. Fanghanel T, Meck V. Pure Appl Chem 2002; 74:1895–1907.
34. Rai D, Serne RJ, Swanson, JL. J Environ Qual 1980; 9:417–420.
35. Rai D. Nucl Technol 1981; 54:107–112.
36. Ockenden DW, Welch GA. J Chem Soc 1956; 3358–3362.
37. Den Auwer C, Simoni E, Conradson S, Madic C. Eur J Inorg Chem 2003; 3843–3859.
38. Hughes MN, Poole RK. Metals and Micro-organisms, London, UK: Chapman and Hall, 1989.
39. Tsezos M, Baird MHI, Shemilt LW. Hydrometallurgy 1987; 17:357–368.
40. Cataldo DA, Garland TR, Wildung RE, Fellows RJ. Health Phys 1989; 57:281–288.
41. Dehut JP, Fosny K, Myttenaere C, Deprins D, Vandecasteele CM. Health Phys 1989; 57:263–267.
42. Lieser KH. Radiochim Acta 1995; 70/71:355–375.
43. Bulman RA. In: Bulman RA, Cooper JR, eds. Speciation of Fission and Activation Products in the Environment. London: Elsevier Applied Science, 1985:213–222.
44. Bate LC, Leddicotte GW. "The radiochemistry of cobalt" Tech. Report No. NAS-NS 3041, USAEC, 1970.
45. Sparkes ST, Long SE. "The chemical speciation of technetium in the environment—a literature survey", Tech. Report No. AERE R12743, UK Atomic Energy Authority, 1988.
46. von Gunten HR, Benes P. Radiochim Acta 1995; 69:1–29.

47. Keepax RE, Jones DM, Pepper, SE, Bryan, ND. In: Keith-Roach MJ, Livens FR, eds. Interactions of Microorganisms with Radionuclides. Amsterdam: Elsevier, 2002:143–177.
48. Elwear S, German KE, Peretrukin VF. J Radioanal Nucl Chem Art 1992; 157:3–14.
49. Kim JI, Zeh P, Delakowitz B. Radiochim Acta 1992; 58/59:147–154.
50. Songkasiri W, Reed DT, Rittmann BE. Radiochim Acta 2002; 90:785–789.
51. Gadd GM, White C. In: Poole RK, Gadd GM, eds. Metal–Microbe Interactions. Oxford, England: Society for General Microbiology, IRL Press, 1989:19–38.
52. Volesky B, May-Phillips HA. Appl Microbiol Biotechnol 1995; 42:797–806.
53. Brady D, Stoll A, Buncan JR. Environ Technol 1994; 15:429–439.
54. Volesky B. Biosorption of Heavy Metals, Boca Raton, Fla.: CRC Press, 1990.
55. Volesky B, Holan ZR. Biotechnol Prog 1995; 11:235–250.
56. deRome L, Gadd GM. J Ind Microbiol 1991; 7:97–104.
57. Higham DP, Sadler PJ, Scawen MD. Science 1984; 225:1043–1046.
58. Turner JS, Robinson NJ. J Ind Microbiol 1995; 14:119–125.
59. Okorov LA, Lichko LP, Kodomtseva VM, Kholodenko VP, Titovsky VT, Kulaev IS. Eur J Biochem 1977; 75:373–377.
60. Jeong BC, Hawes C, Bonthrone KM, Macaskie LE. Microbiology 1997; 143:2497–2507.
61. Marques AM, Roca X, Simon-Pujol MD, Fuste MC, Francisco C. Appl Microbiol Biotechnol 1991; 35:406–410.
62. Strandberg GW, Shumate II SE, Parrott JR. Appl Environ Microbiol 1981; 41:237–245.
63. Suzuki Y, Banfield JF. In: Burns PC, Finch R., eds. Uranium: Mineralogy, Geochemistry and the Environment. Vol. 38. Washington DC, USA: Mineralogical Society of America, 1999:393–432.
64. White C, Gadd GM. Tox Assess 1987; 2:437–447.
65. Lovley DR. Ann Rev Microbiol 1993; 47:263–290.
66. Ehrlich HL. Geomicrobiology. 3rd. ed. New York: Marcel Dekker, Inc, 1996.
67. Lloyd JR, Macaskie LE. Appl Environ Microbiol 1996; 62:578–582.
68. Lloyd JR, Yong P, Macaskie LE. Environ Sci Technol 2000; 34:1297–1301.
69. Lovley DR, Phillips EJP, Gorby YA, Landa E. Nature 1991; 350:413–416.
70. Rusin PA, Quintana L, Brainard JR, Strietelmeier BA, Tait CD, Ekberg SA, Palmer PD, Newton TW, Clark DL. Environ Sci Technol 1994; 28:1686–1690.
71. Barnhart BJ, Campbell EW, Martinez E, Caldwell DE, Hallett R. "Potential microbial impact on transuranic wastes under conditions expected in the waste isolation pilot plant (WIPP)" Tech. Report No. LA-8297-PR, Los Alamos National Laboratory, 1980.
72. Lloyd JR. FEMS Microbiol Rev 2003; 27:411–425.
73. Lovley DR. In: Lovley DR, ed. Environmental Microbe–Metal Interactions. Washington, DC: ASM Press, 2000:3–30.
74. Gadd GM. Adv Microbial Physiol 1999; 41:47–92.
75. Bosecker K. FEMS Microbiol Rev 1997; 20:591–604.
76. Thomas RAP, Beswick AJ, Basnakova G, Moller R, Macaskie LE. J Chem Technol Biotechnol 2000; 75:187–195.
77. Thomas RAP, Lawlor K, Bailey M, Macaskie LE. Appl Environ Microbiol 1998; 64:1319–1322.

78. Satroutdinov AD, Dedyukhina EG, Chistyakova TI, Witschel M, Minkevich IG, Eroshin VK, Egli T. Environ Sci Technol 2000; 34:1715–1720.
79. Macaskie LE, Empson RM, Cheetham AK, Grey CP, Skarnulis, AJ. Science 1992; 257:782–784.
80. Pattanapipitpaisal P, Mabbett AN, Finlay JA, Beswick AJ, Paterson-Beedle M, Essa A, Wright J, Tolley MR, Badar U, Ahmed N, Hobman JL, Brown NL, Macaskie LE. Environ Technol 2002; 23:731–745.
81. Basnakova G, Stephens ER, Thaller MC, Rossolini GM, Macaskie LE. Appl Microbiol Biotechnol 1998; 50:266–272.
82. Barnes LJ, Janssen FJ, Sherren J, Versteegh JH, Koch RO, Scheeren PJH. Chem Eng Res Des 1991; 69A:184–186.
83. Kearney T, Eccles H, Graves D, Gonzalez A. Proceedings of the 18th Annual Conference of the National Low-Level Waste Management Program, Salt Lake City, Utah, U.S.A., May 20–22nd 1996.
84. Postgate JR. The Sulphate Reducing Bacteria. Cambridge, UK: Cambridge University Press, 1979.
85. Diels L, Dong Q, van der Lelie D, Baeyens W, Mergeay M. J Ind Microbiol 1995; 14:142–153.
86. Van Roy S, Peys K, Dresselaers T, Diels L. Res Microbiol 1997; 148:526–528.
87. Anderson S, Apanna VD. FEMS Microbiol Lett 1994; 116:42–48.
88. Ferris FG, Fratton CM, Gertis JP, Schultzelam S, Lollar BS. Geomicrobiol J 1995; 13:57–67.
89. Phillips EJP, Landa ER, Lovley DR. J Ind Microbiol 1995; 14:203–207.
90. Francis CW, Timpson ME, Wilson JH. J Haz Material 1999; 66:67–87.
91. Francis AJ, Dodge CJ, Gillow JB, Papenguth HW. Environ Sci Technol 2000; 34:2311–2317.
92. Macaskie LE, Lloyd JR, Thomas RAP, Tolley MR. Nucl Energy 1996; 35:257–271.
93. Macaskie LE, Jeong BC, Tolley MR. FEMS Microbiol Rev 1994; 14:351–368.
94. Tolley MR, Macaskie LE. UK Patent number GB94/00626, 1994.
95. Clearfield A. Chem Rev 1988; 88:125–148.
96. Hunsberger LR, Ellis AB. Coord Chem Rev 1990; 97:209–224.
97. Pham-Thi M, Columban P. Solid State Ion 1985; 17:295–306.
98. Dorhout PK, Kissane RJ, Abney KD, Avens LR, Eller PG, Ellis AB. Inorg Chem 1989; 28:2926–2930.
99. Watson JHP, Ellwood DC. Minerals Eng 1994; 7:1017–1028.
100. Fendorf S, Li G. Environ Sci Technol 1996; 30:1614–1617.
101. Lloyd JR, Sole VA, Van Praagh CV, Lovley DR. Appl Environ Microbiol 2000; 66:3743–9.
102. Lloyd JR, Chesnes J, Glasauer S, Bunker DJ, Livens FR, Lovley DR. Geomicrobiol J 2002; 19:103–120.
103. Gadd GM. Endeavour 1996; 20:150–156.
104. Lloyd JR, Lovley DR. Curr Opin Biotechnol 2001; 12:248–253.
105. Banaszak JE, Rittman BE, Reed DT. J Radioanal Nucl Chem 1999; 241:385–435.
106. Lloyd JR, Macaskie LE. In: Livens FR, Keith-Roach M, eds. Microbiology and Radioactivity. Elsevier, 2002.
107. Macaskie LE, Lloyd JR. In: Livens FR, Keith-Roach M, eds. Microbiology and Radioactivity. Elsevier, 2002:343–381.

108. Lloyd JR, Anderson RT, Macaskie LE. In: Atlas R, Philp J, eds. Bioremediation. Washington DC: ASM Press, 2004.
109. Thamdrup B. Adv Microbiol Ecol 2000; 16:41–84.
110. Peretrukhin VF, Khizhniak NN, Lyalikova NN, German KE. Radiochem 1996; 38:440–443.
111. Lovley DR, Phillips EJP. Environ Sci Technol 1992; 26:2228–2234.
112. Keith-Roach MJ, Day JP, Fifield LK, Bryan ND, Livens FR. Environ Sci Technol 2000; 34:4273–4277.
113. Fisher NS, Bjerregaard P, Huynh-Ngoc L, Harvey GR. Mar Chem 1983; 13:45–56.
114. Wurtz EA, Sibley TH, Schell WR. Health Phys 1986; 50:79–88.
115. Dhami PS, Gopalakrishnan V, Kannan R, Ramanujam A, Salvi N, Udupa SI. Biotechnol Lett 1998; 20:225–228.
116. Tsezos M, Volesky B. Biotechnol Bioeng 1982; 24:955–969.
117. Gadd GM, White C. Biotechnol Bioeng 1989; 33:592–597.
118. Tsezos M, Volesky B. Biotechnol Bioeng 1981; 23:583–604.
119. Andres Y, MacCordick JH, Hubert JC. Appl Microbiol Biotechnol 1995; 44:271–276.
120. Yong P, Macaskie LE. J Chem Technol Biotechnol 1995; 64:87–95.
121. Yong P, Macaskie LE. J Chem Technol Biotechnol 1998; 71:15–26.
122. Strandberg G, Arnold WD. J Ind Microbiol 1988; 3:329–331.
123. Tolley MR, Macaskie LE. In: Torma AE, Apel ML, Brierley CL, eds. In Biohydro-metallurgical Technologies. The Minerals: Metals and Materials Society, 1993.
124. Yong P. PhD. University of Birmingham, UK, 1996.
125. Wersin P, Hochella MF Jr., Persson P, Redden G, Leckie JO, Harris DW. Geochim Cosmochim Acta 1994; 58:2829–2843.
126. Dick RE, Boswell CD, Macaskie LE. Int. Symp Biohydrometallurgy, Vina del Mar, Chile, Nov. 1995, 1995.
127. Thomas RAP, Morby AP, Macaskie LE. FEMS Microbiol Lett 1997; 155:155–159.
128. Macaskie LE, Bonthrone KM. "Modelling of genetic, biochemical, cellular and microenvironmental parameters determining bacterial sorption and minera-lization processes for recuperation of heavy or precious metals" Tech. Report No. BE 5350, 1996.
129. Jeong BC, Poole PS, Willis AJ, Macaskie, LE. Arch Microbiol 1998; 169:166–173.
130. Tolley MR, Strachan LF, Macaskie LE. J Ind Microbiol 1995; 14:271–280.
131. Macaskie LE, Yong P, Doyle TC, Roig MG, Diaz M, Manzano T. Biotechnol Bioeng 1997; 53:100–109.
132. Lovley D, Phillips EJ. Appl Environ Microbiol 1992; 58:850–856.
133. Lovley DR, Phillips EJP. Appl Environ Microbiol 1994; 60:726–728.
134. Payne RB, Gentry DA, Rapp-Giles BJ, Casalot L, Wall JD. Appl Environ Microbiol 2002; 68:3129–3132.
135. Lloyd JR, Leang C, Hodges Myerson AL, Ciufo S, Sandler SJ, Menthe B, Lovley DR. Biochem J 2003; 369:153–161.
136. Wade R Jr., DiChristina JT. FEMS Microbiol Letts 2000; 184:143–148.
137. Finneran KT, Anderson RT, Nevin KP, Lovley DR. Soil Sed Comtam 2002; 11:339–357.
138. Holmes DE, Finneran KT, Lovley DR. Appl Environ Microbiol 2002; 68:2300–2306.

139. Anderson RT, Vrionis HA, Ortiz-Bernad I, Resch CT, Long PE, Dayvault R, Karp K, Marutzky S, Metzler DR, Peacock A, White DC, Lowe M, Lovley DR. Appl Environ Microbiol 2003; 69:5884–5891.
140. Wahlgren MA, Robbins JA, Edgington DN. In: Hanson WC, ed. Transuranic Elements in the Environment. US Department of Energy, 1980:659–683.
141. Sholkovitz ER, Carey AE, Cochran JK. Nature 1982; 300:159–161.
142. Buesseler KO, Sholkovitz ER. Geochim Cosmochim Acta 1987; 51:2605–2622.
143. Malcolm SJ, Kershaw PJ, Lovett MB, Harvey BR. Geochim Cosmochim Acta 1990; 54:29–35.
144. Morris K, Bryan ND, Livens FR. J Environ Radioact 2001; 56:259–267.
145. Tengerdy RP, Johnson JE, Hollo J, Toth J. Appl Biochem Biotechnol 1981; 6:3–13.
146. Avery SV. J Chem Technol Biotechnol 1995; 62:3–16.
147. Bossemeyer D, Schlosser A, Bakker E. J Bacteriol 1989; 171:2219–2221.
148. Plato P, Denovan JT. Radiat Bot 1974; 14:37–41.
149. Avery SV, Codd GA, Gadd GM. J Gen Microbiol 1991; 137:405–413.
150. Perkins J, Gadd GM. Mycol Res 1993; 97:712–724.
151. Macaskie LE, Dean ACR. Biotechnol Lett 1985; 7:627–630.
152. Zajic JS, Chiu YS. Dev Ind Microbiol 1972; 13:91–100.
153. Watson JS, Scott CD, Faison BD. Appl Biochem Biotechnol 1989; 21/21:201–209.
154. Watson JS, Scott, CD, Faison, BD. Appl Biochem Biotechnol 1989; 20/21:699.
155. Avery SA, Tobin JM. Appl Environ Microbiol 1992; 58:3883–3889.
156. Tsezos M, Keller DM. Biotechnol Bioeng 1983; 25:201–215.
157. Tsezos M, Baird MHI, Shemilt LW. Chem Eng J 1986; 33:B35–B41.
158. Tsezos M, Baird MHI, Shemilt LW. Chem Eng J 1987; 34:B57–B64.
159. Dwivedy KK, Mathur AK. Hydrometallurgy 1995; 38:99–109.
160. Nealson KH, Myers CR. Appl Environ Microbiol 1992; 58:439–443.
161. Lear DW, Oppenheimer CH. Limnol Oceeanogr 19; 7:suppl 44–62.
162. Spooner GM. J Mar Biol Assoc (UK) 1949; 28:587–625.
163. Karavaiko GI, Kareva AS, Avakian ZA, Zakharova VI, Korenevsky AA. Biotechnol Lett 1996; 18:1291–1296.
164. Tsezos M, Remoudaki E, Angelatou V. Int Biodeterior Biodegrad 1996; 38:19–29.
165. Tsezos M, McCready LGR, Bell JP. Biotechnol Bioeng 1989; 34:10–17.
166. Sun H, Wang XR, Wang LS, Dai LM, Li Z, Cheng YJ. Chemosphere 1997; 34:1753–1760.
167. Ghosh S, Sharma A, Talukder G. Biol Trace Element Res 1992; 35:247–271.
168. Garnham GW, Codd GA, Gadd GM. Appl Microbiol Biotechnol 1993; 39:666–672.
169. Wildung RE, McFadden MK, Garland TR. J Environ Qual 1979; 8:156–161.
170. Trabalka JR, Garten CT Jr. In: Lett JT, Ehman UK, Cox AB, eds. Advances in Radiation Biology. Vol. 10. London, UK: Academic Press, 1983:68–73.
171. Garnham GW, Codd GA, Gadd GM. Appl Microbiol Biotechnol 1992; 37:679–684.
172. Henrot J. Health Phys 1989; 57:239–245.
173. Pignolet L, Fonsny K, Capot F, Moureau Z. Health Phys 1989; 57:791–800.
174. Lloyd JR, Cole JA, Macaskie LE. J Bacteriol 1997; 179:2014–2021.
175. Lloyd JR, Mabbett A, Williams DR, Macaskie LE. Hydrometallurgy 2001; 59:327–337.
176. Lloyd JR, Ridley J, Khizniak T, Lyalikova NN, Macaskie LE. Appl Environ Microbiol 1999; 65:2691–2696.

177. Lloyd JR, Nolting H-F, Solé VA, Bosecker K, Macaskie LE. Geomicrobiol J 1998; 15:43–56.
178. Lyalikova NN, Khizhnyak TV. Microbiology 1996; 65:468–473.
179. Kashefi K, Lovley K. Appl Environ Microbiol 2000; 66:1050–1056.
180. Wildung RE, Gorby YA, Krupka KM, Hess NJ, Li SW, Plymale AE, McKinley JP, Fredrickson JK. Appl Environ Microbiol 2000; 66:2451–2460.
181. Lloyd JR, Thomas GH, Finlay JA, Cole JA, Macaskie LE. Biotechnol Bioeng 1999; 66:123–130.
182. Peck HD. In: Odom JM, Singleton R., eds. Sulfate-Reducing Bacteria: Contemporary Perspectives. New York: Springer-Verlag, 1993.
183. De Luca G, Philip P, Dermoun Z, Rousset M, Vermeglio A. Appl Environ Microbiol 2001; 67:4583–4587.
184. Lovley DR, Coates JD. Curr Opin Biotechnol 1997; 8:285–289.
185. Caccavo F Jr, Lonergan DJ, Lovley DR, Davis M, Stolz JF, McInerney JM. Appl Environ Microbiol 1994; 60:3752–3759.
186. Gorby YA, Caccavo F, Bolton H. Environ Sci Technol 1998; 32:244–250.
187. Trollope DR, Evans B. Environ Pollut 1976; 11:109–116.
188. Friedman BA, Dugan PR. Dev Ind Microbiol 1968; 9:381–395.
189. Tsezos M, Remoudaki E, Angelatou V. Int Biodeterior Biodegrad 1996; 35:129–154.
190. Bonthrone KM, Basnakova G, Lin F, Macaskie LE. Nature Biotechnol 1996; 14:635–638.
191. Taghavi S, Mergeay M, Nies D, Van der Lelie D. Res Microbiol 1997; 148:536–551.
192. Amachi S, Kasahara M, Hanada S, Kamagata Y, Shinoyama H, Fujii T, Muramatsu Y. Environ Sci Technol 2003; 37:3885–3890.
193. Councell TB, Landa ER, Lovley DR. Water Air Soil Pollut 1997; 100:99–106.
194. Fuse H, Inoue H, Murakami K, Takimura O, Yamaoka Y. FEMS Microbiol Lett 2003; 229:189–194.
195. Truesdale VW, Watts SF, Rendell AR. Deep-Sea Res Part I-Oceanogr Res Papers 2001; 48:2397–2412.
196. Muramatsu Y, Yoshida S. Geomicrobiol J 1999; 16:85–93.

9

Biogeochemistry of Carbonates: Recorders of Past Oceans and Climate

Rosalind E. M. Rickaby[1] and Daniel P. Schrag[2]
[1]*Department of Earth Sciences, University of Oxford,*
Parks Road, Oxford, OX1 3PR, UK
[2]*Laboratory for Geochemical Oceanography,*
Department of Earth and Planetary Sciences, Harvard University,
20 Oxford Street, Cambridge, Massachusetts 02138, USA

1. INTRODUCTION TO BIOGENIC CARBONATE PROXIES

This chapter will address the use of trace metals in biogenic carbonates as proxies for past ocean conditions with an emphasis on the biogeochemistry of biomineralization. Our aim is to illustrate the utility of trace metal proxies for reconstructing ocean conditions in the past, but also to emphasize how little is known about the detailed mechanisms that underlie these proxies.

In order to probe the record and forcing mechanisms of past climate change beyond the range of direct observation and historical accounts, we are forced to rely on chemical or isotopic proxy records for different ocean properties such as temperature, nutrient abundance, or primary productivity. These proxy records are often in the form of chemical variations encapsulated within exquisitely crafted biomineralized carbonates. The biogenic calcium carbonate minerals, calcite, and aragonite, have an incredibly flexible chemistry. Variations in the abundance of the heavy isotope relative to the light isotope of both O and C (δ^{18}O, and δ^{13}C) in the CO_3^{2-} ion can be frozen within these minerals (e.g., [1–3]). Divalent cations of a similar ionic radius (Mg, Sr, Ba, Mn, Fe, Cu, Zn, Cd) can substitute for Ca^{2+} in the crystal structure (e.g., [4]) and significant levels of U, Li, B, and Na have also been found in carbonates. With the recent advent of multi-collector inductively coupled mass-spectrometry (MC-ICPMS) in the last 10 years, it has become possible to resolve sub per mill (part per thousand) fractionations in the stable isotopes of elements heavier than C and O such as Fe (e.g., [5,6]) or Mo (e.g., [7,8]). A new avenue of research is developing methodologies for the development and application of Ca isotopic variations (δ^{44}Ca) and even isotopic variations in the metals that can substitute for Ca^{2+} (e.g., δ^{26}Mg, δ^{68}Zn, and δ^{57}Fe) as proxies for the past (e.g., [9]).

The information we derive from chemical proxies depends on the residence time of the chemical in the ocean relative to the timescale of interest and the ~ 1.5 kilo year (kyr) mixing time of the ocean. Elements with a relatively short residence time in the ocean have a non-uniform distribution in seawater, which can be indicative of nutrient-like behavior, conservative mixing, or scavenging onto particles. If these elements are then incorporated into biogenic carbonates in direct proportion to their concentration in seawater, they can provide a proxy for reconstructing those processes in the past. For example, the Cd concentration of seawater correlates with phosphate (an essential nutrient) in the ocean with near total depletion in surface waters and enrichment at depth. As a result

Cd/Ca ratios have been used to reconstruct phosphate concentrations in the past both in planktonic foraminifera to measure productivity in surface waters (e.g., [10]), and in benthic foraminifera to monitor changes in the characteristic nutrient signatures of deep water masses and the pattern of ocean circulation (e.g., [11]). For elements where the residence time in the ocean is long relative to the timescale of interest, any variation in the concentration or isotopic value recorded by a biogenic carbonate must be related to factors controlling the partitioning or isotopic fractionation into the carbonate shell such as temperature, salinity, or growth rate. As an example, Mg has a residence time of $\sim$5 million year (Myr), but Mg/Ca varies significantly between glacial and interglacial periods (100 kyr) and is used as a paleothermometer and paleosalinity proxy (in concert with $\delta^{18}O$) on these timescales (e.g., [12–14]). One final application is when the timescale of interest is long relative to the residence time of the element and it is incorporated into the carbonate in direct proportion to its concentration or signature in seawater. In this case, the chemical proxy can reveal changes in the oceanic budget of that element. For example, the Sr isotope curve ($^{87}Sr/^{86}Sr$) measured in marine carbonates on long timescales (e.g., 75 Myr) tells us about the changes in continental weathering processes [15].

With the exception of some systems such as strontium isotopes, the isotopic or chemical fractionation between seawater and the biogenic carbonate almost always involves biological control which is poorly understood. This means that there exists a persistent doubt as to the reliability of these proxies as some have argued that only a full mechanistic understanding of the incorporation of trace metals into biogenic carbonates would allow truly accurate reconstruction of past environments. At the same time, careful calibrations with laboratory or field observations have led to the production of a wide range of proxy records which have provided useful information about past climates and ocean conditions. This highlights the challenge that confronts the paleoceanographer who must work to develop the deepest level of mechanistic understanding of how chemical and isotopic signals are incorporated in biogenic carbonates, but at the same time must continue to use those proxies with the best knowledge available to test hypotheses about ancient climates.

In this chapter, we discuss the comparative trace metal geochemistries of coccolithophores, foraminifera, and corals relative to inorganic calcite and aragonite respectively, in order to define the differing nature of biological selectivity. We then address the biomineralization process for each organism and the biochemical nature of the trace metal selectivity involved at each step of the biomineralization. This overview allows us to develop a mechanistic framework within which to consider trace metal proxies in biogenic carbonate. Notably, the biochemical selectivity is based on similar chemical constraints to those of an inorganic crystal. In each case, the geometry of coordinating oxygens defines a cation specific site with a Ca–O bond length of between 2.2–2.6 Å, but the biochemical process adds a further selectivity due to the energy of dehydration of a cation before binding to the site. We propose that the vital effects of trace metal uptake are related to the varied energy required to dehydrate cations.

2. PARTITION COEFFICIENTS

The origins of chemical proxies for paleoceanography originates from the treatment of carbonate minerals as inorganic materials. The application of thermodynamics to geology in the 1960s created the vision that with a complete understanding of the thermodynamic partitioning of metals into carbonate minerals, the environmental variables (temperature, saturation state) could be calculated from measurements of sediments for various times in the past [16–18].

The true inorganic partition coefficient for trace metal uptake into carbonate, or the isotopic fractionation factor for stable isotopes, is related to the equilibrium partitioning of the particular element or isotope between seawater and calcium carbonate. A partition coefficient greater (less) than one implies that the carbonate is enriched (depleted) in that metal or isotope relative to the seawater content. In inorganic systems, the theoretical partition coefficient is generally agreed to relate to the quotient of the solubility product of $CaCO_3$ and MCO_3 [19–21] and to the activity of the cations in water. In reality, experimental conditions during inorganic precipitation of minerals only approximate equilibrium.

Experimental partition coefficients are affected by kinetic processes relating to the adsorption of the trace metal to kink sites on the growth steps of an actively growing crystal, and also solution boundary related processes. For the incorporation of divalent cations into inorganic calcite, there is a correlation between the experimental partition coefficient and the effective ionic radius in sixfold coordination (Fig. 1). The experimental partitioning of the cations Mg^{2+}, Co^{2+}, Fe^{2+}, Mn^{2+}, Cu^{2+}, Zn^{2+}, and Cd^{2+}, which have ionic radii less than that of Ca^{2+} and form rhombohedral carbonates, increases with increasing effective ionic radius towards that of Ca^{2+}. Cd^{2+} (the element with the most similar ionic radius to Ca) provides a maximum to the partition coefficient, and the trend then decreases with increasing effective ionic radius away from Ca^{2+} to the cations Sr^{2+}, and Ba^{2+} which have ionic radii larger than Ca^{2+}, do not fit the lattice so easily and tend to form orthorhombic carbonates. In general, ions with a similar but smaller radius than Ca^{2+} have the highest partition coefficient and the ions which fit least well into the Ca^{2+} sites have lower partition coefficients.

The original application of equilibrium thermodynamics to biogenic carbonates was done with the awareness of the possible complications from the biomineralization process [1]. These complications were treated by assuming that thermodynamic equilibrium was the underlying physical process, and that any offset from equilibrium was incorporated into a correction that was called the "vital effect" (and was usually assumed to be constant). There was good reason to make this assumption as some experimental data supported this

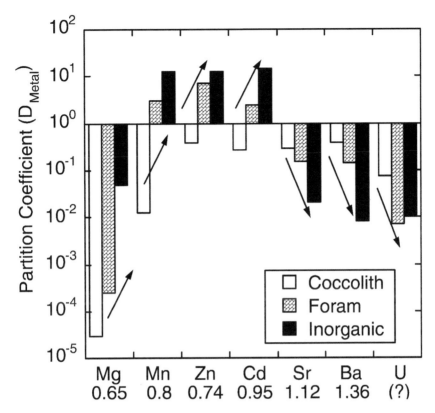

Figure 1 Partition coefficients for a range of metals for inorganic (black bars), foraminifera (brick bars) and coccolith (open bars) calcite. The inorganic calcite and foraminifera partition coefficients are taken from Refs. [22,23]. There is some variance but the graph is not changed significantly as it is plotted on a log scale. The partition coefficients for coccolithophores are based on trace metal analyses of cultured coccolithophores [Rickaby R. E. M., unpublished results]. The ionic radius of each metal in Å is also indicated except for uranium. For most of the metals, we can assume that the metal substitutes for Ca^{2+} in its divalent state. At present there is uncertainty as to the form by which U substitutes into the calcite lattice. There is speculation that it occurs in the form UO_2^+ [24] but this is as yet unproven.

view, such as the variation of oxygen isotopes with temperature in foraminifera that showed constant offsets between species (e.g., [3]). In much current work in paleoceanography, equilibrium thermodynamics in inorganic minerals is still presumed to be the underlying mechanism responsible for the preservation of environmental information (e.g., [25]).

That the pattern of trace metal uptake in biogenic carbonate resembles the inorganic system suggests that either inorganic calcite partitioning is still the dominant control on trace metal uptake into biominerals or that similar kinetic and thermodynamic chemical laws which govern crystal selectivity between Ca^{2+} and a similar trace metal also govern biological selectivity. However, there are differences in selectivity associated with the vitality of these processes which probably arise due to the greater number of steps involved in the biological precipitation process (Fig. 1). For cations which are smaller than Ca^{2+}, the effect of the biology is to discriminate more effectively than inorganic calcite, which implies a greater selectivity during ion transport to the site of nucleation and precipitation. For cations which are larger than Ca^{2+}, the biological discrimination is not as efficient and the biological partition coefficients tend to be greater than the inorganic coefficients. An alternative explanation than reliance on ionic radius alone for these effects is that the smaller ions are "biologically important" metals and often form the metallic co-factors for essential enzymes. Therefore, it could be argued that biological processes are able to recognize and better select amongst the biologically relevant cations.

Further evidence for a non-inorganic imprint on the chemistry of biogenic carbonate is the sensitivity of trace metal uptake to a wider array of environmental factors than one would predict from thermodynamics alone. Not only are the partition coefficients for biological carbonates different than inorganic carbonates by orders of magnitude, but the sensitivity to the environment, e.g., temperature, can be greater or smaller by orders of magnitude or even have an opposite sense (Fig. 2). Most importantly, the investigation of biomineralization over the last two decades [26] has shown a remarkable degree of active transport and control over every step in the biomineralization process. This suggests that whatever fractionations may exist, either in trace metals or isotopes, they are affected by preferential transport, and that the thermodynamic partitioning into inorganic calcite is not the most appropriate context within which to consider trace metal proxies.

3. RECORDERS OF PAST OCEAN CONDITIONS

A comprehensive overview of recent proxies has been provided by Henderson [27]. We will not provide a review here, except to mention briefly some of the applications of the groups of organisms discussed later. However, it is valuable to remember that the motivation in developing new chemical or isotopic proxies is to reconstruct environmental variables in ancient oceans. Therefore, the efforts at exploring the biophysical mechanisms that underlie these proxies have often been pushed aside in favor of simple calibration experiments in which correlations are embraced without a deep understanding of causation. Such efforts have provided a wealth of information about the history of the oceans and climate, but their reliability must ultimately be assessed in the context of a deeper understanding of mechanisms.

Our ultimate goal in reconstructing past climate change is to enhance our understanding of the climate system to create better predictions for the future. From this perspective, we must record changes on timescales which range

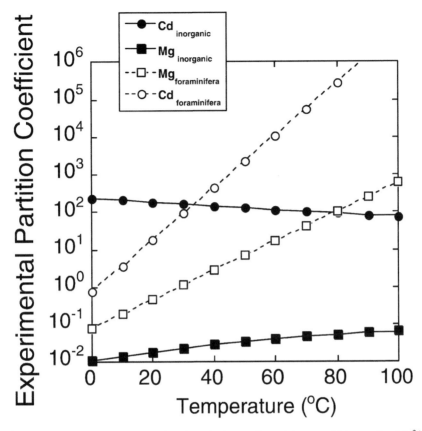

Figure 2 Theoretical and empirical variations of partition coefficients for Mg^{2+} (squares) and Cd^{2+} (circles) into inorganic calcite (open symbols) based on the calculations reported in Ref. [22], and for planktonic foraminifera (closed symbols) [10,12].

across eight orders of magnitude. Later we select just a few examples to illustrate the sort of information we are gaining from biomineral trace metal proxies.

In order to monitor the reaction of our climate system to the exponentially increasing input of greenhouse gases since the industrial revolution, we require information from the subannual, decadal, and century scale. Coral skeletons can grow at rates of up to 10 cm/yr which allow for high resolution annual records, continuous for hundreds of years in some circumstances. Trace metal proxies from coralline aragonite based on Mg/Ca, Cu/Ca, Mn/Ca, and Cd/Ca all yield seasonal information about upwelling and El Niño (e.g., [28–30]), but above all Sr/Ca in coral aragonite is the most extensively used paleothermometer. For example a unique insight of interdecadal variability associated with the industrial revolution within the Pacific has been afforded by records of coral Sr/Ca from across the Pacific which span the last 300 years [31]. These temperature records in the Pacific suggest that the spatial pattern of the inter-decadal Pacific oscillation at least in the South Pacific has varied considerably and undergone a major reorganization at $\sim$1880 AD.

To investigate oceanic conditions associated with abrupt suborbital events, and Milankovitch forced glacial–interglacial cycles which yield insight as to how the climate alters under naturally forcing, we require information on the time-scale of thousands to hundreds of thousands of years. The rain of foraminifera and coccolithophores, amongst other components to ocean sediments accumulate at an average rate of 1–4 cm/kyr, but can be as high as 50–100 cm/kyr so down-core analysis of their chemistry is well suited for reconstructing ocean conditions on these timescales and can be extended into the millions of years by drilling of sedi-ments. Amongst a host of other trace metal proxies in foraminifera, the Mg/Ca paleothermometer and paleonutrient proxy Cd/Ca are perhaps the most prevalent. As an example, Mg/Ca can be used in concert with δ^{18}O to resolve temperature and salinity variations in planktonic forams for surface conditions, and benthic forams for deep water signatures. This combined proxy has been used to focus on the role of thermohaline overturning during millennial scale cooling events which punctuate the last glacial cycle and the most recent transition from glacial to interglacial conditions, $\sim$10 kyr ago. In particular, salinity within the surface of the Caribbean Sea, the main source of surface waters feeding North Atlantic deep water (NADW) formation have been shown to vary in concert with the strength of NADW flow [14]. Furthermore, benthic records across the rapid climate changes of the glacial–interglacial transition from 3146 m in the North Atlantic reveal an oscillation between the changing local dominance of warm, high salinity NADW vs. cold, low-salinity southern-sourced Antarctic bottom water (AABW) associated with stadial and interstadial variations in the Greenland ice core [32]. This reinforces existing thinking that variations in the strength of NADW are intricately linked to millennial scale climate variations as evidenced by benthic foraminiferal Cd/Ca tracing the distinctive nutrient signatures of nutrient poor NADW and nutrient rich AABW (e.g., [33]). Returning to surface waters, planktonic foraminiferal Cd/Ca records from the Southern Ocean refute that increased iron fertilization of productivity in these nutrient rich waters at glacial times could have been responsible for the glacial draw-down of carbon dioxide [10].

The trace metal content of coccolithophores yields information on the same timescales to foraminifera, but about different aspects of the ocean due to the contrasting geochemistry of their biomineralization. The full range of trace metals from coccolithophore calcite in the sediments have not been exploited due to difficulty in separating them from potentially contaminating clays. Nonetheless, Sr/Ca is unaffected by such contamination and correlates with the growth rate of coccolithophores [34–36] and can therefore divulge past pro-ductivity of the oceans (e.g., [37]). We can therefore use proxy records from for-aminifera and coccolithophores to investigate the past when atmospheric carbon dioxide was analogous to our current situation, as the globe oscillated between an icehouse and ice free greenhouse world on timescales of tens of millions of years. A novel methodology has been developed to probe the climate and ocean hundreds of millions of years ago before foraminifera and coccolithophores had evolved. Trace metal ratios (Mg/Ca, Sr/Ca, and Na/Ca) in parallel with stable isotopes from belemnites (cephalopod molluscs which were abundant in the Jurassic ocean) are starting to provide temperature and salinity information (e.g., [38]). These organisms are now extinct and so we have no real calibration data, or information regarding biomineralization mechanisms and as such will not be discussed further.

This is by no means a comprehensive review of how trace metal proxies within biominerals have enhanced our understanding of the past climate, but an illustration of how we can probe a range of climatically important timescales and use different trace metals to explore an assortment of characteristics of the past ocean.

4. BIOMINERALIZATION PROCESSES OF DIFFERENT ORGANISMS

We now seek to address whether the contrast in trace metal geochemistry of biogenic carbonates vs. inorganic carbonates can be understood in terms of the different processes which are important to the biomineralization mechanism of different organisms. First, we shall summarize the biomineralization process of coccolithophores, foraminifera and corals [Fig. 3(a–c)]. The fossil remains of each of these organisms make a significant contribution to our paleoceanographic reconstructions of the past world.

4.1. Coccolithophores

Coccolithophores are single-celled plant plankton which belong to the phylum Haptophyta, and secrete an interlocking sphere of calcite platelets [Fig. 3(a)]. A physical mechanism for the assembly and extrusion of a coccolith has been proposed by Westbroek et al. [39], detailed in Refs. [40,41], and summarized by Young and Henriksen [42].

Growth occurs in a coccolith vesicle derived from the Golgi body and which is supplied with matrix material and calcium via Golgi vesicles. The biomineralization process commences with formation of an organic scale within a vesicle which develops into a complex form, with extensions containing dense particles termed coccolithosomes. The coccolithosomes appear to play a key role in calcification and have been shown to be complexes of acidic polysaccharides with calcium ions [43]. It is thought that they function as calcium vectors during biomineralization and that the polysaccharide phase forms the crystal coatings. Nucleation of a protococcolith ring of alternating orientation simple crystals then occurs around the rim of a precursor base-plate scale followed by crystal growth upward and outward to form the complex crystal

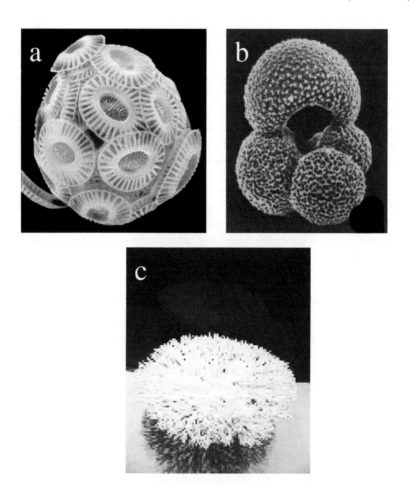

Figure 3 Scanning electron microscopy (SEM) photos of (a) *Emiliania huxleyi* measuring $\sim$10 μm in diameter, (b) *Globigerina bulloides* measuring $\sim$250 μm in diameter, (c) a coral measuring $\sim$10 cm in diameter.

units of the complete coccolith. Assembly of the coccolith, which consists of a cycle of radially and vertically oriented calcite crystals, involves a folded acidic polysaccharide matrix [44]. This polysaccharide matrix is thought to promote and mold calcification by providing uranic acid groups as nucleation sites for Ca^{2+}, but also inhibits crystal growth by adhering to the surface of the coccolith when construction is complete [39]. This organic matrix framework can provide binding sites for the components of a mineral, selectively nucleating specific crystallographic faces, and organic carrier molecules can ensure supersaturation of a phase within mineralizing compartments.

The coccolith grows in an expanding vesicle and much of the morphology is a product of interaction between adjacent crystals. The final structure of the completed coccolith is an emergent result of inorganic growth of crystals defined by the nucleation stage, within a space defined by an expanding organic vesicle. After completion of the coccolith, the vesicle dilates and at this stage, a dense organic coating is visible around and between the coccolith crystals. It is reasonable to infer that the final coating of polysaccharides that can prevent dissolution of the element also serves to inhibit crystal growth. The coccolith is then exocytosed onto the surface to form an interlocking sphere of coccoliths by fusion of the vesicle membrane and cell membrane.

4.2. Perforate Foraminifera

Foraminifera are unicellular calcifying marine amoeba, taxonomically part of the Protista. The most common arrangement is spherical coiling [Fig. 3(b)] with planispiral or low trochispiral tests. Erez [45] summarizes the major steps involved in perforate foraminifera calcification.

In perforate foraminifera, the first step for chamber formation involves delineation of a space using ectoplasm pseudopods. The next step is to define the shape of the newly formed chamber by creating a cytoplasmic bulge that serves as a mold for the organic matrix and as a template for nucleation. The third step is the precipitation of $CaCO_3$ on both sides of a thin organic layer. Radiotracer experiments [46] have confirmed the presence of an internal Ca pool which foraminifera use for this calcification. It is possible that this Ca pool may be connected to small polarizing granules that were recently observed in the endoplasm [47]. Ca is concentrated in the endoplasm in a highly soluble, birefringent mineral phase composed of Ca, Mg, P, and S. The granules are membrane bound and may contain organic matrix or some of its components. The granules provide Ca for the first $CaCO_3$ crystals that precipitate over the newly formed organic matrix. At this stage the chamber consists of a two-dimensional primary wall made of Mg-rich spherulites embedded within the organic matrix.

The second stage of calcification involves massive deposition of a low Mg-calcite wall. This secondary calcite is made of layered crystal aggregates with their *c*-axis perpendicular to the test wall. These units form the secondary lamination and are responsible for the bulk of the skeleton deposition because foraminifera cover their preexisting shell with a new layer of calcite every time a new chamber is built. The biomineralization process forming secondary calcite involves vacuolization of seawater and its modification within the cytoplasm perhaps to reduce the Mg/Ca ratio and elevate the pH. In order to precipitate low Mg-calcite, it is necessary to either concentrate Ca or reduce Mg in the seawater vacuole (as well as increase the pH). The main possibilities considered involve active pumping away of Mg, or alternatively complexation by organic materials perhaps the preudopodial network (bilipid membranes). Towards the end of their life cycles, many planktonic foraminifera deposit several different types of $CaCO_3$ often in the form of a thick crust as is termed gametogenic calcite.

4.3. Corals

Reef corals [Fig. 3(c)] belong to the order Scleractinia, all of which accrete hard exoskeletons. The animal responsible for skeletal formation is the polyp, a double-walled sack of simple design [48].

The calicoblastic layer of the ectoderm, which lies adjacent to the skeletal surface, is considered to be involved in some way in calcification. The basic building blocks of the coral skeleton consist of fine aragonite crystals arranged in three-dimensional fans about a calcification center. Within the calcification centers are submicron sized granular crystals bundled into discrete "nuclear packets". The small size of these granular seed crystal may indicate intracellular mineralization, as suggested by Hayes and Goreau [49]. It is probable that the contents of intracellular vesicles are transported across the apical membrane and exocystosed into the calcifying space. Indeed, this is a likely route for seawater entry. One further suggestion is that the intracellular vesicles in the apical membrane of the calicoblastic ectoderm, with their organic contents, are sites of production and stabilization of amorphous $CaCO_3$ precursors of the granular seed crystals that occupy the centers of calcification. The geometry of the nuclear packets indicates that they are incorporated into the skeleton in a non-rigid state [50].

Calcium ions enter the coral's calcifying space by both passive and active transport [51,52]. Passive entry occurs by way of seawater transported via invaginated vacuoles leaking or diffusing into the calcifying space. Active transcellular transport of both ions occurs enzymatically, via the Ca^{2+} ATPase pump. Regarding the control on precipitation, the origin, physical structure, and function of the putative organic matrix in coral skeletons remains elusive. A model proposed by Barnes [53] argues that fast growing crystals precipitated from a supersaturated solution will compete with each other. The tendency for these crystals to diverge from the optimum axis of growth gives rise to three-dimensional fans. Further compelling evidence for the predominance of physico-chemical factors in the growth of aragonite fibers is the correlation between fiber morphology and coral growth rate [54].

The range of fiber morphologies found amongst the scleractinian taxa could be explained by basic theories of crystal growth in inorganic systems without the need for mediation by an organic macromolecular framework or matrix. Recently however, Cuif et al. [55] have proposed a polycyclic model of crystal growth, involving step-by-step growth of aragonite fibers, each step initiated and guided by a sulfated organic matrix sheet. Alternatively sheets of sulfated organic materials at daily growth boundaries could be inhibitory rather than promotional features.

4.4. Biomineralization and Proxies

It is interesting to note the gradation between the chemistry of inorganic calcite, planktonic foraminifera and the biologically extreme chemistry of coccolith calcite from Fig. 1, i.e., the coccolith calcite experiences the greatest biological influence. Furthermore, all coccolith calcite partition coefficients are less than 1. This corroborates the extreme biological influence on the coccolith calcite and argues for greater selectivity between calcium and trace metals. By contrast, a coral skeleton largely resembles the chemistry that would be expected from inorganic precipitation from seawater. Most trace elements and even small particles occur in the skeleton in proportions reflecting their abundance in seawater, and their tendencies to become incorporated either within or among aragonite crystals. The trace metal geochemistry of corals reflects minimal biological influence. It appears that biomineralization controls the chemistry of biogenic carbonates to differing degrees for different organisms. These differences are easily explained by the relative involvement of biological transport and matrix-mediated precipitation, i.e., coccolithophores experience the most involved intracellular precipitation compared to forams and corals (Fig. 4). There is an importance of seawater vacuolization for both corals and foraminifera, but this process is not thought to occur in coccolithophores. By contrast, organic matrix-mediated precipitation in coccolithophores is key but its

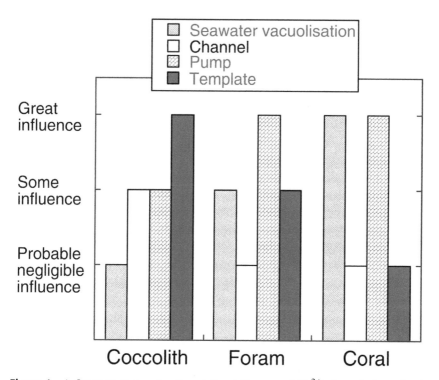

Figure 4 A figure to summarize the relative influences of Ca^{2+} channels (open bars), Ca^{2+} pumps (brick bars), the template (filled bars), and seawater vacuolization (diagonally hatched bars) on the biomineralization process in coccoliths, corals, and foraminifera.

importance is less for foraminifera and is still questioned in corals. The relative trace metal chemistries of coccolithophores, foraminifera, and corals undoubtedly reflect the differing degrees of biological control on the precipitation.

5. BIOLOGICAL DISCRIMINATION BETWEEN CALCIUM AND TRACE METALS

In our quest to understand the mechanisms that account for the different geochemistries of biominerals, we now turn our attention to the trace metal discrimination characteristics of the transport and assemblage processes. Calcium plays a dual role in biomineralizing organisms as both a substrate for calcification and an intracellular regulator.

The cytosol concentration of free calcium is rigorously controlled and maintained at a very low level. For our biological end-member calcite, i.e., the coccolithophores, the Ca^{2+} ions necessary for the formation of calcite diffuse from seawater through Ca^{2+}-selective channels into the cytosol of the coccolithophore driven by a potential difference and by a very low Ca^{2+} activity in the cytosol (0.1 μM) (Fig. 5). This low cytosolic concentration of Ca^{2+} means that Ca^{2+} must be pumped against a concentration gradient at some stage during its transport to the site of precipitation in order to attain saturation. Although this process must be extremely selective for Ca^{2+} ion, the very presence of trace metals in coccolith calcite indicates that the similarly sized trace metal ions must substitute for Ca^{2+} and be transported via the same mechanism but at a different rate as Ca^{2+}.

More generally, the two steps which must impact biogenic calcite chemistry are the transport of ions from seawater across membrane(s) to the site of precipitation, and, the precipitation of the calcite on an organic matrix (Fig. 5). One or both of these processes is common to all biomineralizing organisms. Transport of ions across a membrane is driven by pumps against a concentration gradient or directed through channels when transport is with a concentration gradient. Channels and pumps have very different modes of selectivity and transport. Similarly, the intricate relationship between the molding and precipitation control of the organic template or matrix may control the selection of ions during assemblage of the mineral.

5.1. Selectivity of Ion Channels

Highly specific membrane spanning macromolecular structures, ion channels, serve to facilitate and control the passage of selected charged ions across the hydrocarbon lipid barrier down a concentration gradient. Three divalent ions, Ca^{2+}, Sr^{2+}, and Ba^{2+}, pass readily through all known Ca^{2+} channels. Most other divalent ions act as blockers of Ca^{2+} channels, but in isolated cases, inward currents carried by Mg^{2+}, Mn^{2+}, Co^{2+}, Zn^{2+}, or Ba^{2+} have been demonstrated [56–58]. The selectivity of an ion-selective channel for divalent ions

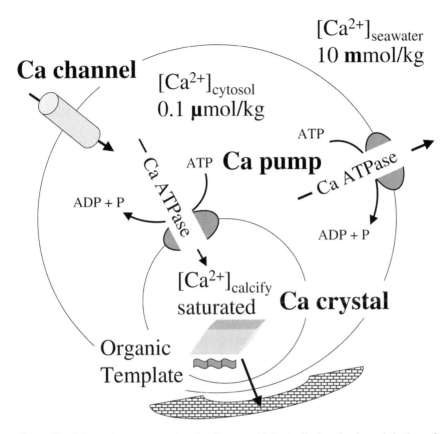

Figure 5 Schematic representation for the most biologically involved precipitation of calcite to show the involvement of Ca^{2+} channels (from high to low Ca^{2+} concentration), Ca^{2+} pumps (from low to high Ca^{2+} concentrations) and the involvement of the template as well as the final stage of crystal precipitation. All calcite biomineralization processes will involve at least one or more of these steps.

follows the sequence $Ca^{2+} > Sr^{2+} > Ba^{2+} \gg Mg^{2+}$ ($Ca^{2+} > Sr^{2+} \sim Ba^{2+} \gg Li^+ > Na^+ > K^+ > Cs^+$) (see Table 1). However, the selectivity of channels is governed by two factors, partitioning into the membrane and mobility once inside. As a result, for some channels the current of Ba^{2+} through the channel can be greater than that of the more selected Ca^{2+} or Sr^{2+}.

Although no detailed molecular structure has been published for a Ca^{2+} channel, many analogies regarding selectivity may be drawn from the study of a K^+ channel by Doyle et al. [59]. If a channel is highly ion-selective, the pore must be narrow enough to force permeating ions into contact with the wall so they can be sensed. These narrow ion selective regions of ion channels are known as the selectivity filter. Doyle et al. [59] showed that a K^+ channel

Table 1 Permeability Ratios P_x/P_{Ca}, for L-Type Ca Channels[a]

Ion	P_x/P_{Ca}	Ion	P_x/P_{Ca}
Ca	1.0	Li	1/424
Sr	0.67	Na	1/1170
Ba	0.40	K	1/3000
		Cs	1/4200

[a]An L-type Ca channel has a large single-channel conductance and a long-lasting current.

begins as a tunnel and then opens into a wide cavity near the middle of the membrane. A K^+ ion would move throughout the internal pore and cavity and still remain mostly hydrated. The chemical composition of the wall lining the pore is predominantly hydrophobic. In contrast, the narrow selectivity filter is lined exclusively by polar main chain atoms belonging to amino acids. So, the selectivity is maintained due to two critical structural factors. When an ion enters, it dehydrates nearly completely. To compensate for the energetic cost of dehydration, the carbonyl oxygen atoms must take the place of the water oxygen atoms, come in very close proximity and coordinate as strongly as the water (Fig. 6).

Secondly, the interactions and hydrogen bonds between surrounding proteins seem to act like a layer of springs stretched radially outwards to hold the pore open at its proper diameter. Smaller or larger ions would distort this structure and disrupt the energy balance. Finally, two K^+ ions at close proximity in the selectivity filter repel each other. The repulsion overcomes the otherwise strong interaction between ion and protein and allows rapid conduction in the setting of high selectivity. This feature is common to Ca^{2+} channels which are highly selective yet capable of high rates of ion transfer. Ca^{2+} channels use diverse mechanisms of gating, but tend to exhibit similar ion permeability characteristics. According to the earlier mechanism, the factors which define the selectivity of a channel are the energy of hydration of a cation, the energy of coordination by carbonyl oxygen, ionic radius, pore radius, and charge.

5.2. Selectivity of Ion Pumps

Ion pumps work in a different way to ion channels. In contrast to ion channels where the high selectivity of binding can slow down the transport of the selected component, i.e., Ca^{2+}, ion pumps will transport most efficiently the highly selected ions. Ca^{2+}, Sr^{2+}, and Mn^{2+} are the only ions that have been demonstrated to be transported by a Ca^{2+}ATP-ase with the formation of concentration gradients. Sumida et al. [60] studied the effects of other divalent cations on the Ca^{2+} uptake by microsomes from bovine aortic smooth muscle and indicated

The chain comprises the signature amino acid sequence from bottom to top: T (Threonine), V (Valine), G (Glycine) and Y (Tyrosine), G (Glycine)

The V and Y side chains are directed away from the ion conduction pathway

The ion conduction pathway is lined by the main chain carbonyl oxygen atoms shown here in red

Two K$^+$ ions (green) located at opposite ends of the selectivity filter with a single water molecule in between. The inner ion is depicted in rapid equilibrium between adjacent coordination sites.

The filter is surrounded by inner and pore helices (white)

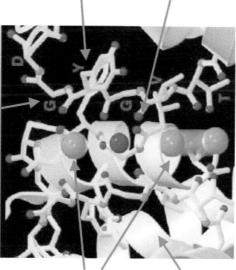

Figure 6 The selectivity filter of a K$^+$ channel shown as a stick representation with the chain closest to the viewer removed. The three chains represented are comprised of signature amino acid sequences threonine, valine, glycine, tyrosine, from bottom to top. The selectivity of the pore is controlled by the coordination of side chain carbonyl oxygen from amino acid groups. Adapted from Ref. [59].

The closest water molecule (red spheres) to the Ca^{2+} ions marked by a star

The coordination of six surrounding O atoms to Ca^{2+} (blue spheres) indicated by white dotted lines

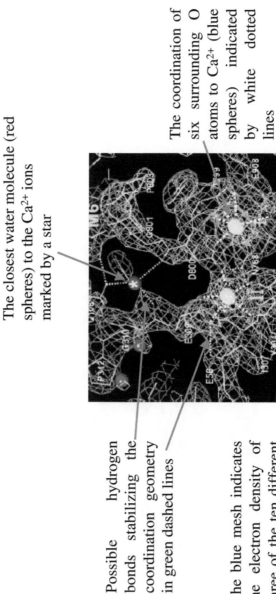

Possible hydrogen bonds stabilizing the coordination geometry in green dashed lines

The blue mesh indicates the electron density of three of the ten different trans-membrane helices, two of which are unwound for efficient coordination geometry (M4 and M6)

that Co^{2+}, Zn^{2+}, Mn^{2+}, Fe^{2+}, and Ni^{2+} did not interfere with Ca^{2+} uptake not the formation of the phosphorylated intermediate and hence, were not in competition with Ca^{2+} transport. Cd^{2+} however, inhibited both Ca^{2+} uptake and the formation of the phosphorylated intermediate but in a non-competitive manner.

The best analogue for Ca^{2+} during transport by Ca^{2+}ATPase is Sr^{2+}, as Sr^{2+} can replace Ca^{2+} at the binding site. The Ca^{2+} ATPase has a higher affinity for Ca^{2+} than for Sr^{2+} [61]. A simple mechanism for catalysis of transport of two calcium ions which is coupled to the hydrolysis of ATP has been investigated for the relative efficiency of transport of Sr^{2+} vs. Ca^{2+}. The first step is the binding of two calcium ions to the exterior side of the vesicles. This binding has a high affinity for Ca with $K_{0.5} \sim 1$ μM; the binding of Sr occurs with a lower affinity of $K_{0.5} \sim 83$ μM [62]. Overall, the mechanism for the transport of Sr^{2+} appears to be the same as that for Ca^{2+} and occurs at a similar rate but with a lower affinity by two orders of magnitude.

The best studied Ca^{2+} ATPase to date is the skeletal muscle sacroplasmic reticulum calcium ATPase (SERCA ATPase) which has been characterized to 2.6 Å detail and shows a high degree of specificity for the transported ion (Fig. 7 [63]). The protein consists of 10 helices which conform to allow the Ca^{2+} ion to enter, be transported across the membrane and then leave. The specificity seems to arise from the coordination geometry of six oxygen atoms which coordinate the Ca^{2+} ion. Interestingly, the six oxygen atoms are located at a distance of 2.2–2.6 Å from the center of each site. This distance can be compared with the Ca–O distance in a calcite crystal of 2.359 Å. This implies that the coordination of the biological cations is controlled in a remarkably similar way to Ca^{2+} in a crystal. Furthermore, rows of main chain carbonyl oxygen within helices point towards the cytoplasm and provide a hydrophilic pathway leading to the Ca^{2+} binding sites. The rows constrict near the Ca^{2+} binding sites, trapping a water molecule. This geometry must be required for displacing water molecules from the Ca^{2+}. In a similar way to the mechanism of selectivity for Ca^{2+} channels, the selectivity of a Ca^{2+} pump is defined by the energy of dehydration and ease to strip water molecules, the energy of binding of the Ca^{2+} to the specific site

---

Figure 7 An extract of the crystal structure of the calcium ATPase of skeletal muscle sarcoplasmic reticulum at 2.6 Å resolution with two calcium ions bound in the transmembrane domain defined by the electron density map. The two calcium ions are located side by side and are surrounded by four transmembrane helices, two of which are unwound for efficient coordination. The binding site has a highly defined geometry which makes it a high-affinity Ca^{2+} binding site due to the coordination of six oxygen atoms from the side chain oxygen atoms of asparagine, glutamate, threonine, aspartic acid, and glycine. This kind of coordination geometry is only possible due to unwinding of the helices. In ATPases that transport heavy metals, the glutamate residue is replaced by cysteine or histidine. The two sites are stabilized by hydrogen-bond networks between the coordinating residues and between residues on other helices. These hydrogen-bond networks must be important for the cooperative binding of two Ca^{2+} ions. Adapted from Ref. [63].

denoted by sixfold oxygen coordination from carbonyl oxygen, ionic radius, and charge.

Ca pumps have been demonstrated to be present in most coccolith membranes [64] and identified in the coccolith vesicle membrane [65]. By association these pumps are inferred to be involved in the calcification process. Al-Horani et al. [66] confirmed that corals pump Ca^{2+} into the calcifying space using the enzyme Ca^{2+} ATPase. No Ca^{2+} pump in biomineralization has been characterized for its characteristics of trace metal selectivity. We can only conclude that it is likely that pumps exhibit a strong selectivity between Ca^{2+} and all trace metals during biomineralization according to the mechanism outlined earlier.

5.3. Selectivity of Acidic Polysaccharide Template

The organic matrix is a preformed insoluble macromoleculear framework that is a key mediator of controlled biomineralization. The matrix subdivides the mineralization spaces, acts as a structural framework for mechanical support, and is interfacially active in nucleation [67]. The matrix is a polymeric framework that consists of a complex assemblage of macromolecules, such as proteins and polysaccharides. In its simplest form the matrix consists of a structural framework of predominantly hydrophobic macromolecules with associated cross-links, onto which are anchored hydrophilic macromolecules that present an active nucleating surface. In many cases, the acidic macromolecules are glycoproteins which are proteins with covalently linked polysaccharide side chains that often contain sulfate and carboxylic acid residues [e.g., Fig. 8(c)].

The central role of the organic matrix in controlling inorganic nucleation is to lower the activation energy by reducing the interfacial energy. Lowering of the activation energy for nucleation is considered to arise from the matching of charge polarity, structure, and stereochemistry at the interface between an inorganic nucleus and an organic macromolecular surface. This leads to control over the rate of nucleation, the number and organization of nucleation sites, polymorph selectivity, and oriented nucleation.

The role of any organic template in the calcification of corals remains questionable, and very little is known about the template used by foraminifera. However, the ease of manipulation of coccolithophores in the laboratory has yielded detailed characterization of the acidic polysaccharides which are involved in nucleation and molding of calcite precipitation. Furthermore, the close relationship between the organic matrix and the coccolith has been demonstrated using the new NanoSIMS technique [Fig. 8(A) and (B)]. Future research is planned to use the high resolution chemical abilities of this technique to map out trace metal distributions in the calcite associated with the organic matrix. Three polysaccharides have been identified as being involved in the precipitation of coccolith calcite. Polysaccharide 1 and 2 (PS1 and PS2) form 20 nm particles with Ca^{2+} ions and attach to the base plate rim [43]. PS2 probably facilitates calcite nucleation whilst PS3 (in *Pleurochrysis*) or coccolith polysaccharide (in *Emiliania huxleyi*) is a sulfated galacturonomannan (Fig. 8(C) [68,69]) which is directly linked to

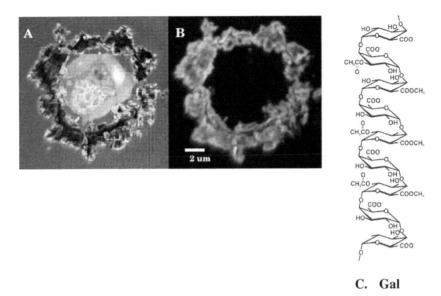

C. Gal

Figure 8 (A) $^{12}C^{14}N^-$ image of a thin (1 μm) section of a cryofixed resin embedded *Coccolithus pelagicus* cell obtained with the Cs^+ ion beam of the Oxford NanoSIMS. The CN^- beam depicts the cell and resolves intracellular compartments and the intricate relationship between the organic template and calcite. The calcified cell has a diameter of 12.5 μm. (B). An $^{16}O^-$ image of the same cell. The $^{16}O^-$ is produced only in the areas where calcite is present and shows the external calcite platelets encircling the cell. (C) The chemical structure of a galactosyluronic acid residue (Gal) which forms part of the acidic polysaccharide involved in calcification separated and identified from coccolithophores.

the growth and shaping of coccolith calcite. PS3 from *Pleurochrysis carterae* and *E. huxleyi* share a similar structure of a backbone of mannosyl residues bearing ester sulfate groups and many galacturonic acid-containing side chains. In a similar manner to the pumps and the channels, the selectivity for nucleation by the template will be controlled by the differential energy of binding of the cations to coordinating oxygen ligands from the acidic polysaccharide.

5.4. Biomineralization Selectivity: A Hypothesis

The biological selectivity of the transporters and matrix is strikingly similar in its base chemistry to the selective assembly of ions into a crystal. In each case the selectivity between Ca^{2+} and trace metals derives from the balance between the energy required for dehydration of the hexaaqua complex of the cation, and the energy released from the new coordination geometry of binding with either carbonyl oxygen from polysaccharides or amino acids, or carbonate oxygen in the crystal. It is remarkable to note that the distance of Ca–O in the site of Ca^{2+} ATPase is 2.2–2.6 Å compared with the Ca–O bond length in

Table 2 Hydration Energy, Electronegativity, and M−O Bond Length for Cations Found Commonly in Biological Calcites

Metal	Gibbs free energy of formation of aqueous ions (kJ/mol) [80]	Electronegativity (Pauling scale)	Bond length M−O in *calcite* or aragonite structure (Å)
Ca²⁺	−553.6	1.0	*2.36*
Sr²⁺	−557.3	1.0	2.57−2.73
Ba²⁺	−560.8	0.9	2.74
Mg²⁺	−454.8	1.2	*2.11*
Mn²⁺	−223.3	1.5	*2.19*
Zn²⁺	−147.2	1.6	*2.11*
Cd²⁺	−77.6	1.7	

calcite of 2.359 Å (other M−O bond lengths are quoted in Table 2) which shows how the size of the crystal binding site is mimicked by biology.

However, the size of the binding site in a biological macromolecule or crystal relative to the ionic radius cannot be the only control on trace metal partitioning. From Fig. 1, it is necessary to explain why biological selectivity appears to be more efficient than crystals for cations of a smaller ionic radius and less efficient than crystals for ions larger than Ca^{2+}. One possibility arises from the different energies of hydration of the cations (Table 2). The Gibbs free energy of formation of aqueous ions of Ca^{2+}, Sr^{2+}, and Ba^{2+} differ by only 1.3% and Sr^{2+} and Ba^{2+} are both incorporated preferentially into biological calcite than inorganic calcite. Mg^{2+} is the next most similar but releases significantly less energy on hydration (18%). Therefore both sides of the energy balance required for Ca^{2+} transport are disrupted be it either in a channel or a pump. Not only will the cations smaller than Ca^{2+} bind less favorably at the oxygen-coordinated specific site, but the energy required for dehydration of the cation before specific binding and transport is significantly different to that for Ca^{2+}. This idea is speculative at the moment, but could be tested using theoretical chemical calculations for the reaction Gibbs free energies. Furthermore, as calcifying organisms rise to the top of the agenda for genetic DNA sequencing, we anticipate identification of the sequence of the Ca transporting proteins and their closest analogies which will enable direct experimentation on selectivity of transporters from the coccolithophores, foraminifera, and corals.

6. BIOLOGICAL ION SELECTIVITY AND THE ENVIRONMENT

All partition coefficients are sensitive to environmental factors including temperature, growth rate or calcification rate and even pH or carbonate ion content of the precipitating media. Whilst we would expect the equilibrium constant of inorganic carbonate equilibria reactions to be affected by the temperature or pressure of the reaction, the sensitivities of trace metal incorporation to these environmental factors departs extensively from inorganic calcite predictions (Fig. 2).

It is worth mentioning here that the biological sensitivity to the environment or "vital effect" is nothing magical. As outlined earlier the dominant controls on ion selectivity by transporters or the matrix hinges on dehydration of the cation and coordination or bonding by oxygen atoms from neighboring amino acids of a macromolecular protein, which is analogous to the binding site within a calcite or aragonite crystal where a cation is dehydrated and then coordinated by six or nine oxygens of neighboring carbonates, respectively. However, it is important to remember that the utility of the chemistry of biogenic carbonate as proxies for past ocean conditions rests on the observed correlations of the chemistry with environmental variables. If such a correlation does not reflect an inorganic mechanism, it must be due to some biochemical process. Here, we explore how environmental factors might control the biochemical discrimination.

6.1. Temperature

Very few studies have investigated the effects of temperature on the selectivity of Ca^{2+} ATPases, or Ca^{2+} channels as the majority of systems of interest are in warm-blooded mammals which maintain a constant body temperature. Nonetheless, we can create a hypothetical model for the selectivity of channels, pumps, and templates based on the activation energy of binding. It should be noted that Ca^{2+} transporters show complex kinetics due to changes in the activation energy of transport, generally characterized by non-linear Arrhenius and van't Hoff plots, but also due to temperature having a great control over the fluidity of the bilayered lipid membrane [70].

If we imagine a Ca^{2+} channel which has high selectivity at binding but as a result inhibits large ion currents of the highly specific ion, then the activation energy for the transport of Ca^{2+} is higher than the similarly sized trace metals. By contrast, if we imagine a Ca^{2+} pump which has high selectivity for both binding and transport, then the activation energy for the transport of Ca^{2+} will be lower than that for the similarly sized trace metals. In each case, as temperature increases, the differential between the two activation energies for transport of Ca^{2+} and a trace metal becomes less important. Therefore, for biomineralization where Ca^{2+} channel processes are dominant we would expect the M/Ca ratio to decrease with increasing temperature, and for biomineralization where Ca^{2+} pumps are more important, we would expect the M/Ca ratio to increase as the selectivity breaks down. So, we propose that positive or negative correlation between trace metal uptake and temperature may be due to the differing importance of channels or pumps in the biomineralization process.

Following the previous logic, we would expect the temperature control on the ordering of ions by the organic matrix or template to be similar to that of a Ca^{2+} pump. The binding is controlled by lowering the activation energy of binding for calcium in particular and would be higher for other trace metals. As temperature increases, we expect the Ca^{2+}-specific ligands to lose their rigidity and as such reduce the selectivity of the nucleation sites such that the M/Ca ratio will increase with increasing temperature.

6.2. Kinetics

Increasingly, growth rate is recognized as controlling the trace metal uptake into biogenic carbonate, and particularly coccolithophores [34–36,71]. Furthermore, the carbonate ion in the surrounding media controls the degree of calcification and mass of carbonate precipitated for planktonic foraminifera, coccolithophores, and corals [72–74]. A further interlinking proposal is that the trace metal uptake is controlled by the carbonate ion, as demonstrated in benthic foraminifera [75,76], and proposed for planktonic foraminifera [77]. As the rate of growth or calcification increases in each case, be it driven by carbonate ion or not, the incorporation of the trace metal increases. We propose that this biological rate or kinetic control on trace metal uptake can be considered as a biological ana-logue of the inorganic model proposed by Lorens [78]. In his model, he proposes a rate-dependent discrimination by the crystal against ions of a different size to Ca^{2+}. We propose that trace metal incorporation into biogenic carbonates is

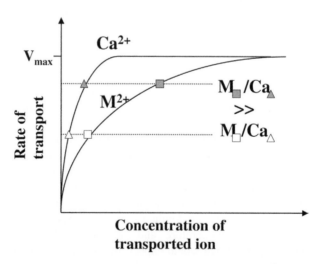

Figure 9 A hypothetical plot of rate of transport of an ion by a Ca^{2+} transporter vs. the concentration of the transported ion for Ca^{2+} (triangles) and M^{2+} (squares). The slightly higher charge density of the calcium ion relative to the metal ion leads to stronger bonding of the calcium ion by the transporter and more efficient transport at lower concentrations. As the rate of transport of ions increases from the open shapes to the gray shapes, the M/Ca rate of the transported ions will also increase as marked by the dotted lines.

controlled by a rate-dependent discrimination of binding by the transport and template macromolecules.

In channels, pumps or template, there is a binding site specific to the transported ion, namely calcium. In each case, Ca^{2+} will be bound more strongly than the trace metal ions [79]. This means that the maximal rate of transport by a Ca^{2+} pump or channel (V_{max}) would be attained at a lower concentration of Ca^{2+} than of the trace metal (Fig. 9). As the rate of transport increases, the concentration of the transported trace metal increases proportionally more than the Ca^{2+}. The M/Ca rate transported to the vesicle and available for precipitation will increase with increased rates of pumping. At higher rates, the discrimination between Ca^{2+} and trace metals to the nucleation sites of the organic template will be less efficient. In essence, at higher rates of reaction, more mistakes will be made and each process in the biomineralization process is likely to enhance incorporation of trace metals at higher rates of growth or precipitation.

7. SUMMARY

Trace metal proxies bound within the calcium carbonate tests of oceanic organisms provide a unique insight into how the climate system works on timescales which span eight orders of magnitude, from annual to hundreds of millions of years. Whilst the motivation for developing these proxies was the idea that thermodynamic equilibria control the chemistry during precipitation, in reality the application of trace metal proxies relies upon empirical calibration. Such calibration can be applied to a wide range of environmental reconstructions, but more accurate application of proxies requires a mechanistic understanding of the biomineralization process.

The partitioning of trace metals into biogenic carbonates reflects to some extent the same pattern as an inorganic crystal, but there is an additional selectivity and differing environmental sensitivity to, e.g., temperature, which confirms that biochemical processes also play a role in the uptake and assembly of ions into a crystal. Different organisms display differing degrees of biological control on their carbonate chemistry. Aragonitic coral chemistry is most similar to inorganic precipitation from seawater whilst coccolithophores are most different, and these contrasts correlate with the degree of control of the organism over its biomineralization.

Selectivity between Ca and trace metals during biomineralization arises during transport by pumps, channels, or nucleation upon an organic matrix. The biological selectivity of the transporters and matrix is strikingly similar in its base chemistry to the selective assembly of ions into a crystal. In each case, the selectivity between Ca^{2+} and trace metals derives from the balance between the energy required for dehydration of the hexaaqua complex of the cation, and the energy released from the new coordination geometry of binding with either carbonyl oxygen from polysaccharides or amino acids, or carbonate oxygen in the crystal. This is a speculative idea, but with some careful chemical calculations based on the energy of binding of Ca^{2+} or the trace metal ions to these macromolecular structures, it provides an alternative thermodynamic framework within which to consider the application of trace metal proxies.

ACKNOWLEDGMENTS

R. Rickaby acknowledges the support of the Natural Environmental Research Council grant number: NER/M/S/2002/00123 and is also grateful for constructive discussions with Sam Shaw, Nick Belshaw, and Don Fraser.

ABBREVIATIONS

AABW	Antarctic bottom water
$\delta^{no}E$	isotopic variation of the element ^{no}E, e.g., of ^{18}O, relative to an internationally accepted standards in parts per thousand or per mill
kyr	kilo years
MC-ICPMS	multi-collector inductively coupled plasma mass-spectrometry
Myr	million years
NADW	North Atlantic deep water
PS	polysaccharide
SEM	scanning electron microscopy
SERCA	skeletal muscle sarcoplasmic reticulum calcium ATPase
SI	International System of Units
SIMS	secondary ion mass spectrometry

REFERENCES

1. Urey HC. J Chem Soc 1947; 562–581.
2. Epstein S, Buchsbaum HA, Lowenstam HA, Urey HC. Bull Geol Soc Am 1953; 64:1315–1326.
3. Bemis BE, Spero HJ, Bijma J, Lea DW. Paleoceanography 1998; 13:150–160.
4. Boyle EA. Earth Planet Sci Lett 1981; 53:11–35.
5. Croal LR, Johnson CM, Beard BL, Newman DK. Geochim Cosmochim Acta 2004; 68:1227–1242.
6. Icopini GA, Anbar AD, Ruebush SS, Tien M, Brantley SL. Geology 2004; 32:205–208.
7. Arnold GL, Anbar AD, Barling J, Lyons TW. Science 2004; 304:87–90.
8. Siebert C, Nagler TF, von Blanckenburg F, Kramers JD. Earth Planet Sci Lett 2003; 211:159–171.
9. Anbar AD. Earth Planet Sci Lett 2004; 217:223–236.
10. Elderfield H, Rickaby REM. Nature 2000; 405:305–310.
11. Boyle EA, Keigwin LD. Earth Planet Sci Lett 1985/86; 76:135–150.
12. Anand P, Elderfield H, Conte MH. Paleoceanography 2003; 18:art. no. 1050.

13. Rosenthal Y, Oppo DW, Linsley BK. Geophys Res Lett 2003; 30:art. no. 1428.
14. Schmidt MW, Spero HJ, Lea DW. Nature 2004; 428:160–163.
15. Palmer MR, Elderfield H. Nature 1985; 314:526–528.
16. Holland HD. Prog. Rep. U.S. AEC Contact No. AT (30-1), 2266 (1960).
17. Holland HD, Borcsik M, Munoz J, Oxburgh UM. Geochim Cosmochim Acta 1963; 27:957–977.
18. Kinsman DJJ, Holland HD. Geochim Cosmochim Acta 1969; 33:1–17.
19. Driessens FCM. ACS Symp Ser 1986; 323:524–560.
20. Sverjensky DA. Geochim Cosmochim Acta 1984; 48:1127–1134.
21. Sverjensky DA. Geochim Cosmochim Acta 1985; 49:853–864.
22. Rimstidt JD, Balog A, Webb J. Geochim Cosmochim Acta 1998; 62:1851–1863.
23. Lea DW. Trace Elements in Foraminiferal Calcite. In: Sen Gupta B, ed. Modern Foraminifera. Dordrecht: Kluwer, 1999:259–277.
24. Shen GT, Dunbar RB. Geochim Cosmochim Acta 1995; 59:2009–2024.
25. Lea DW. Elemental and Isotopic Proxies of Marine Temperatures. In: Elderfield H, ed. The Oceans and Marine Geochemistry. Vol. 6. Treatise on Geochemistry. Holland HD and Tuerekian KK, eds. Oxford: Elsevier-Pergamon, 2003: 365–390.
26. Weiner S, Dove PM. Rev Minera Geochem 2003; 54:1–29.
27. Henderson GM. Earth Planet Sci Lett 2002; 203:1–13
28. Linn LJ, Delaney ML, Druffel ERM. Geochim Cosmochim Acta 1990; 54:387–394.
29. Delaney ML, Linn LJ, Druffel ERM. Geochim Cosmochim Acta 1993; 57:347–354.
30. Reuer MK, Boyle EA, Cole JE. Earth Planet Sci Lett 2003; 210:437–452.
31. Linsley BK, Wellington GM, Schrag DP, Ren L, Salinger MJ, Tudhope AW. Clim Dyn 2004; 22:1–11.
32. Skinner LC, Shackleton NJ, Elderfield H. Geochem Geophys Geosys 2003; 4:art. no. 1098.
33. Keigwin LD, Boyle EA. Paleoceanography 1999; 14:164–170.
34. Stoll HM, Klaas CM, Probert I. Glob Planet Change 2002; 34:153–171.
35. Stoll HM, Rosenthal Y, Falkowski P. Geochim Cosmochim Acta 2002; 66:927–936.
36. Rickaby REM, Schrag DP, Zondervan I, Riebesell U. Glob Biogeochem Cycles 2002; 16:art. no. 1006.
37. Stoll HM, Bains S. Paleoceanography 2003; 18:art. no. 1049.
38. Bailey TR, Rosenthal Y, McArthur JM, van de Schootbrugge B, Thirlwall MF. Earth Planet Sci Lett 2003; 212:307–320.
39. Westbroek P, Vanderwal P, Borman AH, Devrind JPM, Kok D, Debruijn WC, Parker SB. Phil Trans R Soc Lond B 1984; 304:435–444.
40. Simkiss K, Wilbur KM. Biomineralization: Cell Biology and Mineral Deposition, San Diego, CA: Academic Press, 1989.
41. Lowenstam HA, Weiner S. On Biomineralization, New York: Oxford University Press, 1989.
42. Young JR, Henriksen K. Rev Mineral Geochem 2003; 54:189–215.
43. Marsh ME. Protoplasma 1994; 177:108–122.
44. Young JR, Didymus JM, Bown PR, Prins B, Mann S. Nature 1992; 356:516–518.
45. Erez J. Rev Mineral Geochem 2003; 54:115–149.
46. Anderson OR, Faber WW. J Foram Res 1984; 14:303–308.
47. Erez J, Bentov S, Brownlee C, Raz M, Rinkevich B. Geochim Cosmochim Acta 2002; 66(suppl 1): A216-A216.

48. Cohen AL, McConnaughey TA. Rev Mineral Geochem 2003; 54:151–187.
49. Hayes RL, Goreau NI. Biol Bull 1977; 152:26–40.
50. Constanz BR. Skeletal organisation in Acropora, In: Crick RE, ed. Origin, Evolution and Modern Aspects of Biomineralization in Plants and Animals. New York: Plenum Press, 1989:175–200.
51. Ip YK, Krishnaveni P. J Exp Zool 1991; 258:273–276.
52. Ferrier-Pages C, Boisson F, Allemand D, Tambutte E. Mar Ecol-Progr Ser 2002; 245:93–100.
53. Barnes DJ. Science 1970; 170:1305–1308.
54. Constanz BR. Palaios 1986; 1:52–157.
55. Cuif JP, Lecointre G, Perrin C, Tillier A, Tillier S. Zoologica Scripta 2003; 32:459–473.
56. Hagiwara S, Byerly L. Ann Rev Neurosci 1981; 4:69–125.
57. Almers W, Palade PT. J Physiol-Lond 1981; 312:159–176.
58. Hess P, Lansman JB, Tsien RW. J Gen Physiol 1986; 88:293–319.
59. Doyle DA, Cabral JM, Pfuetzner RA, Kuo AL, Gulbis JM, Cohen SL, Chait BT, MacKinnon R. Science 1998; 280:69–77.
60. Sumida M, Hamada M, Takenaka H, Hirata Y, Nishigauchi K, Okuda H. J Biochem 1986; 100:765–772.
61. Yu X, Inesi G. J Biol Chem 1995; 270:4361–4367.
62. Fujimori T, Jencks WP. J Biol Chem 1992; 267:18466–18474.
63. Toyoshima C, Nakasako M, Nomura H, Ogawa H. Nature 2000; 405:647–655.
64. Kwon DK, Gonzalez EL. J Phycol 1994; 30:689–695.
65. Araki Y, Gonzalez EL. J Phycol 1998; 34:79–88.
66. Al-Horani FA, Al-Moghrabi SM, de Beer D. Mar Biol 2003; 142:419–426.
67. Mann S. Biomineralization: Principles and Concepts in Bioinorganic Materials Chemistry. New York: Oxford University Press, 2001.
68. Fichtinger-Schepman AMJ, Kamerling JP, Versluis C, Vliegenthart JFG. Carbohydr Res 1981; 93:105–123.
69. Marsh ME, Ridall AL, Azadi P, Duke P. J Struct Biol 2002; 139:39–45.
70. Caldwell CR, Haug A. Physiol Plant 1981; 53:117–124.
71. Stoll HM, Schrag DP. Geochem Geophys Geosys 2000; 1:1999GC000015.
72. Kleypas JA, Buddemeier RW, Archer D, Gattuso JP, Langdon C, Opdyke BN. Science 1999; 284:118–120.
73. Barker S, Elderfield H. Science 2002; 297:833–836.
74. Riebesell U, Zondervan I, Rost B, Tortell PD, Zeebe RE, Morel FMM. Nature 2000; 407:364–367.
75. McCorkle DC, Martin PA, Lea DW, Klinkhammer GP. Paleoceanography 1995; 10:699–714.
76. Marchitto TM, Curry WB, Oppo DW. Paleoceanography 2000; 15:299–306.
77. Boyle EA, Erez J. EOS Trans. AGU 84 (52) Ocean Sci Meet Suppl 2004; Abstract OS21G-01.
78. Lorens RB. Geochim Cosmochim Acta 1981; 45:553–561.
79. Stephan S, Hasselbach W. Eur J Biochem 1991; 196:231–237.
80. Stark JG, Wallace HG. Chemistry Data Book, in International System of Units (SI). 2d ed. London: John Murray Ltd., 1990.

Subject Index

T - #0109 - 111024 - C88 - 229/152/16 - PB - 9780367454210 - Gloss Lamination